Lecture Notes in Mathematics

Edited by A. Dold, F. Takens and B. Teissier

Editorial Policy
for the publication of monographs

1. Lecture Notes aim to report new developments in all areas of mathematics – quickly, informally and at a high level. Monograph manuscripts should be reasonably self-contained and rounded off. Thus they may, and often will, present not only results of the author but also related work by other people. They may be based on specialized lecture courses. Furthermore, the manuscripts should provide sufficient motivation, examples and applications. This clearly distinguishes Lecture Notes from journal articles or technical reports which normally are very concise. Articles intended for a journal but too long to be accepted by most journals, usually do not have this "lecture notes" character. For similar reasons it is unusual for doctoral theses to be accepted for the Lecture Notes series.

2. Manuscripts should be submitted (preferably in duplicate) either to one of the series editors or to Springer-Verlag, Heidelberg. In general, manuscripts will be sent out to 2 external referees for evaluation. If a decision cannot yet be reached on the basis of the first 2 reports, further referees may be contacted: the author will be informed of this. A final decision to publish can be made only on the basis of the complete manuscript, however a refereeing process leading to a preliminary decision can be based on a pre-final or incomplete manuscript. The strict minimum amount of material that will be considered should include a detailed outline describing the planned contents of each chapter, a bibliography and several sample chapters.
Authors should be aware that incomplete or insufficiently close to final manuscripts almost always result in longer refereeing times and nevertheless unclear referees' recommendations, making further refereeing of a final draft necessary.
Authors should also be aware that parallel submission of their manuscript to another publisher while under consideration for LNM will in general lead to immediate rejection.

3. Manuscripts should in general be submitted in English.
Final manuscripts should contain at least 100 pages of mathematical text and should include
– a table of contents;
– an informative introduction, with adequate motivation and perhaps some
   historical remarks: it should be accessible to a reader not intimately familiar
   with the topic treated;
– a subject index: as a rule this is genuinely helpful for the reader.

Continued on back inside cover

Lecture Notes in Mathematics 1742

Editors:
A. Dold, Heidelberg
F. Takens, Groningen
B. Teissier, Paris

Springer
Berlin
Heidelberg
New York
Barcelona
Hong Kong
London
Milan
Paris
Singapore
Tokyo

André Unterberger

Quantization and Non-holomorphic Modular Forms

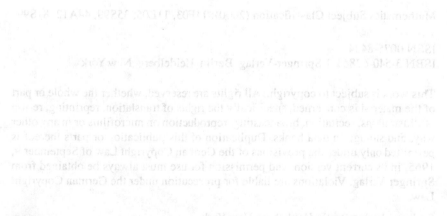

Springer

Author

André Unterberger
Mathématiques (UPRESA 6056)
Université de Reims
Moulin de la Housse, BP 1039
51687 Reims Cedex 2, France

E-mail: andre.unterberger@univ-reims.fr

Cataloging-in-Publication Data applied for

Die Deutsche Bibliothek - CIP-Einheitsaufnahme

Unterberger, André:
Quantization and non-holomorphic modular forms / André Unterberger. -
Berlin ; Heidelberg ; New York ; Barcelona ; Hong Kong ; London ;
Milan ; Paris ; Singapore ; Tokyo : Springer, 2000
 (Lecture notes in mathematics ; 1742)
 ISBN 3-540-67861-1

Mathematics Subject Classification (2000): 11F03, 11L05, 35S99, 44A12, 81S99

ISSN 0075-8434
ISBN 3-540-67861-1 Springer-Verlag Berlin Heidelberg New York

Springer-Verlag Berlin Heidelberg New York
a member of BertelsmannSpringer Science+Business Media GmbH

© Springer-Verlag Berlin Heidelberg 2000
Printed in Germany

Typesetting: Camera-ready TeX output by the author
SPIN: 10724305   41/3142-543210 - Printed on acid-free paper

Foreword: instructions for use

The aim of this manuscript is to bring together quantization theory and the theory of non–holomorphic modular forms. It depends on a certain number of ideas from quantization theory, pseudodifferential analysis, partial differential equations and elementary harmonic analysis on one hand, from modular form theory on the other.

As it adresses itself to two rather distinct possible audiences, we include the present foreword, as an answer to the question "who might be interested in reading what ?". Still, let us stress that, from our point of view, trade between mathematical disciplines should be conducted on a reciprocal basis: we thus hope that some number theorists may view our present investigations not only, or even mostly, as a — well–founded or not — claim that pseudodifferential analysis has something to contribute to modular form theory but, also, as an invitation to join a possibly unfamiliar playground. That the game is far from being over will be shown at the end of this foreword.

The reader interested in modular form theory, but not in motivations from quantization theory, might be advised to jump from the introduction to section 7, in which the Rankin–Selberg unfolding method is extended to some fair degree: thanks to the hyperfunction concept, the extended method permits to recover the Roelcke–Selberg coefficients of any $f \in L^2(\Gamma \backslash \Pi)$ with $\Delta f \in L^2(\Gamma \backslash \Pi)$, without any decrease at infinity being required; Π is the upper half–plane, Δ is the Laplace–Beltrami operator on Π and $\Gamma = SL(2, \mathbf{Z})$. The method is then applied (sections 9, 11 and 12) to the case when f is the product, or the Poisson bracket, of two Eisenstein series $E_{\frac{1-\nu_1}{2}}$ and $E_{\frac{1-\nu_2}{2}}$. This brings (section 10) certain Dirichlet series in two variables $\zeta_n(s,t)$ and $\zeta_n^-(s,t)$, related to Kloosterman–Selberg series, into the picture. In section 14, the preceding results are generalized so far as the admissible values of ν_1 and ν_2 are concerned: also, the coefficients of the discrete terms in the decomposition of $f_1 f_2$ or $\{f_1, f_2\}$ are made explicit in terms of L–functions and Hecke's theory. In retrospect, this permits to analyze the complex continuation of the series $\zeta_n^\pm(s,t)$. A central section of this work is section 15, which describes some rather simple Dirichlet series in one variable (theorems 15.2 and 15.3), the poles of which yield the eigenvalues of Δ. It also lets (theorems 15.6 and 15.7) the Maass cusp–forms appear as the residues with respect to some complex parameter μ of some simple series, generalizing Eisenstein's. Now it is not a novel thing to describe the eigenvalues of Δ and the Maass cusp–forms (or their Fourier coefficients) as poles and residues of appropriate series: but our series $F_{\mu,\nu}$, closely modelled after Eisenstein's, look as simple as one might wish; also, our method, based on the idea that, *in some algebraic sense*, Eisenstein series alone should generate all non-holomorphic modular forms, may be new too. To our knowledge, previous results of a related nature regarding Maass cusp-forms were based either on the study of the Green's kernel of the automorphic Laplacian [28, 20], or on Selberg's non-holomorphic Poincaré series [33] (which occur here in section 20, for a quite distinct purpose). For the benefit of readers who would insist on delimiting exactly what is new in our results, we show (in the remark which follows the proof of proposition 14.6) that when $|\text{Re} \, (s - t)| > \frac{3}{2}$, the complex continuation of $\zeta_n(s, t)$ could be obtained as a consequence of known results concerning the Kloosterman-Selberg series: but this does not hold when (s, t) lies in the domain $|\text{Re} \, (s - t)| < 1$ crucial for our investigations in section 15.

What precedes covers most of what is contained in the present monograph concerning non-holomorphic modular forms viewed independently from other subjects, though sections 19-20 could also be viewed in this context. Actually, these are in part the result of our desire to understand, from an analyst's point of view, why quadratic numbers enter the Maass construction of cusp-forms for congruence subgroups of Γ, but they also owe part of their inspiration to classical potential theory, namely to the notion of single layer potential. Though it is difficult to get one's hands on cusp-forms in any concrete way, it is easy to construct automorphic functions which are regular and modular of the non-holomorphic type almost everywhere. For instance, Eisenstein series arise as such a kind of object, with singularities at infinity (and all Γ-equivalent points): substituting a finite point z^o for ∞, one is led to Hejhal-type pseudo-cusp-forms. This is especially interesting, number-theoretically, when z^o is taken as an imaginary quadratic point. We here show how to construct automorphic, almost modular, functions on Π, with mild singularities (discontinuities in the normal derivative) spread out on the (locally finite) collection of Γ-transforms of a line with extremities conjugate in some real quadratic extension of $\mathbb{Q}$.

One last construction which, we hope, can interest number theorists, is our generalization of the Rankin-Cohen products [5] of modular forms to the non-holomorphic case: this was a major starting point of the present work, and it is explained in the introduction which follows.

All the rest of the monograph is concerned with quantization theory: this may start as a combination of representation theory (a group G, a Hilbert space $\mathcal{H}$, a unitary representation π of G into $\mathcal{H}$) and pseudodifferential analysis (the prescription of a covariant *quantizing map* from some *phase space* X on which G acts to the space of operators on $\mathcal{H}$, or its inverse *symbol map* [42]). Arithmetic comes into play as soon as some arithmetic subgroup Γ of G has been fixed, and one tries to analyze the space, or hopefully algebra, of operators whose symbols are Γ–invariant "functions" on X.

This very rich structure is interesting for several reasons, including the following: two quite distinct phase spaces X and $\tilde{X}$ may carry symbolic calculi related to the same representation π [40]. In this way, identifying the two species of symbols of the same operator commuting with the group of unitaries $\pi(\Gamma)$ leads to an identification of automorphic objects living on X or $\tilde{X}$. For instance, $\mathbb{R}^2\backslash\{0\}$ (with ξ and $-\xi$ identified) and the upper half–plane Π are so related since (section 17) one can relate (parts of) the Weyl calculus on $L^2(\mathbb{R})$ to some of the symbolic calculi available with Π as a phase space. In this way, non–holomorphic modular forms on Π are identified with homogeneous Γ–invariant distributions on $\mathbb{R}^2\backslash\{0\}$ (on which Γ acts in a linear way).

This can also be done with the help of the Radon transformation (section 4) or elementary representation theory (sections 2, 3). However, as shown in section 13, only the point of view of pseudodifferential analysis permits a true understanding of the "correct" Hilbert space of automorphic (*i.e.,* Γ–invariant) distributions on $\mathbb{R}^2$: this is not a trivial problem because there is no fundamental domain for the action of Γ on $\mathbb{R}^2\backslash\{0\}$. Still, a number of explicit automorphic distributions are nice objects to play with: this is described in section 16, where attention is also paid, in this classical distribution setting, to constructions with a more adelic taste.

On the other hand, something which we had not expected beforehand occurs, and is described in section 18: namely, it is possible to identify automorphic distributions with Cauchy data for the Lax–Phillips scattering theory for the automorphic wave equation. This identification has two versions: a first one in which both species appear as boundary values for two related boundary–value problems in the three–dimensional light–cone, and a deeper one in which the connection is analyzed through quantization theory. It is part of our projects (actually well under way) to develop the automorphic Weyl calculus, in which automorphic distributions act as symbols: this calculus is very exotic in comparison to all known pseudodifferential analyses and, in preparation for it, section 5 contains a composition formula for Weyl symbols of the usual type which, we hope, readers well versed in pseudodifferential analysis may find interesting.

Considering the quite impressive point to which the scope of modular form theory has been extended in recent years, it is hoped, but we have not even started working on this, that a comparable analysis may be developed on $\mathbb{R}^n$, in connection with general arithmetic subgroups of the symplectic group.

CONTENTS

1. Introduction

Like most mathematicians, the present author — usually a devotee of pseudodifferential analysis or quantization theory — has been secretly in love with number theory, especially the theory of modular forms, for years. To declare himself, he waited for some opportunity, which finally presented itself when he first came across H.Cohen's bilinear products [5] of (holomorphic) modular forms, in the nice expository paper by Zagier [49]. These are expressions

$$F_j^{k_1, k_2}(f_1, f_2) = \sum_{l=0}^{j} (-1)^l \binom{k_1 + j - 1}{l} \binom{k_2 + j - 1}{j - l} f_1^{(j-l)} f_2^{(l)}$$

$$(1.1)$$

in which f_1 and f_2 are holomorphic modular forms of weights k_1 and k_2 and j is a non–negative integer: the expression reduces to the pointwise product of f_1 and f_2 in the case when $j = 0$; it always produces a cusp–form, of weight $k_1 + k_2 + 2j$, when $j \geq 1$. Now, if you are familiar with pseudodifferential analysis, you cannot fail to recognize at once the characteristic scent of some kind of composition rule for symbols. As luck would have it, in collaboration with J.Unterberger, we had just completed some work [44] on quantization theory in which the one–sheeted hyperboloid acted as a phase space: the Cohen products were then shown [45] to be just the various terms (under some disguise) in the expansion of the composition, according to this symbolic calculus, of appropriate symbols. A related, though not truly similar — since no genuine operators are present there — point of view was developed independently in [6].

The present work started with a view towards completing the same program in a way that would enable one to consider non–holomorphic modular forms instead. Of course, as soon as the deep relationship between the Cohen products and symbolic calculi had been established, the way the problem had to be considered presented itself in a natural way. But it turned out to be considerably more difficult technically, in that integral bilinear operators with very singular kernels had to be tackled with in lieu of differential operators.

Our main incentive had to do with the production — not quite the same as a construction, *cf. infra* — of Maass cusp–forms, the definition of which will be recalled shortly in this introduction. If, in the holomorphic case, it is nice, for instance, to watch how the two basic Eisenstein series G_4 and G_6 can join forces to produce the Ramanujan Δ–function (from

(1.1), with $j = 1$: *cf.* [49], p.250), nothing identical could happen in the non–holomorphic case. For, on one hand, the discrete parameter j from holomorphic theory had to be replaced by a continuous one, and cusp–forms do exist only for exceptional values of the parameter; next, experience has shown how difficult it is to actually construct just one Maass cusp–form (to our knowledge, though this has of course been done by Maass and others (*cf.* for instance [9, 3]) in the case of appropriate congruence subgroups of $\Gamma = SL(2, \mathbb{Z})$, using algebraic number theory, such a construction is still waiting in the case of the group Γ itself).

From the very beginning, we thus envisioned the non–holomorphic ana-logue of Cohen's products primarily as a machine for producing Maass cusp–forms, and have kept this point of view in the present work: other applica-tions will wait. But, though we indeed explain the link between our bilinear products and quantization theory (in sections 5, 8, 13, 17), we made our best to present most of our results without appealing to the latter. This change of emphasis was due on one hand to a change in our own interests while this work developed; next, to the fact — unexpected beforehand — that these bilinear products also result from a certain decomposition of pointwise prod-ucts or Poisson brackets of pairs of functions on the Poincaré half–plane; finally to the hope that some readers might be interested in the modular function – theoretic aspects of these lecture notes without feeling the need for a quantization approach.

We can now proceed towards a description of the constructions and results contained in the present work. The first main ingredient is the full non–unitary principal series [22] of representations of $G = SL(2, \mathbb{R})$, to wit the family (π_ν), $\nu \in \mathbb{C}$, of representations of G in some appropriate Hilbert space H of functions on the real line ($L^2(\mathbb{R})$ will do if Re $\nu = 0$, *i.e.*, in the unitary case), defined by the formula

$$(\pi_\nu(g)u)(s) = |-cs + a|^{-1-\nu} u(\frac{ds - b}{-cs + a}), \qquad g = \begin{pmatrix} a & b \\ c & d \end{pmatrix}.$$

$$(1.2)$$

There is another classical realization of the same representation (*cf.* [10], p.37), in which the Hilbert space H consists of functions on the Poincaré half–plane $\Pi = G/K$, $K = SO(2)$, actually eigenfunctions of the Laplacian

$$\Delta = -y^2 \left(\frac{\partial^2}{\partial x^2} + \frac{\partial^2}{\partial y^2} \right):$$

$$(1.3)$$

the parameter ν in (1.2) corresponds to the eigenvalue $\frac{1-\nu^2}{4}$ of Δ. The ad-vantages of working with one or the other of the two realizations compensate to the point that it is very handy to work with both simultaneously. This is why, in section 3, we describe, in an arithmetic setting, how the Γ–invariant vectors from the two realizations should be identified with one another. In the realization by means of functions on Π, these Γ–invariant vectors are none other than the classical notion of non–holomorphic modular forms: of course, one should realize at once that these functions are very far from lying in H. They can, however, lie in $L^2(\Gamma\backslash\Pi)$, *i.e.*, be square–integrable when restricted to a fundamental domain (an open subset of Π, essentially a set of representatives of the set of Γ–orbits in Π): they are then called (Maass) cusp–forms. In the first setting, one is not even dealing with bona fide functions of s (*cf.* (1.2)), and one has to use an appropriate distribution concept, which is described in section 2. There is a nice dictionary, in which, for instance, *Eisenstein distributions* living on $\mathbb{R}$ give rise, on Π, to the usual Eisenstein series (1.8), and the functional equation of Eisenstein series is put (*cf.* (3.33)) into another perspective.

But it is not for these frivolous satisfactions that we advocate the use of both realizations of the principal series of representations of G. For it is only when using the more singular realization that one can convince oneself (this will be done in section 8) that one is in possession of all possible bilinear products that would make a "composition" of non–holomorphic modular forms possible. The first of these products is the rule $\mathbf{L}_{\nu_1,\nu_2;\nu}$ defined by the equation

$$\mathbf{L}_{\nu_1,\nu_2;\nu}(u_1,u_2)(s) = \int \chi_{\nu_1,\nu_2;\nu}(s_1,s_2;s)\, u_1(s_1)\, u_2(s_2)\, ds_1\, ds_2\,,$$

$$(1.4)$$

where

$$\chi_{\nu_1,\nu_2;\nu}(s_1,s_2;s) =$$

$$|s_1-s_2|^{\frac{1}{2}(-1+\nu+\nu_1+\nu_2)}|s_1-s|^{\frac{1}{2}(-1-\nu+\nu_1-\nu_2)}|s_2-s|^{\frac{1}{2}(-1-\nu-\nu_1+\nu_2)}\,. \quad (1.5)$$

It satisfies the *covariance rule*

$$\pi_\nu(g)\mathbf{L}_{\nu_1,\nu_2;\nu}(u_1,u_2) = \mathbf{L}_{\nu_1,\nu_2;\nu}(\pi_{\nu_1}(g)u_1,\pi_{\nu_2}(g)u_2),\qquad g\in G,$$

$$(1.6)$$

which expresses in particular that if u_1 (*resp.* u_2) is a modular distribution with parameter ν_1 (*resp.* ν_2), then $\mathbf{L}_{\nu_1,\nu_2;\nu}(u_1,u_2)$ is a modular distribution with parameter ν. This gives half the bilinear products we had been looking after: the missing ones are obtained by inserting appropriate signs in the kernel defined in (1.5). The reason why things are not so clear when working on Π is that, there, one considers only eigenfunctions of Δ, not arbitrary functions as one does on $\mathbb{R}$. However, to be truthful, expressions like (1.4) are hardly manageable in view of their singularities, especially when distributions are substituted for u_1 and u_2, as will be seen in section 8: this is why we must now transfer (1.4) to Π.

The answer is surprisingly simple (section 8), and can be expressed through the following recipe (out of the arithmetic environment to start with, *i.e.*, for functions on Π corresponding to more regular functions u_1 and u_2): start from two eigenfunctions of the Laplacian, take their pointwise product, finally extract from the result the spectral density at some given eigenvalue of Δ. The same would work with the second family of bilinear products we have been hinting at above, substituting Poisson brackets for pointwise products. When this has been done, coming back to automorphic funstions, we can essentially forget that the whole project started with quantization theory, and pretend that, for some reason, we are interested in decomposing (according to the spectral theory of Δ) products or Poisson brackets of non–holomorphic modular forms. It will be nice to observe that, once we have enriched the total linear space of modular forms with this extra algebraic structure, we can claim that it is generated by Eisenstein series alone.

At this point, we wish to indicate why bilinear expressions like (1.4) impose themselves when viewed in connection with symbolic calculi. In section 5, we look at the Weyl calculus of operators on $L^2(\mathbb{R})$ from a novel point of view. Namely, we decompose symbols — they live on $\mathbb{R}^2$ — as sums of homogeneous ones: no, not the polynomial symbols of integral homogeneous degree the reader may be familiar with, which are sometimes suitable in asymptotic expansions. Rather, we consider the continuous decomposition, provided by a Fourier transformation, of a symbol as a sum of symbols homogeneous of degree $-1 - i\lambda$, $\lambda \in \mathbb{R}$. We then show that, if one starts from two homogeneous symbols h_1 and h_2, and decompose again their composition (in the sense of the Weyl calculus) $h_1 \# h_2$, what one gets is precisely expressions that reduce to (1.4), up to some explicit coefficients. Had we started from some much more exotic symbolic calculi (the ones associated [39, 42] with the projective discrete series $(\mathcal{D}_{\tau+1})_{\tau>-1}$ of representations of G: they essentially reduce to the Weyl calculus when $\tau + 1 = \frac{1}{2}$ or $\frac{3}{2}$), we would have obtained the same result; only the coefficients would have changed. We shall relegate this to the end (section 17) of this work, and mention here that it explains why (*cf. supra*) our bilinear products are related to the unquantized structure on the half–plane as described by pointwise products or Poisson brackets: these occur, as the computation shows, if one considers the asymptotics, as $\tau \to \infty$, of the composition formula associated with $\mathcal{D}_{\tau+1}$.

In section 4, we briefly describe another essential tool, in terms suitable to our task, namely the Radon transform, which has been studied in a much more general context by Helgason [17]. It connects analysis on two homogeneous spaces of G: the Poincaré half–plane G/K and the horocyclic space $\Xi = G/MN$, where $M = \{\pm I\}$, $N = \{\left(\begin{smallmatrix} 1 & b \\ 0 & 1 \end{smallmatrix}\right), b \in \mathbb{R}\}$. One can view the second of these spaces as the space $\mathbb{R}^2 \backslash \{0\}$ (on which G acts in a linear way), in which any two points the negative of each other have been identified: thus functions on Ξ are just even functions on $\mathbb{R}^2$. The Radon transform is a map from functions on Π to functions on Ξ that intertwines the two quasi–regular actions of G on the two spaces of functions. Its main advantage is that, when viewed as an operator on $L^2(\Xi) = L^2_{\text{even}}(\mathbb{R}^2)$ through this transfer, the operator $\pi^{-2}(\Delta - \frac{1}{4})$ reduces to the square of the Euler operator

$$\mathcal{E} = \frac{1}{2i\pi}(\xi_1 \frac{\partial}{\partial\xi_1} + \xi_2 \frac{\partial}{\partial\xi_2} + 1): \tag{1.7}$$

another way to see the Radon transform is as a global correspondence that reduces, when the degree of homogeneity $-1 - i\lambda$ (the eigenvalue $\frac{1}{4}(1 + \lambda^2)$ of Δ) is fixed, to the intertwining operator described above, connecting the two realizations of $\pi_{i\lambda}$.

Serious matters start in sections 6 and 7, in which we extend the Radon transform to an arithmetic setting. What we get is a new way to handle the Roelcke–Selberg decomposition of an arbitrary function in the space $L^2(\Gamma\backslash\Pi)$. Recall that, on this space, the Laplacian (some self-adjoint extension of it) has both a continuous spectrum $[\frac{1}{4}, \infty[$ and a (mysterious) discrete spectrum. Recall the definition

$$E_{\frac{1-\nu}{2}}(z): = \frac{1}{2} \sum_{\substack{|m|+|n|\neq 0 \\ (n,m)=1}} \left(\frac{|mz - n|^2}{\operatorname{Im} z}\right)^{\frac{\nu-1}{2}} \tag{1.8}$$

of Eisenstein series. The Roelcke–Selberg theorem states that any function $f \in L^2(\Gamma\backslash\Pi)$ can be expanded as

$$f(z) = \Phi^0 + \frac{1}{8\pi} \int_{-\infty}^{\infty} \Phi(\lambda) E_{\frac{1-i\lambda}{2}}(z) \, d\lambda + \sum_{j\geq 1} \Phi^j \mathcal{M}_j(z), \tag{1.9}$$

where Φ^0 is a constant, the function Φ lies in $L^2(\mathbb{R})$ and satisfies the symmetry property that the function $\lambda \mapsto (\zeta^*(i\lambda))^{-1} \Phi(\lambda)$, with

$$\zeta^*(s): = \pi^{-\frac{s}{2}} \Gamma(\frac{s}{2}) \zeta(s), \tag{1.10}$$

should be even: finally, the functions $\mathcal{M}_j$ are Maass cusp-forms which may be chosen as an orthonormal family for the Petersson scalar product (the canonical scalar product on $L^2(\Gamma\backslash\Pi)$), and the Φ^j's are constants.

One problem we take another look at is how to effectively get the coefficients Φ^0, $\Phi(\lambda)$ of this decomposition, as well as the sum $f_{\lambda_k} = \sum \Phi^j \mathcal{M}_j(z)$ of all terms corresponding to some discrete eigenvalue $\frac{1+\lambda_k^2}{4}$ of the Laplacian, starting from an arbitrary f. The obvious answer to the first question (to wit, $\Phi(\lambda) = (E_{\frac{1-i\lambda}{2}}|f))$, even when completed by the Rankin–Selberg trick (*cf. infra*) is not satisfactory, since the integral usually diverges. To get a more efficient answer, we take advantage of the Radon transform, and analyze it in the Γ-invariant setting. Our result expresses the function $\Phi(\lambda)$ as a hyperfunction: recall that hyperfunction theory starts when you wish to add analytic functions with disjoint domains! This is exactly what is needed here. It should be pointed out (we realized this much later) that our method actually extends the Rankin–Selberg celebrated *unfolding* method, as it reduces to it in the case when the latter one is applicable (*i.e.*, when f goes to zero at infinity, in the fundamental domain, rather rapidly). The orthogonal projection f_{λ_k} of f onto the eigenspace of Δ corresponding to the eigenvalue $\frac{1+\lambda_k^2}{4}$ is also obtained in a related way, through its Fourier coefficients.

This method is then applied (sections 9 and 11: the most technical of the paper) to the case when $f = f_{\nu_1,\nu_2}$ is the product of two Eisenstein series $E_{\frac{1-\nu_1}{2}}$ and $E_{\frac{1-\nu_2}{2}}$: now, such a function is never in $L^2(\Gamma\backslash\Pi))$, but, *provided* $\mathrm{Re}\,(\nu_1 \pm \nu_2) \neq \pm 1$, you can get such a function if you first substract a small number of well–chosen Eisenstein series. One can then explicitly express the coefficient $\Phi(\lambda)$ of the decomposition (1.9) as

$$\Phi(\lambda) = \frac{\zeta^*\left(\frac{1+i\lambda-\nu_1+\nu_2}{2}\right)\zeta^*\left(\frac{1+i\lambda+\nu_1-\nu_2}{2}\right)\zeta^*\left(\frac{1-i\lambda-\nu_1-\nu_2}{2}\right)\zeta^*\left(\frac{1+i\lambda-\nu_1-\nu_2}{2}\right)}{\zeta^*(\nu_1)\,\zeta^*(\nu_2)\,\zeta^*(-i\lambda)}$$

(1.11)

(theorem 9.6). The Fourier coefficients of the function f_{λ_k} can be computed as residues of a certain Dirichlet series, a one–variable version of the Dirichlet series in two variables

$$\zeta_n(s,t) = \frac{1}{4} \sum_{\substack{m_1 m_2 \neq 0 \\ (m_1,m_2)=1}} |m_1|^{-s}\,|m_2|^{-t}\,e^{2i\pi n \frac{\overline{m}_2}{m_1}},$$

$$(\overline{m}_2 m_2 \equiv 1 \bmod m_1). \quad (1.12)$$

These series are introduced in section 10 and their basic properties are established. In a certain domain of the variables (s,t), which has no intersection with that in which the expansion (1.12) is valid (an intermediary is required), the continuation of $\zeta_n(s,t)$ can be expressed *as a series of* so-called Kloosterman–Selberg *series*: the latter ones were introduced by Selberg [33] and considered also by Deshouillers–Iwaniec [7], Goldfeld–Sarnak [12], Iwaniec [20]. The treatment by Iwaniec — based on the analysis of the automorphic Green function — is most precise so far as the continuation properties of the Kloosterman–Selberg series is concerned and we have done our best to see whether the part of our preceding result dealing with the Fourier coefficients of the discrete part f_{λ_k} could be derived from the results there. The discussion following proposition 14.6 shows that this is the case when $|\mathrm{Re}\,(\nu_1 + \nu_2)| > 1$ and $|\mathrm{Re}\,(\nu_1 - \nu_2)| > 1$, not in the other cases: no analytic continuation is possible across the lines $\mathrm{Re}\,(\nu_1 \pm \nu_2) = \pm 1$.

It is the case when $|\mathrm{Re}\,(\nu_1 - \nu_2)| < 1$, however, that will be needed in the next topic discussed in this introduction, namely the construction of generating series *of sorts* of Maass cusp–forms. We say "of sorts" since locating the eigenvalues of the modular Laplacian as the poles of some analytic function is not quite as nice, e.g. so far as numerical computation is concerned, as identifying them with the zeroes of some function would be. A similar remark applies to the polar parts of our function, which do not quite qualify as explicit expressions of Maass cusp–forms.

Assume $|\operatorname{Re} \nu| < 1$ and consider the following series, to be compared to the Eisenstein series (1.8):

$$F_{\mu,\nu}(z) = \frac{1}{2} \sum_{\substack{n \in \mathbf{Z},\, m \in \mathbf{Z}^{\times} \\ m|n(n-1)}} (\frac{m_1}{m_2})^{\frac{\nu}{2}} |m|^{\frac{1-i\mu}{2}} \left(\frac{|mz - n|^2}{\operatorname{Im} z} \right)^{\frac{i\mu-1}{2}}, \tag{1.13}$$

where the pair m_1, m_2 is the pair of positive integers characterized by the conditions $|m| = m_1 m_2$, $m_1|n-1$, $m_2|n$. When meaningful, it is $\mathbf{Z}$–periodic as a function of z and satisfies the equation $\Delta F_{\mu,\nu} = \frac{1+\mu^2}{4} F_{\mu,\nu}$. It is not Γ–invariant, which, in some sense, will turn out to our advantage! Indeed, as a function of μ, it is analytic in the half–plane $\operatorname{Im} \mu > 1 + |\operatorname{Re} \nu|$, and the function

$$F_{\mu,\nu}(z) - F_{\mu,\nu}(-\frac{1}{z})$$

extends to the half–plane $\operatorname{Im} \mu > -1 + |\operatorname{Re} \nu|$. Thus, when we have proved that the function $\mu \mapsto F_{\mu,\nu}(z)$ extends as a meromorphic function to this larger half–plane, the coefficients of the terms with negative orders in the Laurent expansion at each pole will have to be modular forms! The poles show up in two families: all numbers $-i\omega$ with $\zeta^*(\omega) = 0$, and most λ_k's with $\frac{1}{4}(1+\lambda_k^2)$ in the even part of the discrete spectrum of Δ. "Even" means that we retain only those eigenvalues for which there exists an eigenfunction invariant under the symmetry $z \mapsto -\bar{z}$.

To understand exactly which λ_k's are actually poles requires some deeper analysis (section 14) of the relationship between the Roelcke–Selberg decomposition of products of Eisenstein series and Hecke's theory. Recall that the Hecke operators (cf. (8.8) if you are not familiar with the concept) constitute a certain sequence of operators — not differential operators, though: they are more in the nature of a collection of magnifying glasses — acting on spaces of (non–holomorphic, in our case) modular forms, which commute with the modular Laplacian. It is then possible, for each eigenvalue $\frac{1}{4}(1 + \lambda_k^2)$ of the Laplacian, to define an orthonormal basis $(\mathcal{M}_{k,\ell})_\ell$ of the (finite–dimensional) associated eigenspace with the following property: each $\mathcal{M}_{k,\ell}$, besides being an eigenfunction of Δ, is a joint eigenfunction of the family of all Hecke operators. A fundamental notion, especially in connection with Hecke's theory, is that of the L–function $L(.,\mathcal{M})$ associated with any Maass cusp–form $\mathcal{M}$ (cf. for instance [3], p. 107). We can then make our description of the poles of the function $\mu \mapsto F_{\mu,\nu}(z)$ fully precise as follows: a number λ_k with $\frac{1}{4}(1 + \lambda_k^2)$ in the even part of the discrete spectrum of Δ is a pole if and only if at least one of the functions $L(s, \mathcal{M}_{k,\ell})$ (where ℓ can take any value) does not vanish at $s = \frac{1 \pm \nu}{2}$: moreover, the residue of the function under examination at $\mu = \lambda_k$ is a (non–zero) multiple of the cusp–form $\sum_\ell L(\frac{1 \pm \nu}{2}, \mathcal{M}_{k,\ell}) \mathcal{M}_{k,\ell}(z)$.

Needless to say, there are corresponding results for the odd eigenfunctions and eigenvalues of Δ: the situation is simpler then, since the non-trivial zeros of the zeta function do not contribute. The starting point, this time, is the Roelcke–Selberg decomposition of the Poisson bracket — instead of the product — of two Eisenstein series. In both cases, the λ_k's with $\frac{1}{4}(1 + \lambda_k^2)$ in the even, or odd, part of the discrete spectrum of Δ can be located as the poles of certain Dirichlet series: one of the simplest, which yields at least all the λ_k's with $\frac{1}{4}(1 + \lambda_k^2)$ even and simple, and $L(\frac{1}{2}, \mathcal{M}_k) \neq 0$, is the series $\sum_{m \geq 1} a_m m^{-s}$, with

$$a_m = \sum_{\substack{q \bmod 2m \\ q^2 \equiv 1 \bmod 4m}} e^{i\pi \frac{q}{m}}. \tag{1.14}$$

In section 13, we analyze automorphic distributions on $\mathbb{R}^2$, i.e., Γ-invariant tempered distributions: note that there is no fundamental domain for the linear action of Γ in $\mathbb{R}^2$ and that no non–constant automorphic *function* can exist in this sense. Still, with the help of the Weyl calculus of operators, one can define in a natural way a Hilbert space of such distributions: as it turns out, it is essentially a rephrased version of $L^2(\Gamma \backslash \Pi) \oplus L^2(\Gamma \backslash \Pi)$, with the advantage that the spectral decomposition with respect to Δ has to be replaced by the decomposition of automorphic distributions into their homogeneous components. Moving from $\mathbb{R}^2$ to the $(1 + 2)$–dimensional forward light–cone, one can identify the Hilbert space of automorphic distributions with the Hilbert space of Cauchy data for the Lax–Phillips scattering theory [25]. This identification can be made in two ways: the deeper one fully uses the resources of quantization theory, as described in section 17. In section 16, we familiarize ourselves with examples of automorphic distributions and we show how a number of calculations with an adelic taste can be carried within a classical distribution setting.

Finally, in sections 19 and 20, we try to answer the following question: can one associate, in a natural way, classes of automorphic functions with quadratic orbits of $PSL(2, \mathbb{R})$ in $P_1(\mathbb{R})$, in a way comparable with that in which Eisenstein series are associated to the orbit $P_1(\mathbb{Q})$? The positive answer to this question, and the fact that this leads also to some approach of the Maass cusp–forms (again in a hyperfunction setting), depend on the analytic continuation of interesting Dirichlet series, generalizing Hecke's, such as

$$
\sum_{\left(\begin{smallmatrix} n_1 & n_2 \\ m_1 & m_2 \end{smallmatrix}\right) \in \Gamma_\infty \backslash \Gamma / \Gamma_\rho} |N(m_1\rho + m_2)|^{\frac{\nu-1}{2}} e^{i\pi m \, \mathrm{Tr}(\frac{n_1\rho+n_2}{m_1\rho+m_2})} \left| \frac{m_1\rho + m_2}{m_1\bar\rho + m_2} \right|^{\frac{i\pi q}{\kappa \log \eta}},
$$
(1.15)

in which a quadratic irrational ρ, together with the norm and trace from the field $\mathbb{Q}$ to the field $\mathbb{Q}(\rho)$, occurs, Γ_ρ is the stabilizer of ρ in Γ, finally η is an appropriate unit of the field $\mathbb{Q}(\rho)$.

We take this opportunity to thank J.M. Deshouillers for an interesting discussion, and references as well, concerning Kloosterman sums. We are very grateful to the French C.N.R.S. which, by offering us two years without teaching duties, facilitated our completion of this work. Finally, we wish to thank for its kind hospitality the Erwin Schrödinger Institute where some of the latter parts of this project were completed.

2. Distributions associated with the non–unitary principal series

The representation π_ν defined in (1.2) is generally not unitary, but it belongs to the full non–unitary principal series of representations of G as defined in [22], p.38. The Hilbert space to be chosen is $L^2(\mathbb{R}, (1+s^2)^{\mathrm{Re}\ \nu} ds)$, and each operator $\pi_\nu(g)$ is bounded since

$$\int |(\pi_\nu(g)u)(s)|^2 (1 + s^2)^{\mathrm{Re}\ \nu} ds$$

$$= \int |-cs + a|^{-2 - 2\,\mathrm{Re}\ \nu} |u(\frac{ds - b}{-cs + a})|^2 (1 + s^2)^{\mathrm{Re}\ \nu} ds$$

$$= \int |u(t)|^2 [(a^2 + c^2)t^2 + 2(ab + cd)t + b^2 + d^2]^{\mathrm{Re}\ \nu} dt$$

$$= \int |u(t)|^2 \left(\frac{|z - t|^2}{\mathrm{Im}\ z}\right)^{\mathrm{Re}\ \nu} dt \tag{2.1}$$

with $z = g^{-1}.i \in \Pi$.

Recall that the space $C^\infty_{-\nu}$ of C^∞–vectors of the representation $\pi_{-\nu}$ can be defined ([22], p.51) in the usual way, even though $\pi_{-\nu}$ may not be unitary: $u \in L^2(\mathbb{R}, (1 + s^2)^{-\mathrm{Re}\ \nu} ds)$ will lie in $C^\infty_{-\nu}$ if the map $g \mapsto \pi_{-\nu}(g)u$, valued in this Hilbert space, is C^∞ on G. A basis of the space of infinitesimal operators of the representation $\pi_{-\nu}$, defined by the identity $\pi_{-\nu}(\exp t\epsilon_j) = \exp(2i\pi t e_j)$, $t \in \mathbb{R}$, with

$$\epsilon_0 = \begin{pmatrix} 0 & \frac{1}{2} \\ -\frac{1}{2} & 0 \end{pmatrix}, \quad \epsilon_1 = \begin{pmatrix} 0 & -\frac{1}{2} \\ -\frac{1}{2} & 0 \end{pmatrix}, \quad \epsilon_2 = \begin{pmatrix} \frac{1}{2} & 0 \\ 0 & -\frac{1}{2} \end{pmatrix}, \tag{2.2}$$

is given, as is easily shown, as

$$e_0 = (-4i\pi)^{-1}[(s^2 + 1)\frac{d}{ds} + (1 - \nu)s]$$

$$e_1 = (-4i\pi)^{-1}[(s^2 - 1)\frac{d}{ds} + (1 - \nu)s]$$

$$e_2 = (-2i\pi)^{-1}[s\frac{d}{ds} + \frac{1 - \nu}{2}]. \tag{2.3}$$

Since e_0 is an elliptic operator, the space $C^\infty_{-\nu}$ is contained in $C^\infty(\mathbb{R})$. Moreover, if $u \in C^\infty_{-\nu}$, then Du lies in $L^2(\mathbb{R}, (1 + s^2)^{-\mathrm{Re}\ \nu} ds)$ for all D in the algebra generated by e_0, e_1, e_2. However, this condition is not sufficient (since the representation, not of class one, associated as in (1.2) with the function $s \mapsto |s|^{-1-\nu}\mathrm{sign}(s)$ rather than $|s|^{-1-\nu}$, would have exactly the same *formal* infinitesimal generators). In particular, if $u \in C^\infty_{-\nu}$, the function $|s|^{1-\nu}u(s)$ has a limit at ∞, *the same at $\pm\infty$*: we denote it as u_∞.

Either $u_\infty = 0$ or $u(s) \sim u_\infty |s|^{-1+\nu}$, $|s| \to \infty$:

$$u \in C^\infty_{-\nu} \Rightarrow u(s) = u_\infty |s|^{-1+\nu} + O(|s|^{-2+\text{Re }\nu}), \qquad |s| \to \infty.$$

$$(2.4)$$

Note that the notation u_∞ always makes implicit reference to the fact that u is considered as an element of the space $C^\infty_{-\nu}$: which ν is to be thought of will always be clear from the context.

In order to connect the two classical realizations of the representation $\pi_{-\nu}$, let us introduce (following [10], p.37), for $\nu \neq 1, 2, \ldots$, the function

$$\phi^{-\nu}(s) = \pi^{\frac{\nu-1}{2}} \Gamma(\frac{1-\nu}{2})(1+s^2)^{\frac{\nu-1}{2}},$$

$$(2.5)$$

invariant under all operators $\pi_{-\nu}(g)$, $g \in SO(2) \subset SL(2, \mathbb{R})$ (in other words, a K–invariant vector of the representation $\pi_{-\nu}$). The normalization coefficient in (2.5) has been chosen for arithmetic reasons (*cf.* end of proof of proposition 3.8). More generally, set

$$\phi_z^{-\nu}(s) = \pi^{\frac{\nu-1}{2}} \Gamma(\frac{1-\nu}{2}) \left(\frac{|z-s|^2}{\text{Im } z}\right)^{\frac{\nu-1}{2}}$$

$$(2.6)$$

for $z = x + iy \in \Pi$, so that $\phi_z^{-\nu} = \pi_{-\nu}(g)\phi^{-\nu}$ if $z = g.i$. Then the function $\phi^{-\mu}$ will belong to $C^\infty_{-\nu}$ if and only if $\mu = \nu - 2k$ for some non–negative integer k. This is necessary since

$$\frac{\pi^{\frac{1-\mu}{2}}}{\Gamma(\frac{1-\mu}{2})} (\pi_{-\nu}(\begin{pmatrix} 0 & -1 \\ 1 & 0 \end{pmatrix})\phi^{-\mu})(s) = |s|^{\nu-\mu}(1+s^2)^{\frac{\mu-1}{2}}$$

$$(2.7)$$

has to be a C^∞ function of s near 0; the converse comes from an expansion of the first factor on the right-hand side in the equation

$$\frac{\pi^{\frac{1-\nu+2k}{2}}}{\Gamma(\frac{1-\nu+2k}{2})} (\pi_{-\nu}(\begin{pmatrix} a & b \\ c & d \end{pmatrix})\phi^{-\nu+2k})(s) = (-cs+a)^{2k} \left(\frac{|z-s|^2}{\text{Im } z}\right)^{\frac{\nu-1}{2}-k}, \qquad z = g.i.$$

$$(2.8)$$

Also, note for immediate reference that if k is a positive integer, the function $s \mapsto s\phi^{-\nu+2k}(s)$ also belongs to $C^\infty_{-\nu}$.

The following elementary proposition is quite helpful:

PROPOSITION 2.1. *Assume* Re $\nu < 1$, $\frac{\nu}{2} \notin \mathbb{Z}$. *Then the Fourier transformation* $u \mapsto \hat{u}$, $\hat{u}(\sigma) = \int u(s) \exp(-2i\pi s\sigma) ds$, *is an isomorphism from the space* $C^\infty_{-\nu}$ *onto the space of functions* $v = v(\sigma)$ *on* $\mathbb{R}$ *which satisfy the following properties:*

(i) v *is* C^∞ *outside 0;*

(ii) v *and its derivatives are rapidly decreasing at infinity;*

(iii) *near 0,* v *admits an expansion*

$$v(\sigma) \sim a_0 + a_1\sigma + a_2\sigma^2 + \ldots$$
$$+ |\sigma|^{-\nu}(b_0 + b_1\sigma + b_2\sigma^2 + \ldots)$$

$$(2.9)$$

where any specified derivative of the remainder shall be $O(|\sigma|^N)$ with N as large as one pleases provided the expansion is pushed far enough.

PROOF. Assume that $u \in C^\infty_{-\nu}$: then the C^∞ function $\pi_{-\nu}(\begin{pmatrix} 0 & -1 \\ 1 & 0 \end{pmatrix})u$ satisfies near 0 an expansion of the kind

$$(\pi_{-\nu}(\begin{pmatrix} 0 & -1 \\ 1 & 0 \end{pmatrix})u)(s) \sim \beta_0(1+s^2)^{\frac{\nu-1}{2}} + \beta_1 s^2 (1+s^2)^{\frac{\nu-3}{2}} + \dots$$
$$- \gamma_0 s (1+s^2)^{\frac{\nu-3}{2}} - \gamma_1 s^3 (1+s^2)^{\frac{\nu-5}{2}} - \dots , \quad (2.10)$$

from which one gets the expansion at infinity

$$\pi^{\frac{\nu-1}{2}}\Gamma(\frac{1-\nu}{2}) u(s) \sim \beta_0\phi^{-\nu} + \beta_1\phi^{-\nu+2} + \dots$$
$$+ \gamma_0 s \phi^{-\nu+2} + \gamma_1 s \phi^{-\nu+4} + \dots ; \quad (2.11)$$

the properties of $\hat{u}$ are then a consequence of the formula ([**27**], p.401)

$$\widehat{\phi^{-\mu}}(\sigma) = 2|\sigma|^{-\frac{\mu}{2}}K_{\frac{\mu}{2}}(2\pi|\sigma|) = \pi^{\frac{\mu}{2}}\Gamma(\frac{\mu}{2})\Gamma(\frac{2-\mu}{2}) \times$$
$$\left[-\sum_{m\geq 0} \frac{\pi^{2m}\sigma^{2m}}{m!\,\Gamma(\frac{\mu}{2}+2m+1)} + |\sigma|^{-\mu}\sum_{m\geq 0}\frac{\pi^{-\mu+2m}\sigma^{2m}}{m!\,\Gamma(-\frac{\mu}{2}+2m+1)} \right]. \quad (2.12)$$

The remainder term, on the other hand, shall only contribute to $\hat{u}$ some extra term differentiable near 0 as many times as one pleases. In the reverse direction, in view of the relations

$$\frac{d}{ds}\phi^{-\mu} = -2\pi s \phi^{-\mu+2},$$
$$\frac{d^2}{ds^2}\phi^{-\mu} = -2\pi \phi^{-\mu+2} - 4\pi^2 \phi^{-\mu+4}, \quad (2.13)$$

everything boils down to showing that any sum

$$S_N(\sigma) = a_0 + a_1\sigma + \dots + a_N\sigma^N + |\sigma|^{-\nu}(b_0 + b_1\sigma + \dots + b_N\sigma^N) \quad (2.14)$$

can be written, up to error terms that have the same kind of expansion as the right-hand side of (2.9) and start with terms in σ^{N+1} or $|\sigma|^{-\nu}\sigma^{N+1}$, as linear combinations of $\widehat{\phi^{-\nu}},\dots,\sigma^N\widehat{\phi^{-\nu}},\widehat{\phi^{-\nu+2}},\dots,\sigma^N\widehat{\phi^{-\nu+2}}$. Now this comes essentially from the facts that

$$\frac{\pi^{\frac{1-\nu}{2}}}{\Gamma(-\frac{\nu}{2})}\widehat{\phi^{-\nu}} - \frac{\pi^{\frac{3-\nu}{2}}}{\Gamma(\frac{2-\nu}{2})}\widehat{\phi^{-\nu+2}}$$

has no constant term, and that $\widehat{\phi^{-\nu+2}}$ has no $|\sigma|^{-\nu}$ term.   □

PROPOSITION 2.2. *Assume $-1 <$ Re $\nu < 1$. The map $\theta_{-\nu}$, defined through*

$$\widehat{\theta_{-\nu}u}(\sigma) = |\sigma|^\nu\hat{u}(\sigma), \qquad u \in C^\infty_{-\nu}, \quad (2.15)$$

maps the space $C_{-\nu}^{\infty}$ to the space C_{ν}^{∞}; it intertwines the representation $\pi_{-\nu}$ with the representation π_{ν}. Observe that $\theta_{\nu}\theta_{-\nu} = I$.

PROOF. The first point is a consequence of proposition 2.1. For the second one, first assume that $-1 < \mathrm{Re}\,\nu < 0$. Then, with

$$\omega_{-\nu} = \pi^{-\nu-\frac{1}{2}} \frac{\Gamma(\frac{\nu+1}{2})}{\Gamma(-\frac{\nu}{2})},$$ (2.16)

one may write, if $u \in C_{-\nu}^{\infty}$,

$$(\theta_{-\nu}u)(s) = \omega_{-\nu} \int |s - t|^{-\nu-1} u(t)\, dt,$$ (2.17)

where the integral converges absolutely: indeed, $|u(t)| \le C\,(1+|t|)^{\mathrm{Re}\,\nu-1}$ as a consequence of (2.4). That $\theta_{-\nu}$ acts as an intertwining operator from $C_{-\nu}^{\infty}$ to C_{ν}^{∞} is then the consequence of a formal computation: the general case can be obtained by analytic continuation, noting also that the map $u \mapsto u_1$, with $u_1(s) = (1 + s^2)^{\frac{\mu-\nu}{2}} u(s)$, is an isomorphism from $C_{-\nu}^{\infty}$ to $C_{-\mu}^{\infty}$, so as to give a meaning to the analytic dependence on ν of a function mapped into a ν–dependent space.

$\square$

DEFINITION 2.3. Endow the linear space $C_{-\nu}^{\infty}$ with the Fréchet structure associated with the collection of norms $u \mapsto \|Du\|_{L^2(\mathbb{R},(1+s^2)^{-\mathrm{Re}\,\nu}ds)}$, where D varies in the algebra generated by the operators e_0, e_1, e_2 defined in (2.3). We then define the space $\mathcal{D}_{\nu}'$ as the topological dual of $C_{-\nu}^{\infty}$, and denote as π_{ν} the representation of G on this space defined by the equation

$$< \pi_{\nu}(g)\mathfrak{T}, u > = < \mathfrak{T}, \pi_{-\nu}(g^{-1})u >, \qquad \mathfrak{T} \in \mathcal{D}_{\nu}', \quad u \in C_{-\nu}^{\infty}, \quad g \in G.$$ (2.18)

In the way familiar when dealing with distributions, the two possible definitions of $\pi_{\nu}(g)\mathfrak{T} = \pi_{\nu}(g)f$ would agree in the case when $\mathfrak{T}$ is associated with a function f on $\mathbb{R}$ in the natural way.

DEFINITION 2.4. Given $\mathfrak{T} \in \mathcal{D}_{\nu}'$, we shall denote as $\Theta_{\nu}\mathfrak{T}$ the function on the upper half–plane Π defined as

$$(\Theta_{\nu}\mathfrak{T})(z) = < \mathfrak{T}, \phi_z^{-\nu} >,$$ (2.19)

where $\phi_z^{-\nu}$ was defined in (2.6).

Note that, obviously,

$$(\Theta_{\nu}\pi_{\nu}(g)\mathfrak{T})(z) = (\Theta_{\nu}\mathfrak{T})(g^{-1}.z)$$ (2.20)

for all $g \in G$. The definition (2.19) is only a slight generalization of formula (2) in [10], p.37.

THEOREM 2.5. For any $\mathfrak{T} \in \mathcal{D}_{\nu}'$, there exists N and C such that

$$|(\Theta_{\nu}\mathfrak{T})(z)| \le C(\cosh d(i,z))^N, \qquad z \in \Pi.$$ (2.21)

The function $\Theta_\nu \mathfrak{T}$ is an eigenfunction of Δ (cf. (1.3)) for the eigenvalue $\frac{1-\nu^2}{4}$.

PROOF. Computing the eigenvalues of the matrix $\begin{pmatrix} \frac{1}{y} & -\frac{x}{y} \\ -\frac{x}{y} & \frac{x^2+y^2}{y} \end{pmatrix}$, we get, since $\cosh d(i,z) = \frac{1+y^2+x^2}{2y}$, the inequality

$$(4\cosh d(i,z))^{-1}(1+t^2) \leq \frac{(x-t)^2+y^2}{y} \leq 4\cosh d(i,z)(1+t^2),$$
$$(2.22)$$

from which it follows, applying (2.1), that $\pi_{-\nu}(g)$ has a norm less than $2^{|\operatorname{Re}\nu|}(\cosh d(i,g.i))^{\frac{|\operatorname{Re}\nu|}{2}}$ as an endomorphism of the space $L^2(\mathbb{R},(1+s^2)^{-\operatorname{Re}\nu}ds)$. In particular,

$$\|\phi_z^{-\nu}\| \leq C\,(\cosh d(i,z))^{\frac{|\operatorname{Re}\nu|}{2}} \qquad (2.23)$$

for some constant C, and all $z \in \Pi$. Here and in the rest of the present proof, the norm considered is always that in $L^2(\mathbb{R},(1+s^2)^{-\operatorname{Re}\nu}ds)$. More generally, let D be an element in the algebra referred to in definition 2.3, a linear combination of products of at most k factors, each of which is one of the e_j's. Then $\|D\phi^{-\nu}\| \leq C$ for some C, and (with another constant C)

$$\|\pi_{-\nu}(g)D\pi_{-\nu}(g^{-1})\phi_z^{-\nu}\| = \|\pi_{-\nu}(g)D\phi^{-\nu}\| \leq C(\cosh d(i,z))^{\frac{|\operatorname{Re}\nu|}{2}} \qquad (2.24)$$

if $z = g.i$. Next

$$\pi_{-\nu}(g)e_j\pi_{-\nu}(g^{-1}) = \frac{1}{2i\pi}\frac{d}{dt}\bigg|_{t=0}(\pi_{-\nu}(g)\pi_{-\nu}(\exp t\,\epsilon_j)\pi_{-\nu}(g^{-1})) \quad (2.25)$$

$$= \frac{1}{2i\pi}\frac{d}{dt}\bigg|_{t=0}\pi_{-\nu}(\exp t\,Ad(g)\epsilon_j) \qquad (2.26)$$

is the infinitesimal operator of the representation $\pi_{-\nu}$ associated with the element $Ad(g)\epsilon_j \in \mathfrak{g}$. Finally, with respect to any linear basis of $\mathfrak{g}$, $Ad(g^{-1})$ is represented by a matrix the entries of which are quadratic in terms of the matrix $g = \begin{pmatrix} a & b \\ c & d \end{pmatrix}$: thus, as an endomorphism of $\mathfrak{g}$, $Ad(g^{-1})$ has norm $\leq C(a^2+b^2+c^2+d^2) = 2\,C\cosh d(i,z)$. As a consequence,

$$\|D\phi_z^{-\nu}\| \leq C\,(\cosh d(i,z))^{\frac{|\operatorname{Re}\nu|}{2}+k}, \qquad (2.27)$$

from which the first part of theorem 2.5 follows: the second follows from an examination of the function $\phi_z^{-\nu}$ in (2.6). □

Remark. In a more general context, this correspondence from distributions to joint eigenfunctions of the algebra $D(G/K)$ associated with some symmetric space G/K was studied in [26].

3. Modular distributions

In the present work, we shall set

$$\Gamma = SL(2, \mathbf{Z}) \tag{3.1}$$

and study Γ–invariant distributions. Much of what follows in this section may be considered as just a rephrasing of well–known facts regarding automorphic functions. It is nevertheless unavoidable as has been explained in the introduction, where we noted that switching from automorphic functions to modular distributions and way back is a necessity. One may also note, as a by–product, that statements in terms of modular distributions may involve nicer sets of coefficients (*cf.* propositions 3.8 and 3.9 below) than the usual formulations.

DEFINITION 3.1. *Given $\nu \in \mathbf{C}$, a ν–modular distribution shall be any $\mathfrak{T} \in \mathcal{D}'_\nu$ satisfying the equation*

$$\pi_\nu(g)\mathfrak{T} = \mathfrak{T} \tag{3.2}$$

for all $g \in \Gamma$.

THEOREM 3.2. *Let $\mathfrak{T}$ be a ν–modular distribution. Then the function $\Theta_\nu \mathfrak{T}$, the definition of which is given in definition 2.4, is an eigenfunction of Δ for the eigenvalue $\frac{1-\nu^2}{2}$. It is automorphic, i.e., Γ–invariant on Π. Last, it satisfies the estimate*

$$|(\Theta_\nu \mathfrak{T})(z)| \leq C\,(\mathrm{Im}\ y)^N \tag{3.3}$$

for some N as $|z|$ goes to infinity while z stays within the fundamental domain $|z| \geq 1$, $-\frac{1}{2} \leq \mathrm{Re}\ z \leq \frac{1}{2}$ of Γ.

PROOF. This is a corollary of (2.20) and of theorem 2.5. □

PROPOSITION 3.3. *Let $\nu \in \mathbf{C}$ be given, with $\mathrm{Re}\ \nu < -1$. Consider the distribution $\mathfrak{E}_\nu$ such that*

$$< \mathfrak{E}_\nu, u > \ = \frac{1}{2} \sum_{\substack{|m|+|n|\neq 0}} |m|^{\nu-1} u(\frac{n}{m}) \tag{3.4}$$

for every $u \in C^\infty_{-\nu}$: here, relying on (2.4), we have set, by convention,

$$|m|^{\nu-1}u(\frac{n}{m}) := u_\infty \,|n|^{\nu-1} \qquad \text{in the case when } m = 0,\ n \neq 0. \tag{3.5}$$

It is ν-modular. We shall call it the Eisenstein ν-distribution. One may also write

$$< \mathfrak{E}_\nu, u >= \zeta(1-\nu)[\sum_{m\geq1,(n,m)=1} m^{-1+\nu}u(\frac{n}{m}) + u_\infty]. \qquad (3.6)$$

PROOF. The series converges because, from

$$|u(s)| \leq C|s|^{-1+\text{Re }\nu}, \qquad |s| \to \infty, \qquad (3.7)$$

we get, if $m \geq 1$,

$$\sum_{n\in\mathbf{Z}} |u(\frac{n}{m})| = \sum_{q\in\mathbf{Z}}\sum_{n=0}^{m-1} |u(q+\frac{n}{m})|$$

$$\leq m \sum_{q\in\mathbf{Z}} \sup\{|u(s)| : q \leq s < q+1\}$$

$$\leq C\,m\,[1+\sum_{q\geq1} q^{-1+\text{Re }\nu}] \leq Cm. \qquad (3.8)$$

Also note that the linear form $u \mapsto u_\infty$ is continuous on $C^\infty_{-\nu}$ since $u_\infty = (\pi_{-\nu}(\left(\begin{smallmatrix}0&-1\\1&0\end{smallmatrix}\right))u)(0)$ and the embedding $C^\infty_{-\nu} \subset C^\infty(\mathbf{R})$ is continuous.

Next, observe that, if $\text{Re }\nu < -1$ and $u \in C^\infty_{-\nu}$, the series on the right-hand side of (3.6) can be written as

$$\sum_{d\geq1} d^{\nu-1}[\sum_{m\geq1,(n,m)=1} m^{-1+\nu}u(\frac{n}{m}) + u_\infty]$$

$$= \frac{1}{2}[\sum_{m\neq0,n\in\mathbf{Z}} |m|^{\nu-1}u(\frac{n}{m}) + \sum_{n\neq0,m=0} |n|^{\nu-1}u_\infty]$$

$$= \frac{1}{2}\sum_{|m|+|n|\neq0} |m|^{\nu-1}u(\frac{n}{m})$$

$$=< \mathfrak{E}_\nu, u > . \qquad (3.9)$$

To check that $\mathfrak{E}_\nu$ is ν-modular, we may now use (3.6) as a definition of $< \mathfrak{E}_\nu, u >$. Let $g = \left(\begin{smallmatrix}a&b\\c&d\end{smallmatrix}\right) \in \Gamma$, with $c \neq 0$, and set $v = \pi_{-\nu}(g)u$. Then, assuming $m \geq 1$ and $(n,m) = 1$, one has

$$v(\frac{n}{m}) = m^{1-\nu}|-cn+am|^{-1+\nu}u(\frac{dn-bm}{-cn+am}) \qquad (3.10)$$

in the case when $am - cn \neq 0$; also, when $am - cn = 0$, i.e., $\left(\begin{smallmatrix}m\\n\end{smallmatrix}\right) = \pm\left(\begin{smallmatrix}c\\a\end{smallmatrix}\right)$,

$$v(\frac{n}{m}) = \lim_{s\to\frac{n}{m}} |-cs+a|^{-1+\nu}u_\infty|\frac{ds-b}{-cs+a}|^{-1+\nu}$$

$$= u_\infty m^{1-\nu}|dn-bm|^{-1+\nu}$$

$$= u_\infty |c|^{1-\nu}; \qquad (3.11)$$

finally

$$v_\infty = |c|^{-1+\nu} u(-\frac{d}{c}). \tag{3.12}$$

The first part of the proposition easily follows from the last three equations.

$\square$

PROPOSITION 3.4. *One has*

$$(\Theta_\nu \mathfrak{E}_\nu)(z) = \pi^{\frac{\nu-1}{2}} \Gamma(\frac{1-\nu}{2}) \zeta(1-\nu) E_{\frac{1-\nu}{2}}(z), \qquad z \in \Pi, \tag{3.13}$$

where the right–hand side stands for the classical Eisenstein series so denoted (e.g. [36], p.207).

PROOF. Indeed, as a consequence of (3.6), (2.6) and (2.19), we get, with $z = x + iy$,

$$\frac{\pi^{\frac{1-\nu}{2}}}{\Gamma(\frac{1-\nu}{2}) \zeta(1-\nu)}(\Theta_\nu \mathfrak{E}_\nu)(z) = y^{\frac{1-\nu}{2}} \left[1 + \sum_{m\geq 1, (n,m)=1} m^{-1+\nu} |z - \frac{n}{m}|^{\nu-1}\right]$$

$$= y^{\frac{1-\nu}{2}} \left[1 + \sum_{m\geq 1, (n,m)=1} |mz - n|^{\nu-1}\right]$$

$$= \frac{1}{2} y^{\frac{1-\nu}{2}} \sum_{\substack{m,n\in\mathbf{Z} \\ (n,m)=1}} |mz - n|^{\nu-1}. \tag{3.14}$$

$\square$

DEFINITION 3.5. Given $\nu \in \mathbb{C}$ with $\mathrm{Re}\,\nu < 1$, $\frac{\nu}{2} \notin \mathbf{Z}$, and $u \in C^\infty_{-\nu}$, we shall denote as $\mathrm{res}_0(\hat{u})$ and $\mathrm{res}_{-\nu}(\hat{u})$ the coefficients denoted respectively as a_0 and b_0 in the expansion (2.9) of $\hat{u}$ near 0.

Remark. One may remember $\mathrm{res}_0(\hat{u})$ and $\mathrm{res}_{-\nu}(\hat{u})$ as the "residues" of $\hat{u}$ at 0 of respective weights 0 and $-\nu$. The following proposition will be useful later.

PROPOSITION 3.6. *Let* $g = \left(\begin{smallmatrix} a & b \\ c & d \end{smallmatrix}\right) \in G$, *assume that* $-1 < \mathrm{Re}\,\nu < 1$, *and let* $u \in C^\infty_{-\nu}$. *Let* $v = \pi_{-\nu}(g)u$. *If* $c \neq 0$, *one has*

$$\mathrm{res}_{-\nu}(\hat{v}) = \pi^{\frac{1}{2}-\nu} \frac{\Gamma(\frac{\nu}{2})}{\Gamma(\frac{1-\nu}{2})} |c|^{-1+\nu} u(-\frac{d}{c}),$$

$$\mathrm{res}_0(\hat{v}) = \pi^{\frac{1}{2}+\nu} \frac{\Gamma(-\frac{\nu}{2})}{\Gamma(\frac{1+\nu}{2})} |c|^{-1-\nu} (\theta_{-\nu}u)(-\frac{d}{c}). \tag{3.15}$$

If $c = 0$, one has

$$\mathrm{res}_{-\nu}(\hat{v}) = \pi^{\frac{1}{2}-\nu} \frac{\Gamma(\frac{\nu}{2})}{\Gamma(\frac{1-\nu}{2})} |d|^{-1+\nu} u_\infty \,,$$

$$\mathrm{res}_0(\hat{v}) = \pi^{\frac{1}{2}+\nu} \frac{\Gamma(-\frac{\nu}{2})}{\Gamma(\frac{1+\nu}{2})} |d|^{-1-\nu} (\theta_{-\nu}u)_\infty \,. \tag{3.16}$$

Remark. Observe that the residues at zero of $\hat{v}$ only depend on the class $(\dot{c}\ \dot{d})$ of g in $N \backslash G$.

PROOF. From (2.12) it follows that if $\mu = \nu - 2k$ where k is a positive integer, neither $\hat{\phi}^{-\mu}$ nor $\frac{d}{d\sigma}\hat{\phi}^{-\mu}$ can have any term in $|\sigma|^{-\nu}$ in its expansion near 0. Looking at (2.11), one sees that the coefficient denoted as β_0 there is nothing but u_∞ as defined in (2.4): thus, from (2.11) and (2.12) again (the latter one applied in the case when $\mu = \nu$), one gets

$$\mathrm{res}_{-\nu}(\hat{u}) = \pi^{\frac{1}{2}-\nu} \frac{\Gamma(\frac{\nu}{2})}{\Gamma(\frac{1-\nu}{2})} u_\infty \,. \tag{3.17}$$

Estimating

$$v(s) = |-cs + a|^{-1+\nu} u\left(\frac{ds - b}{-cs + a}\right) \tag{3.18}$$

as $|s|$ goes to infinity, one gets

$$v_\infty = |c|^{\nu-1} u\left(-\frac{d}{c}\right) \qquad \text{if } c \neq 0 \,,$$

$$v_\infty = |a|^{\nu-1} |\frac{d}{a}|^{\nu-1} u_\infty = |d|^{\nu-1} u_\infty \qquad \text{if } c = 0 \,. \tag{3.19}$$

From the definition of $\theta_{-\nu}$ in proposition 2.2, it follows that

$$\mathrm{res}_0(\hat{u}) = \mathrm{res}_\nu(\widehat{\theta_{-\nu}u}) \,. \tag{3.20}$$

Thus

$$\mathrm{res}_0(\hat{v}) = \mathrm{res}_0(\widehat{\pi_{-\nu}(g)u}) = \mathrm{res}_\nu(\widehat{\theta_{-\nu}\pi_{-\nu}(g)u})$$

$$= \mathrm{res}_\nu(\widehat{\pi_\nu(g)\theta_{-\nu}u}) \tag{3.21}$$

since $\theta_{-\nu}$ is an intertwining operator. This makes it possible to reduce the computation of $\mathrm{res}_0(\hat{v})$ to that already made. □

PROPOSITION 3.7. *Given* $\nu \in \mathbf{C}$ *with* $\mathrm{Re}\ \nu < 1$, $\frac{\nu}{2} \notin \mathbf{Z}$, *and a* ν–*modular distribution* $\mathfrak{T}$, *there exists a unique collection of coefficients* $\{C_0, C_\infty, (c_n)_{n \in \mathbf{Z}^\times}\}$, *with the sequence* (c_n) *having at most polynomial increase as* $|n| \to \infty$, *such that*

$$< \mathfrak{T}, u >= C_0 \mathrm{res}_0(\hat{u}) + C_\infty \mathrm{res}_{-\nu}(\hat{u}) + \sum_{n \neq 0} c_n \hat{u}(n) \tag{3.22}$$

for all $u \in C^\infty_{-\nu}$. *The function* $\Theta_\nu \mathfrak{T}$ *is a (Maass) cusp-form if and only if the "cuspidal" coefficients* C_0 *and* C_∞ *vanish.*

PROOF. The series on the right–hand side makes sense, and defines a continuous linear form on $C_{-\nu}^{\infty}$, in view of proposition 2.1: by a density argument, it would then be sufficient to check that the identity above (for some fixed set of coefficients) is indeed true whenever $u = \phi_z^{-\nu}$ for some $z \in \Pi$. From (2.12), it follows that

$$\mathrm{res}_0(\hat{\phi}^{-\nu}) = \pi^{\frac{\nu}{2}} \Gamma(-\frac{\nu}{2}), \qquad \mathrm{res}_{-\nu}(\hat{\phi}^{-\nu}) = \pi^{-\frac{\nu}{2}} \Gamma(\frac{\nu}{2}). \tag{3.23}$$

Since

$$\phi_z^{-\nu}(s) = y^{\frac{\nu-1}{2}} \phi^{-\nu}(\frac{s-x}{y}), \tag{3.24}$$

(2.12) yields

$$\hat{\phi}_z^{-\nu}(\sigma) = 2 y^{\frac{1}{2}} e^{-2i\pi x\sigma} |\sigma|^{-\frac{\nu}{2}} K_{\frac{\nu}{2}}(2\pi|\sigma|y) \tag{3.25}$$

and

$$\mathrm{res}_0(\hat{\phi}_z^{-\nu}) = \pi^{\frac{\nu}{2}} \Gamma(-\frac{\nu}{2}) y^{\frac{\nu+1}{2}}, \qquad \mathrm{res}_{-\nu}(\hat{\phi}_z^{-\nu}) = \pi^{-\frac{\nu}{2}} \Gamma(\frac{\nu}{2}) y^{\frac{-\nu+1}{2}}. \tag{3.26}$$

What we are looking for is thus an expansion

$$< \mathfrak{T}, \phi_z^{-\nu} > = C_0 \, \pi^{\frac{\nu}{2}} \Gamma(-\frac{\nu}{2}) y^{\frac{\nu+1}{2}} + C_\infty \, \pi^{-\frac{\nu}{2}} \Gamma(\frac{\nu}{2}) y^{\frac{-\nu+1}{2}}$$
$$+ 2 \sum_{n \neq 0} c_n e^{-2i\pi nx} y^{\frac{1}{2}} |n|^{-\frac{\nu}{2}} K_{\frac{\nu}{2}}(2\pi|n|y). \tag{3.27}$$

In view of theorem 3.2, which states that $(\Theta_\nu \mathfrak{T})(z) = < \mathfrak{T}, \phi_z^{-\nu} >$ is an automorphic function with a decent behaviour at infinity, one sees that this expansion is a very classical one (cf. [36], p.208). The last part follows from (3.26) as well.   □

PROPOSITION 3.8. *Let* Re $\nu < -1$. *The Eisenstein distribution* $\mathfrak{E}_\nu$ (*cf.* (3.4) *or* (3.6)) *can be defined, in terms of the Fourier transform of* $u \in C_{-\nu}^{\infty}$, *as*

$$< \mathfrak{E}_\nu, u > = \zeta(-\nu) \mathrm{res}_0(\hat{u}) + \zeta(\nu) \mathrm{res}_{-\nu}(\hat{u}) + \sum_{n \neq 0} \sigma_\nu(|n|) \hat{u}(n), \tag{3.28}$$

where $\sigma_\nu(|n|)$ *denotes, as usually, the sum of the ν–th powers of all positive divisors of* $|n|$.

PROOF. We may assume Re $\nu < -1$, $\nu \notin \mathbf{Z}$. It suffices to prove (3.28) in the following two cases: (i) when $u = \phi^{-\nu}$; (ii) when $u_\infty = 0$. In view of (2.19), (3.13), the first case follows from the classical expansion

$$\pi^{-s} \Gamma(s) \zeta(2s) E_s(z) = y^s \pi^{-s} \Gamma(s)\zeta(2s) + y^{1-s}\pi^{s-1}\Gamma(1-s)\zeta(2-2s)$$
$$+ 2 \sum_{n \neq 0} |n|^{s-\frac{1}{2}} \sigma_{1-2s}(|n|) y^{\frac{1}{2}} K_{s-\frac{1}{2}}(2\pi|n|y) e^{2i\pi nx}. \tag{3.29}$$

In the case when $u_\infty = 0$, *i.e.*, $\mathrm{res}_{-\nu}(\hat{u}) = 0$ (*cf.* (3.17)), one has

$$(\zeta(1-\nu))^{-1} < \mathfrak{E}_\nu, u > = \sum_{\substack{d \geq 1 \\ (c,d)=1}} d^{\nu-1} u\left(\frac{c}{d}\right)$$

$$= \sum_{c \in \mathbf{Z}} u(c) + \sum_{d \geq 2} d^{\nu-1} \sum_{\substack{0 < c < d \\ (c,d)=1}} \sum_{m \in \mathbf{Z}} u\left(m + \frac{c}{d}\right)$$

$$= \sum_{n \in \mathbf{Z}} \hat{u}(n) \left[1 + \sum_{d \geq 2} d^{\nu-1} \sum_{\substack{0 < c < d \\ (c,d)=1}} e^{2i\pi n \frac{c}{d}} \right]$$

$$= \sum_{n \in \mathbf{Z}} \hat{u}(n) \sum_{d \geq 1} d^{\nu-1} c_d(|n|), \qquad (3.30)$$

where we have used Ramanujan's symbol ([**16**, p. 56]). Denoting as ϕ the Euler function and using [**16**], theorem 292, we get

$$(\zeta(1-\nu))^{-1} < \mathfrak{E}_\nu, u > = \hat{u}(0) \sum_{d \geq 1} d^{\nu-1} \phi(d) + (\zeta(1-\nu))^{-1} \sum_{n \in \mathbf{Z}^\times} \hat{u}(n) \, \sigma_\nu(|n|), \qquad (3.31)$$

and the first term on the right–hand side reduces to $\hat{u}(0) \frac{\zeta(-\nu)}{\zeta(1-\nu)}$ according to [**16**], theorem 288. □

One may observe that (3.28) and (3.29) are essentially equivalent. Still, the coefficients in (3.28) look simpler, which is due to the fact that, by concentrating on the distribution $\mathfrak{E}_\nu$ rather than on its image under Θ_ν, what we have done is tantamount to what would be, in an adelic setting, isolating the archimedean place from the other ones. This is why the two factors $\zeta(1-\nu)$ and $\pi^{\frac{\nu-1}{2}} \Gamma(\frac{1-\nu}{2})$ appear in separate ways, the first one in the definition of $\mathfrak{E}_\nu$, the second one in that of $\phi_z^{-\nu}$.

Proposition 3.8, together with proposition 2.1, permits to define, in particular when $\mathrm{Re}\ \nu < 1$, the analytic continuation of the distribution-valued function $\nu \mapsto \mathfrak{E}_\nu$. One may note that an extension of proposition 2.1, disregarding the assumption $\mathrm{Re}\ \nu < 1$, would be an easy matter if one would agree to consider distributions rather than locally summable functions v in the range of the Fourier transformation.

PROPOSITION 3.9. *Assuming* $-1 < \mathrm{Re}\ \nu < 1$, *define the intertwining operator* $\theta_\nu : \mathcal{D}'_\nu \to \mathcal{D}'_{-\nu}$ *by the equation*

$$< \theta_\nu \mathfrak{T}, u > = \ < \mathfrak{T}, \theta_\nu u >, \qquad u \in C^\infty_\nu, \qquad (3.32)$$

where, in its occurrence on the right–hand side, $\theta_\nu : C^\infty_\nu \to C^\infty_{-\nu}$ *is defined by means of proposition 2.2. Then the* ν-modular distribution $\mathfrak{E}_\nu$ satisfies *the equation*

$$\theta_\nu \mathfrak{E}_\nu = \mathfrak{E}_{-\nu} . \tag{3.33}$$

PROOF. This is a consequence of (3.28), together with the equations $\widehat{\theta_\nu u}(\sigma) = |\sigma|^{-\nu} \hat{u}(\sigma)$ and $\sigma_\nu(|n|) = |n|^\nu \sigma_{-\nu}(|n|)$. $\qquad\square$

Remark. Since $\theta_\nu \phi_z^\nu = \phi_z^{-\nu}$ as a consequence of (2.15) and (2.12), one can see, with the help of proposition 3.4, that the usual functional equation of Eisenstein's series is an immediate consequence of the way $\mathfrak{E}_\nu$ and ϕ_z^ν transform under the intertwining operator θ_ν.

4. The principal series of $SL(2,\mathbf{R})$ and the Radon transform

We parametrize the generic elements of the subroups N, A, K occurring in the Iwasawa decomposition of $G = SL(2,\mathbf{R})$ as

$$n = \begin{pmatrix} 1 & b \\ 0 & 1 \end{pmatrix}, \quad a = \begin{pmatrix} e^{\frac{r}{2}} & 0 \\ 0 & e^{-\frac{r}{2}} \end{pmatrix}, \quad k = \begin{pmatrix} \cos\frac{\theta}{2} & \sin\frac{\theta}{2} \\ -\sin\frac{\theta}{2} & \cos\frac{\theta}{2} \end{pmatrix}. \tag{4.1}$$

Following the normalizations in ([17],ch.II,§3), we set $dn = \pi^{-1}db$, $da = \pi\,dr$, $dk = (4\pi)^{-1}d\theta$, which corresponds to the choice of the Haar measure

$$dg = e^{-2\rho(\log a)}\,dn\,da\,dk \tag{4.2}$$

on $G = NAK$. Recall that ρ, the positive half-root, is the element of $\mathfrak{a}^*$ characterized by $\rho\left(\begin{pmatrix} 1 & 0 \\ 0 & -1 \end{pmatrix}\right) = 1$.

The homogeneous space G/K is identified with the Poincaré upper half-plane Π in the standard way: its base point is i. Since the class of $g \in G$ in G/N is characterized by the left column of the matrix g, the space $\Xi = G/MN$, with $M = \{\pm I\}$, can be identified with the quotient of $\mathbf{R}^2\backslash\{0\}$ by the equivalence relation which identifies two points the negative of each other: we denote the generic point of Ξ as $\xi = \pm\begin{pmatrix} \xi_1 \\ \xi_2 \end{pmatrix}$. Choosing $\pm\begin{pmatrix} 1 \\ 0 \end{pmatrix}$ as the base-point ξ^o in Ξ, we have $g.\xi^o = \pm\begin{pmatrix} a \\ c \end{pmatrix}$ if $g = \begin{pmatrix} a & b \\ c & d \end{pmatrix}$. Still following Helgason's normalizations, one can see that the measures $d\mu(z)$ and $d\xi$ on Π and Ξ respectively have to be normalized in the standard way (i.e., $d\mu(x + iy) = y^{-2}dx\,dy$) for the first one, and by

$$\int_\Xi h(\xi)\,d\xi = \int_{\mathbf{R}^2\backslash\{0\}} h\left(\pm\begin{pmatrix} \xi_1 \\ \xi_2 \end{pmatrix}\right)\,d\xi_1\,d\xi_2 \tag{4.3}$$

for the second one. One then defines the Radon transform V from functions f on Π to functions on Ξ by the formula

$$(Vf)(g.\xi^o) = \int_N f((gn).i)\,dn \tag{4.4}$$

and its dual transform V^* (from functions h on Ξ to functions on Π) as

$$(V^*h)(g.i) = \int_K h((gk).\xi^o)\,dk. \tag{4.5}$$

Recalling that, on Π, the hyperbolic distance is characterized by $\cosh d(i, x + iy) = \frac{1+x^2+y^2}{2y}$, it is easily seen [41] that the integral defining $(Vf)(g.\xi^o)$ converges pointwise if, say, $|f(z)| \leq C(\cosh d(i, z))^{-\mu}$ for some $\mu > \frac{1}{2}$ and $C > 0$, yielding then a continuous function $Vf \in L^2(\Xi)$; the integral (4.5) converges pointwise as soon as h is continuous.

The two operators V and V^* are formally adjoint of each other under the given normalizations of the measures on Π and Ξ. With the choice of coordinates on Π and Ξ made above, one has

$$(Vf)\left(\pm\begin{pmatrix}\xi_1\\\xi_2\end{pmatrix}\right) = \frac{1}{\pi}\int_{-\infty}^{\infty} f\left(\frac{\xi_1^2(i+b)}{\xi_1\xi_2(i+b)+1}\right) db \qquad (4.6)$$

and

$$(V^*h)(z) = \frac{1}{2\pi}\int_0^{2\pi} h\left(\pm\begin{pmatrix} y^{\frac{1}{2}}\cos\frac{\theta}{2} - xy^{-\frac{1}{2}}\sin\frac{\theta}{2} \\ -y^{-\frac{1}{2}}\sin\frac{\theta}{2}\end{pmatrix}\right) d\theta. \qquad (4.7)$$

Identifying $L^2(\Xi)$ with $L_{\text{even}}^2(\mathbf{R}^2\setminus\{0\}) = L_{\text{even}}^2(\mathbf{R}^2)$, we define on this space the essentially self–adjoint Euler operator (with initial domain $C_0^\infty(\mathbf{R}^2\setminus\{0\})$)

$$\mathcal{E} = \frac{1}{2i\pi}(\xi_1\frac{\partial}{\partial\xi_1} + \xi_2\frac{\partial}{\partial\xi_2} + 1): \qquad (4.8)$$

one then has

$$(t^{2i\pi\mathcal{E}}h)(\xi) = t\, h(t\xi) \qquad (4.9)$$

for every $h \in L^2(\Xi)$ and $t > 0$, which makes it possible to define in particular the operator

$$T = \left(\frac{\pi}{2}\right)^{\frac{1}{2}} \frac{\Gamma(\frac{1}{2} - i\pi\mathcal{E})}{\Gamma(-i\pi\mathcal{E})} = \pi^{-\frac{1}{2}}(-i\pi\mathcal{E})\int_0^{\infty} t^{-\frac{1}{2}}(1+t)^{-1+i\pi\mathcal{E}} dt. \qquad (4.10)$$

On $L^2(\mathbf{R}^2\setminus\{0\})$, we shall always use the *symplectic* Fourier transformation $\mathcal{F}$ characterized by

$$(\mathcal{F}h)(\xi) = \int_{\mathbf{R}^2} h(\eta)e^{-2i\pi[\xi,\eta]} d\eta \qquad (4.11)$$

with $[\xi,\eta] = -\xi_1\eta_2 + \xi_2\eta_1$: it commutes with the quasi–regular action of G on $L^2(\mathbf{R}^2\setminus\{0\})$ and satisfies $\mathcal{F}^2 = I$. Also note that $\mathcal{F}\mathcal{E} = -\mathcal{E}\mathcal{F}$. We shall also need (especially in sections 5 and 13, in relation with the Weyl calculus) the transformation $\mathcal{G} = 2^{2i\pi\mathcal{E}}\mathcal{F}$, i.e.,

$$(\mathcal{G}h)(\xi) = 2\int_{\mathbf{R}^2} h(\eta)e^{-4i\pi[\xi,\eta]} d\eta. \qquad (4.12)$$

We then set, as an operator on $L^2(\mathbf{R}^2\setminus\{0\})$,

$$\kappa = \frac{\Gamma(\frac{1}{2} + i\pi\mathcal{E})}{\Gamma(\frac{1}{2} - i\pi\mathcal{E})}(2\pi)^{-2i\pi\mathcal{E}}\mathcal{G}, \qquad (4.13)$$

an involutive unitary transformation. To understand κ, one may note ([41], prop. 4.1) that if $h(\xi) = h_0(|\xi|^2)$, then $(\kappa h)(\xi) = |\xi|^{-2}h_0(|\xi|^{-2})$.

The slight difference between the definition of κ in (*loc.cit.*) and our present one, *to wit* the substitution of $\pi^{-2i\pi\mathcal{E}}\mathcal{G} = \pi^{-i\pi\mathcal{E}}\mathcal{G}\pi^{i\pi\mathcal{E}}$ for $\mathcal{G}$, may be accounted for by the fact that our new coordinates $\pm\xi$ on Ξ are linked to the coordinates $\pm x$ used in the former reference by the equation $\xi = \pi^{\frac{1}{2}}x$: this change has been made to agree with Helgason's normalization, and also explains the difference, by factors of powers of π, in the definition of T above, or in that of $\mathcal{R}_{\tau+1}$ (*cf.infra*, (17.19)). Recall the definition (1.3) of the hyperbolic Laplacian Δ on Π.

THEOREM 4.1. *The unitary transformation TV, initially defined on the space of continuous functions on Π with a compact support, extends as an isometry from $L^2(\Pi)$ onto the subspace $\mathrm{Ran}(TV)$ of $L^2_{\mathrm{even}}(\mathbb{R}^2\backslash\{0\})$ consisting of all functions invariant under the symmetry $T\kappa T^{-1}$. The operator V^*T^* extends on $\mathrm{Ran}(TV)$ as the inverse of TV, and is zero on the subspace $(\mathrm{Ran}(TV))^\perp$ of $L^2_{\mathrm{even}}(\mathbb{R}^2\backslash\{0\})$ consisting of all functions h with $T\kappa T^{-1}h = -h$. Moreover, the isometry TV intertwines the two quasi-regular actions of G on $L^2(\Pi)$ and $L^2(\Xi)$ respectively, and transforms the operator $\Delta - \frac{1}{4}$ on $L^2(\Pi)$ into the operator $\pi^2 \mathcal{E}^2$ on $L^2(\Xi)$.*

PROOF. The computation of the norm of TVf, $f \in L^2(\Pi)$, can be taken from ([17], ch.II, §3), and the last part is a precise way to state the well-known isomorphism $D(G/K) \simeq D_W(A)$ (*loc.cit.*, p.87). The full theorem, including this characterization of the range of TV, is just theorem 5.1 in [41].

$\square$

As a consequence of spherical representation theory, if $f \in C_0^\infty(\Pi)$, one may write (Mehler's formula)

$$f(z) = \int_0^\infty f_\lambda(z) \left(\frac{\pi\lambda}{2} \tanh \frac{\pi\lambda}{2}\right) d\lambda \qquad (4.14)$$

with

$$f_\lambda(z) = \frac{1}{4\pi^2} \int_\Pi f(w)\,\mathfrak{P}_{-\frac{1}{2}+\frac{i\lambda}{2}}(\cosh d(z,w))\,d\mu(w), \qquad (4.15)$$

where $\mathfrak{P}_{-\frac{1}{2}+\frac{i\lambda}{2}}$ is the usual Legendre function, d stands for the geodesic distance on Π, and the numerical factor on the right-hand side of (4.14) is $|c(\frac{i\lambda}{2})|^{-2}$ in terms of Harish-Chandra's c-function. Let $\mathcal{H}_{i\lambda}$ be the completion of the space of all f_λ ($f \in C_0^\infty(\Pi)$) under the norm

$$\|f_\lambda\|^2_{\mathcal{H}_{i\lambda}} = (4\pi^2)^{-2} \int_{\Pi\times\Pi} f(z)\,\bar{f}(w)\,\mathfrak{P}_{-\frac{1}{2}+\frac{i\lambda}{2}}(\cosh d(z,w))\,d\mu(z)\,d\mu(w)$$

$$= (4\pi^2)^{-1}\,(f, f_\lambda)_{L^2(\Pi)}. \qquad (4.16)$$

Then

$$\|f\|^2_{L^2(\Pi)} = 4\pi^2 \int_0^\infty \|f_\lambda\|^2_{\mathcal{H}_{i\lambda}} \left(\frac{\pi\lambda}{2} \tanh \frac{\pi\lambda}{2}\right) d\lambda, \qquad (4.17)$$

and since

$$\Delta f_\lambda = \frac{1}{4}(1 + \lambda^2)f_\lambda, \qquad (4.18)$$

(4.14) and (4.17) express the decomposition of $L^2(\Pi)$ as a Hilbert direct integral of eigenspaces of Δ.

On the other hand, given $h \in L^2(\Xi) = L^2_{\text{even}}(\mathbb{R}^2\backslash\{0\})$, one has

$$h = \int_{-\infty}^{\infty} h_\lambda \, d\lambda \tag{4.19}$$

with

$$h_\lambda(\xi) = \frac{1}{2\pi} \int_0^{\infty} t^{i\lambda} h(t\xi) \, dt, \tag{4.20}$$

where, for every $\lambda \in \mathbb{R}$, h_λ belongs to the space $\text{Hom}_{\text{even}}^{-1-i\lambda}$ of (even) homogeneous functions of degree $-1 - i\lambda$. Note that the subscript λ refers to the spectral decomposition of $\mathcal{E}$ in (4.19), to that of Δ in (4.14): no confusion should arise since one is dealing with functions on $\mathbb{R}^2\backslash\{0\}$ in the first case, with functions on Π in the second one. Even functions, homogeneous of a given degree, can be identified with functions on the real line through the correspondence $h_\lambda \mapsto h_\lambda^\flat$, where

$$h_\lambda^\flat(s) = h_\lambda(s, 1). \tag{4.21}$$

Then, under the decomposition (4.19), one has

$$\|h\|_{L^2(\mathbb{R}^2\backslash\{0\})}^2 = 4\pi \int_{-\infty}^{\infty} \|h_\lambda^\flat\|_{L^2(\mathbb{R})}^2 \, d\lambda. \tag{4.22}$$

The quasi–regular representation π of G in $L^2_{\text{even}}(\mathbb{R}^2\backslash\{0\})$ defined by

$$(\pi(g)h)(\xi) = h(g^{-1}\xi) \tag{4.23}$$

decomposes under (4.19) as the sum of the representations $\pi_{i\lambda}$, $\lambda \in \mathbb{R}$, defined in (1.2). Indeed, for every λ,

$$(\pi(g)h)_\lambda^\flat(s) = |-cs + a|^{-1-i\lambda} h_\lambda^\flat \left(\frac{ds - b}{-cs + a} \right), \quad g = \begin{pmatrix} a & b \\ c & d \end{pmatrix} \tag{4.24}$$

as is readily seen from (4.21). Also, $(\mathcal{F}h)_{-\lambda} = \mathcal{F}h_\lambda$, from which it is immediate (inverting (4.21)) that

$$(\mathcal{F}h)_{-\lambda}^\flat(s) = \int_{\mathbb{R}^2\backslash\{0\}} |\eta_2|^{-i\lambda} h_\lambda^\flat(t) e^{2i\pi(s-t)\eta_2} \, dt \, d\eta_2. \tag{4.25}$$

Now, denote as $u \mapsto \hat{u}$ the usual Fourier transformation on $L^2(\mathbb{R})$, normalized as

$$\hat{u}(\sigma) = \int u(s) e^{-2i\pi s\sigma} \, ds. \tag{4.26}$$

Given $\lambda \in \mathbb{R}$, the unitary map $\theta_{i\lambda} \colon L^2(\mathbb{R}) \to L^2(\mathbb{R})$, defined by

$$\widehat{\theta_{i\lambda}u}(\sigma) = |\sigma|^{-i\lambda} \hat{u}(\sigma), \tag{4.27}$$

is the canonical intertwining operator from the Hilbert space of the representation $\pi_{i\lambda}$ to that of the representation $\pi_{-i\lambda}$. Thus (4.25) just expresses that

$$(\mathcal{F}h)^b_{-\lambda} = \theta_{i\lambda} h^b_\lambda. \tag{4.28}$$

Let us finally link the decompositions (4.14) and (4.19) on one hand, (4.17) and (4.22) on the other, by means of theorem 4.1: making this link explicit will bring to light a few formulas that will be needed in the sequel. First observe that the interval of integration is not the same in (4.14), (4.17) as in (4.19), (4.22): this is related, as indicated by the last part of theorem 4.1, to the Weyl group of G having two elements. Let $f \in C_0^\infty(\Pi)$ and let $h = TVf$. Since, as a consequence of (4.10) and (4.13), the involutive transformation $T\kappa T^{-1}$ exchanges the spaces $\mathrm{Hom}_{\mathrm{even}}^{-1-i\lambda}$ and $\mathrm{Hom}_{\mathrm{even}}^{-1+i\lambda}$, one has

$$T\kappa T^{-1}(h_\lambda - h_{-\lambda}) = (T\kappa T^{-1}h)_{-\lambda} - (T\kappa T^{-1}h)_\lambda = h_{-\lambda} - h_\lambda,$$

so that $h_\lambda - h_{-\lambda}$ belongs to the kernel of V^*T^* in view of theorem 4.1. Thus

$$f = V^*T^*TVf = \int_{-\infty}^\infty V^*T^*(TVf)_\lambda \, d\lambda = 2\int_0^\infty V^*T^*(TVf)_\lambda \, d\lambda. \tag{4.29}$$

Now, on the space $\mathrm{Hom}_{\mathrm{even}}^{-1-i\lambda}$, T^*T acts as the scalar

$$\left(\frac{\pi}{2}\right)^{\frac12}\frac{\Gamma\left(\frac{1+i\lambda}{2}\right)}{\Gamma\left(\frac{i\lambda}{2}\right)} \times \left(\frac{\pi}{2}\right)^{\frac12}\frac{\Gamma\left(\frac{1-i\lambda}{2}\right)}{\Gamma\left(\frac{-i\lambda}{2}\right)} = \frac{\pi\lambda}{4}\tanh\frac{\pi\lambda}{2}, \tag{4.30}$$

hence

$$f = \int_0^\infty \left(\frac{\pi\lambda}{2}\tanh\frac{\pi\lambda}{2}\right) V^*(Vf)_\lambda \, d\lambda. \tag{4.31}$$

Also, noting as a consequence of (4.10) that T acts on $(Vf)_\lambda$ as the scalar $\left(\frac{\pi}{2}\right)^{\frac12}\frac{\Gamma(\frac12+\frac{i\lambda}{2})}{\Gamma(\frac{i\lambda}{2})}$ and using (4.21), (4.20) and (4.6) as well, one gets for almost all λ that

$$(TVf)^b_\lambda(s) = (2\pi)^{-\frac32}\frac{\Gamma(\frac12+\frac{i\lambda}{2})}{\Gamma(\frac{i\lambda}{2})}\int_0^\infty t^{i\lambda-2}\, dt \int_{-\infty}^\infty f(\frac{s^2(i+b)}{s(i+b)+t^2})\, db: \tag{4.32}$$

performing the change of variables characterized by

$$z = \frac{s^2(i+b)}{s(i+b)+t^2}, \qquad d\mu(z) = \frac{2\, dt\, db}{t} \tag{4.33}$$

so that $t^2 = \frac{|z-s|^2}{y}$, one gets

$$(TVf)^b_\lambda(s) = \frac12(2\pi)^{-\frac32}\frac{\Gamma(\frac12+\frac{i\lambda}{2})}{\Gamma(\frac{i\lambda}{2})}\int_\Pi \left(\frac{|z-s|^2}{y}\right)^{-\frac12-\frac{i\lambda}{2}} f(z)\, d\mu(z). \tag{4.34}$$

In the reverse direction, invert for fixed λ the map $h_\lambda \mapsto h_\lambda^\flat$ defined in (4.21) as $u_\lambda \mapsto u_\lambda^\sharp$ with

$$u_\lambda^\sharp(\xi) = |\xi_2|^{-1-i\lambda} \, u_\lambda(\frac{\xi_1}{\xi_2}). \tag{4.35}$$

From (4.10) and (4.7) it then follows (setting $s = -y \cot\frac{\theta}{2} + x$ in the latter formula) that

$$(V^* T^* u_\lambda^\sharp)(z) = (2\pi)^{-\frac{1}{2}} \frac{\Gamma(\frac{1}{2} - \frac{i\lambda}{2})}{\Gamma(-\frac{i\lambda}{2})} \int_{-\infty}^{\infty} u_\lambda(s) \left(\frac{|z-s|^2}{y} \right)^{-\frac{1}{2} + \frac{i\lambda}{2}} ds. \tag{4.36}$$

Observe that, if $u_\lambda \in L^2(\mathbb{R})$, the integral on the right–hand side is at most $\pi^{\frac{1}{2}} \|u_\lambda\|$ in view of the Cauchy-Schwarz inequality.

From (4.34) and (4.36), still under the assumption that $f \in C_0^\infty(\Pi)$, one gets that

$$(V^*(Vf)_\lambda)(z) = \frac{1}{4\pi^3} \int_\Pi f(w) \, d\mu(w) \int_{-\infty}^{\infty} \left(\frac{|z-s|^2}{\text{Im } z} \right)^{-\frac{1}{2}+\frac{i\lambda}{2}} \left(\frac{|w-s|^2}{\text{Im } w} \right)^{-\frac{1}{2}-\frac{i\lambda}{2}} ds. \tag{4.37}$$

Now

$$\frac{1}{\pi} \int_{-\infty}^{\infty} \left(\frac{|z-s|^2}{\text{Im } z} \right)^{-\frac{1}{2}+\frac{i\lambda}{2}} \left(\frac{|w-s|^2}{\text{Im } w} \right)^{-\frac{1}{2}-\frac{i\lambda}{2}} ds = \mathfrak{P}_{-\frac{1}{2}+\frac{i\lambda}{2}}(\cosh d(z,w)) \tag{4.38}$$

as a consequence of Plancherel's formula together with the identities ([27], p.401)

$$\pi^{-\frac{1}{2}} \int_{-\infty}^{\infty} \left(\frac{|z-s|^2}{y} \right)^{-\frac{1}{2}+\frac{i\lambda}{2}} e^{-2i\pi s\sigma} \, ds$$

$$= y^{\frac{1}{2}} e^{-2i\pi\sigma x} \frac{2\pi^{-\frac{i\lambda}{2}}}{\Gamma(\frac{1}{2} - \frac{i\lambda}{2})} |\sigma|^{-\frac{i\lambda}{2}} K_{\frac{i\lambda}{2}}(2\pi|\sigma|y) \tag{4.39}$$

and ([27], p.413)

$$\int_0^{\infty} K_{\frac{i\lambda}{2}}(2\pi\sigma \, \text{Im } z) \, K_{\frac{i\lambda}{2}}(2\pi\sigma \, \text{Im } w) \cos(2\pi\sigma \, \text{Re }(z-w)) \, d\sigma$$

$$= \frac{1}{8}(\text{Im } z \, \text{Im } w)^{-\frac{1}{2}} \, \Gamma(\frac{1}{2} + \frac{i\lambda}{2}) \Gamma(\frac{1}{2} - \frac{i\lambda}{2}) \, \mathfrak{P}_{-\frac{1}{2}+\frac{i\lambda}{2}}(\cosh d(z,w)). \tag{4.40}$$

Thus

$$(V^*(Vf)_\lambda)(z) = \frac{1}{4\pi^2} \int_\Pi f(w) \, \mathfrak{P}_{-\frac{1}{2}+\frac{i\lambda}{2}}(\cosh d(z,w)) \, d\mu(w)$$

$$= f_\lambda(z) \tag{4.41}$$

so that (4.31) (which is nothing but the image of the decomposition (4.19) under TV) is not distinct from (4.14).

We finally prove that, for arbitrary $\lambda \in \mathbf{R}$ and $f \in C_0^\infty(\Pi)$, one has

$$\|(TVf)_\lambda^\flat\|_{L^2(\mathbf{R})}^2 = \frac{\pi}{2}\left(\frac{\pi\lambda}{2}\tanh\frac{\pi\lambda}{2}\right)\|f_\lambda\|_{\mathcal{H}_{i\lambda}}^2 \qquad (4.42)$$

from which it follows that theorem 4.1, again, connects (4.17) and (4.22). Indeed, from (4.34) and (4.30),

$$\|(TVf)_\lambda^\flat\|_{L^2(\mathbf{R})}^2 = \frac{1}{32\pi^3}\left(\frac{\lambda}{2}\tanh\frac{\lambda}{2}\right)$$

$$\int f(z)\bar{f}(w)\,d\mu(z)\,d\mu(w)\int_{-\infty}^\infty \left(\frac{|z-s|^2}{\operatorname{Im} z}\right)^{-\frac{1}{2}-\frac{i\lambda}{2}}\left(\frac{|w-s|^2}{\operatorname{Im} w}\right)^{-\frac{1}{2}+\frac{i\lambda}{2}} ds, \qquad (4.43)$$

a formula that only needs being compared to (4.16) with the help of (4.38).

Using (4.36) together with (2.19) and (2.6), one sees that

$$(V^*T^*u_\lambda^\natural)(z) = \frac{2^{-\frac{1}{2}}\pi^{-\frac{i\lambda}{2}}}{\Gamma(-\frac{i\lambda}{2})}(\Theta_{i\lambda}u_\lambda)(z) \qquad (4.44)$$

so that $\Theta_{i\lambda}$ has a (one-sided) inverse $\Theta_{i\lambda}^{-1}$ defined as

$$\Theta_{i\lambda}^{-1}f = (SVf)_\lambda^\flat, \qquad (4.45)$$

where

$$S: = 2^{-\frac{1}{2}}\pi^{i\pi\mathcal{E}}(\Gamma(i\pi\mathcal{E}))^{-1}T. \qquad (4.46)$$

Actually, the map in (4.45) has a much more important role than that of a simple inverse of $\Theta_{i\lambda}$ since f need not be an eigenfunction of Δ: this map really provides the spectral decomposition of f, viewed through the Radon transformation. Now theorem 2.5 or theorem 3.2 permits to give a meaning to the right–hand side of (4.45) in the case when u_λ is replaced by an $i\lambda$–modular distribution. It is more difficult to extend the meaning of (4.34) (or (4.45)) to the case when $f \in L^2(\Gamma\backslash\Pi)$: this is the object of section 6.

5. Another look at the composition of Weyl symbols

Recall that the Weyl calculus of operators on $L^2(\mathbb{R})$ is the rule that associates with a function h on $\mathbb{R}^2$ (a *symbol*) the operator $\mathrm{Op}(h)$ defined as

$$(\mathrm{Op}(h)u)(x) = \int h(\frac{x+y}{2}, \eta)\, u(y)\, e^{2i\pi(x-y)\eta}\, dy\, d\eta, \qquad u \in L^2(\mathbb{R}). \tag{5.1}$$

The problem we solve in this section is how to make the composition $(h_1, h_2) \mapsto h_1 \# h_2$ of two Weyl symbols (the result is by definition the Weyl symbol of the product of the two operators with symbols h_1 and h_2) fit with the decomposition of symbols into their homogeneous parts. We first need, however, to slightly generalize $(4.19), (4.22)$ since we have to deal with symbols which are not necessarily even functions on $\mathbb{R}^2$. The result of the present section will make it possible for us to understand the role of the integral kernels introduced in (1.5). In this very computational section, we found it convenient to list separately, towards the end, a certain number of elementary formulas.

If $\delta = 0$ or 1 and $\mu \in \mathbb{C}$, set

$$|s|_\delta^\mu = |s|^\mu (\mathrm{sign}\, s)^\delta; \tag{5.2}$$

also abbreviate $|s|_1^\mu$ as $\langle s \rangle^\mu$. Given $\nu \in \mathbb{C}$ and $u \in L^2(\mathbb{R}, (1+s^2)^{\mathrm{Re}\, \nu}\, ds)$, one may then define [22]

$$(\pi_{\nu,\delta}(g)u)(s) = |-cs + a|_\delta^{-1-\nu}\, u(\frac{ds-b}{-cs+a}) \tag{5.3}$$

for every matrix $g = \left(\begin{smallmatrix} a & b \\ c & d \end{smallmatrix}\right) \in SL(2, \mathbb{R})$. The function $\pi_{\nu,\delta}(g)u$ satisfies the same integrability condition as u, and $\pi_{\nu,\delta}$ is a bounded representation: it is unitary in the case when $\nu = i\lambda$ is pure imaginary. Depending on whether $\delta = 0$ (this is the case already considered in section 2) or 1, it is of class one (*i.e.*, it admits a K-invariant vector) or not.

In this section which deals with the Weyl calculus, it is convenient to change our notations from the preceding section, now denoting as $X = (x, \xi)$ the generic point of $\mathbb{R}^2$. Just as in (4.19), the Fourier inversion formula applied to the function $f(\tau) = e^\tau h(e^\tau X), h \in \mathcal{S}(\mathbb{R})$, yields the continuous decomposition

$$h = \int_{-\infty}^{\infty} h_\lambda \, d\lambda, \tag{5.4}$$

$$h_\lambda(X) = \frac{1}{2\pi} \int_0^\infty t^{i\lambda} \, h(tX) \, dt \tag{5.5}$$

of h as a sum of functions homogeneous of degree $-1 - i\lambda$. Observe that $h_\lambda(X)$ also makes sense for a complex λ with Im $\lambda < 1$. We may then, in (5.4), make a contour deformation from the real line to the line Im $\lambda = a$, $a > 0$: with $\nu = i\lambda$, one then has Re $\nu < 0$, a situation that will make it possible to consider genuine (convergent) integrals in lemma 5.2.

In the case when h is even or odd, one can recover h_λ from the function $h_\lambda^\flat$ on the real line defined as

$$h_\lambda^\flat(s) = h_\lambda(s, 1) \tag{5.6}$$

since one may write

$$h_\lambda(x, \xi) = |\xi|_\delta^{-1-i\lambda} h_\lambda^\flat\left(\frac{x}{\xi}\right) \tag{5.7}$$

with $\delta = 0$ or 1 according to the parity of h_λ. Splitting h, in general, into its even and odd parts, one may introduce the two functions

$$h_{\lambda,0}^\flat(s) = \frac{1}{2} [h_\lambda(s, 1) + h_\lambda(-s, -1)], \qquad h_{\lambda,1}^\flat(s) = \frac{1}{2} [h_\lambda(s, 1) - h_\lambda(-s, -1)]. \tag{5.8}$$

One then always has

$$\|h\|_{L^2(\mathbf{R}^2)}^2 = 4\pi \int_{-\infty}^{\infty} \left[\|h_{\lambda,0}^\flat\|_{L^2(\mathbf{R})}^2 + \|h_{\lambda,1}^\flat\|_{L^2(\mathbf{R})}^2 \right] d\lambda. \tag{5.9}$$

The group $G = SL(2, \mathbf{R})$ acts on $L^2(\mathbf{R}^2)$ under the (quasi)–regular representation π defined as

$$(\pi(g)h)(X) = h(g^{-1}X) : \tag{5.10}$$

the formulas (5.4) to (5.8) provide the (continuous) decomposition of π as a sum of irreducible representations since, as is immediate,

$$(\pi(g)h)_{\lambda,\delta}^\flat(s) = |-cs + a|_\delta^{-1-i\lambda} h_{\lambda,\delta}^\flat\left(\frac{ds - b}{-cs + a}\right) \tag{5.11}$$

if $g = \left(\begin{smallmatrix} a & b \\ c & d \end{smallmatrix}\right)$.

Even (*resp.* odd) symbols on $\mathbf{R}^2$ are those of operators that commute (*resp.* anticommute) with the operator C defined as $(Cu)(x) = u(-x)$. In particular, the composition $h_1 \# h_2$ of two symbols in $L^2(\mathbf{R}^2)$ with given parities is even (*resp.* odd) according to whether the two parities are the same or not. Recall that the Weyl calculus provides an isometry from the space $L^2(\mathbf{R}^2)$ onto the space of Hilbert–Schmidt operators on $L^2(\mathbf{R})$, which makes such a composition meaningful. From the decomposition of h_1, h_2 given by the formulas (5.4)–(5.8), one gets a decomposition

$$(h_1 \# h_2)^\flat_{\lambda,\delta}(s) = \sum_{\substack{\delta_1 + \delta_2 \\ \equiv \delta \bmod 2}} \int_{-\infty}^\infty \int_{-\infty}^\infty d\lambda_1 \, d\lambda_2$$

$$\int_{\mathbf{R}^2} K^{\delta_1,\delta_2;\delta}_{i\lambda_1,i\lambda_2;i\lambda}(s_1, s_2; s) \, (h_1)^\flat_{\lambda_1,\delta_1}(s_1) \, (h_2)^\flat_{\lambda_2,\delta_2}(s_2) \, ds_1 \, ds_2 . \quad (5.12)$$

To convince oneself that, at least for almost all sets $(\delta_1, \delta_2; \delta; \lambda_1, \lambda_2; \lambda)$, the kernel $K^{\delta_1,\delta_2;\delta}_{i\lambda_1,i\lambda_2;i\lambda}(s_1, s_2; s)$ is well-defined (almost everywhere) as a *function* of $(s_1, s_2; s)$, it suffices to integrate the left-hand side against $(\bar{h}_3)^\flat_{\lambda,\delta}(s) \, ds \, d\lambda$, where $(h_3)^\flat_{\lambda,\delta}$ arises from the decomposition (5.4)–(5.8) of an arbitrary function $h_3 \in L^2(\mathbf{R}^2)$, and to use the polarized form of (5.9). We shall not make too much fuss in what follows about these justifications, which are quite easy but notationally burdening. Our problem is computing the integral kernel $K^{\delta_1,\delta_2;\delta}_{i\lambda_1,i\lambda_2;i\lambda}(s_1, s_2; s)$.

Because of the covariance of the Weyl calculus under the metaplectic representation ($cf.$ (13.1)), one has the identity

$$(h_1 \# h_2)(dx - b\xi, -cx + a\xi) = (\tilde{h}_1 \# \tilde{h}_2)(x, \xi) \quad (5.13)$$

if $\tilde{h}_j(x, \xi) = h_j(dx - b\xi, -cx + a\xi)$. From (5.10) it thus follows that if $\mathbf{K}^{\delta_1,\delta_2;\delta}_{i\lambda_1,i\lambda_2;i\lambda}$ denotes the (bilinear) operator with integral kernel $K^{\delta_1,\delta_2;\delta}_{i\lambda_1,i\lambda_2;i\lambda}$, one has

$$\pi_{i\lambda,\delta}(g) \, \mathbf{K}^{\delta_1,\delta_2;\delta}_{i\lambda_1,i\lambda_2;i\lambda}(u_1, u_2) = \mathbf{K}^{\delta_1,\delta_2;\delta}_{i\lambda_1,i\lambda_2;i\lambda}(\pi_{i\lambda_1,\delta_1}(g)u_1, \pi_{i\lambda_2,\delta_2}(g)u_2)$$

$$(5.14)$$

for an arbitrary pair (u_1, u_2), from which (performing a change of variables in the unwritten explicit integral form of the right–hand side) one gets the identity

$$K^{\delta_1,\delta_2;\delta}_{i\lambda_1,i\lambda_2;i\lambda}\left(\frac{as_1 + b_1}{cs_1 + d_1}, \frac{as_2 + b_2}{cs_2 + d_2}; \frac{as + b}{cs + d}\right) =$$

$$|cs_1 + d|^{1-i\lambda_1}_{\delta_1} \, |cs_2 + d|^{1-i\lambda_2}_{\delta_2} \, |cs + d|^{1+i\lambda}_{\delta} \, K^{\delta_1,\delta_2;\delta}_{i\lambda_1,i\lambda_2;i\lambda}(s_1, s_2; s). \quad (5.15)$$

Consider now, for complex ν_1, ν_2, ν and $\varepsilon, \varepsilon_1$ and ε all set to the value 0 or 1, the integral kernels (containing those in (1.4)–(1.5) as a special case)

$$\chi^{\varepsilon_1,\varepsilon_2;\varepsilon}_{\nu_1,\nu_2;\nu}(s_1, s_2; s) = |s_1 - s_2|^{\frac{-1+\nu+\nu_1+\nu_2}{2}}_{\varepsilon} \, |s_2 - s|^{\frac{-1-\nu-\nu_1+\nu_2}{2}}_{\varepsilon_1} \, |s - s_1|^{\frac{-1-\nu+\nu_1-\nu_2}{2}}_{\varepsilon_2}.$$

$$(5.16)$$

It is immediate to check that

$$\chi^{\varepsilon_1,\varepsilon_2;\varepsilon}_{\nu_1,\nu_2;\nu}\left(\frac{as_1 + b_1}{cs_1 + d_1}, \frac{as_2 + b_2}{cs_2 + d_2}; \frac{as + b}{cs + d}\right) =$$

$$|cs_1 + d|^{1-\nu_1}_{\delta_1} \, |cs_2 + d|^{1-\nu_2}_{\delta_2} \, |cs + d|^{1+\nu}_{\delta} \, \chi^{\varepsilon_1,\varepsilon_2;\varepsilon}_{\nu_1,\nu_2;\nu}(s_1, s_2; s) \quad (5.17)$$

provided that $\delta_1, \delta_2, \delta = 0$ or 1 satisfy

$$\delta_1 \equiv \varepsilon_2 + \varepsilon, \quad \delta_2 \equiv \varepsilon + \varepsilon_1, \quad \delta \equiv \varepsilon_1 + \varepsilon_2 \quad \text{mod } 2. \quad (5.18)$$

Coming back to our kernel $K^{\delta_1,\delta_2;\delta}_{i\lambda_1,i\lambda_2;i\lambda}$, and fixing $\varepsilon, \varepsilon_1$ and ε satisfying the relations (5.18) (since $\delta \equiv \delta_1 + \delta_2$ mod 2, it suffices to choose $j = 0$ or 1 and to set $\varepsilon \equiv j + \delta$, $\varepsilon_1 \equiv j + \delta_1$, $\varepsilon_2 \equiv j + \delta_2$), one sees that the ratio

$$(\chi^{\varepsilon_1,\varepsilon_2;\varepsilon}_{i\lambda_1,i\lambda_2;i\lambda}(s_1, s_2; s))^{-1} \times (K^{\delta_1,\delta_2;\delta}_{i\lambda_1,i\lambda_2;i\lambda})(s_1, s_2; s)$$

is invariant under all substitutions $(s_1, s_2; s) \mapsto (\frac{as_1+b_1}{cs_1+d_1}, \frac{as_2+b_2}{cs_2+d_2}; \frac{as+b}{cs+d})$, $g = \begin{pmatrix} a & b \\ c & d \end{pmatrix} \in SL(2, \mathbb{R})$. Now the orbit under $SL(2, \mathbb{R})$ of a point $(s_1, s_2, s) \in (\dot{\mathbb{R}})^3$ with distinct coordinates (where $\dot{\mathbb{R}}$ denotes the projective line) is characterized by the sign of $(s_1 - s_2)(s_2 - s)^{-1}(s - s_1)^{-1}$.

As a consequence, one has an identity

$$K^{\delta_1,\delta_2;\delta}_{i\lambda_1,i\lambda_2;i\lambda}(s_1, s_2; s) = \sum_{j=0}^{1} a^{\delta_1,\delta_2;\delta}_{\varepsilon_1,\varepsilon_2;\varepsilon}(i\lambda_1, i\lambda_2; i\lambda) \, \chi^{\varepsilon_1,\varepsilon_2;\varepsilon}_{i\lambda_1,i\lambda_2;i\lambda}(s_1, s_2; s),$$
$$(5.19)$$

where

$$\varepsilon \equiv j + \delta, \quad \varepsilon_1 \equiv j + \delta_1, \quad \varepsilon_2 \equiv j + \delta_2, \quad (5.20)$$

and our remaining problem in this section is to determine the coefficients $a^{\delta_1,\delta_2;\delta}_{\varepsilon_1,\varepsilon_2;\varepsilon}(i\lambda_1, i\lambda_2; i\lambda)$.

To that effect we shall compute both sides of equation (5.12) for the simplest available pair (h_1, h_2) of symbols with non–commuting associated operators.

LEMMA 5.1. *Let ν_1 and ν_2 be complex numbers with $|\text{Re }(\nu_1 \pm \nu_2)| < 1$, $\nu_1 \neq 0$, $\nu_2 \neq 0$, so that in particular $|\text{Re } \nu_1| < 1$, $|\text{Re } \nu_2| < 1$. Set $h_1(x, \xi) = |x|_{\delta_1}^{-1-\nu_1}$ for some $\delta_1 = 0$ or 1: concerning the meaning of h_1 as a distribution, cf. the comment right after (5.52); with $\delta_2 = 0$ or 1 too, set $h_2(x, \xi) = |\xi|_{\delta_2}^{-1-\nu_2}$. Let $h = h_1 \# h_2$, a tempered distribution on $\mathbb{R}^2$. In the space $\mathcal{S}'(\mathbb{R}^2)$, it admits the decomposition*

$$h = \int_{-\infty}^{\infty} h_{\lambda,\delta} \, d\lambda, \quad (5.21)$$

with

$$h^b_{\lambda,\delta}(s) = 2^{\frac{\nu_1+\nu_2-i\lambda-5}{2}} \, \pi^{\frac{\nu_1+\nu_2-i\lambda}{2}} \, \frac{\Gamma(\frac{-\nu_1+\delta_1}{2})\Gamma(\frac{-\nu_2+\delta_2}{2})}{\Gamma(\frac{\nu_1+\delta_1+1}{2})\Gamma(\frac{\nu_2+\delta_2+1}{2})} \times$$

$$\left[i^{\delta_2-\delta} \frac{\Gamma(\frac{1+\nu_1-\nu_2+i\lambda}{4})\Gamma(\frac{1+\nu_1+\nu_2-i\lambda+2\delta_1}{4})\Gamma(\frac{1-\nu_1+\nu_2+i\lambda+2\delta}{4})}{\Gamma(\frac{1-\nu_1+\nu_2-i\lambda}{4})\Gamma(\frac{1-\nu_1-\nu_2+i\lambda+2\delta_1}{4})\Gamma(\frac{1+\nu_1-\nu_2-i\lambda+2\delta}{4})} \, |s|^{\frac{-1-\nu_1+\nu_2-i\lambda}{2}} \right.$$

$$\left. + i^{-\delta_2-\delta+1} \frac{\Gamma(\frac{3+\nu_1-\nu_2+i\lambda}{4})\Gamma(\frac{3+\nu_1+\nu_2-i\lambda-2\delta_1}{4})\Gamma(\frac{3-\nu_1+\nu_2+i\lambda-2\delta}{4})}{\Gamma(\frac{3-\nu_1+\nu_2-i\lambda}{4})\Gamma(\frac{3-\nu_1-\nu_2+i\lambda-2\delta_1}{4})\Gamma(\frac{3+\nu_1-\nu_2-i\lambda-2\delta}{4})} \, \langle s \rangle^{\frac{-1-\nu_1+\nu_2-i\lambda}{2}} \right],$$

$$\tag{5.22}$$

where $\delta = 0$ or 1 is characterized by $\delta \equiv \delta_1 + \delta_2 \bmod 2$. Recall that $\langle s \rangle^\mu = |s|_1^\mu$.

PROOF. First observe that the function $s \mapsto |s|_\epsilon^{\frac{-1-\nu_1+\nu_2-i\lambda}{2}}$ should be lifted, under (5.7), to the function (homogeneous of degree $-1 - i\lambda$, with parity related to δ)

$$(x,\xi) \mapsto |\xi|_{\epsilon'}^{\frac{-1+\nu_1-\nu_2-i\lambda}{2}} \, |x|_\epsilon^{\frac{-1-\nu_1+\nu_2-i\lambda}{2}}, \qquad \epsilon' \equiv \epsilon + \delta \bmod 2. \tag{5.23}$$

Now

$$(x\frac{\partial}{\partial x})|x|_\epsilon^{\frac{-1-\nu_1+\nu_2-i\lambda}{2}} = (\frac{-1-\nu_1+\nu_2-i\lambda}{2})|x|_\epsilon^{\frac{-1-\nu_1+\nu_2-i\lambda}{2}}, \tag{5.24}$$

so that integrations by parts, together with asymptotics for the Gamma function on vertical lines ([27], p.13), here reduced to the observation that there are as many Gamma factors upstairs as downstairs, give a meaning to the right-hand side of (5.21) as a weakly convergent integral in $S'(\mathbb{R}^2)$.

The *standard* (*i.e.*, convolutions first) symbol of the operator $\mathrm{Op}(h_1) \circ \mathrm{Op}(h_2)$ is of course, simply, the function

$$g(x,\xi) = |x|_{\delta_1}^{-1-\nu_1} \, |\xi|_{\delta_2}^{-1-\nu_2}. \tag{5.25}$$

The Weyl symbol h of the same operator is ([38], p.15) linked to g by the equation $h = \exp(\frac{1}{4i\pi}\frac{\partial^2}{\partial x \partial \xi})g$, *i.e.*, denoting as $\hat{g}^1$ the Fourier transform of g with respect to its first variable,

$$\hat{h}^1(\eta,\xi) = \hat{g}^1(\eta, \xi - \frac{\eta}{2}). \tag{5.26}$$

From (5.53), one first gets

$$h(x,\xi) = (-i)^{\delta_1} \, \pi^{\nu_1+\frac{1}{2}} \, \frac{\Gamma(\frac{-\nu_1+\delta_1}{2})}{\Gamma(\frac{\nu_1+\delta_1+1}{2})} \int_{-\infty}^{\infty} e^{2i\pi x\eta} \, |\eta|_{\delta_1}^{\nu_1} \, |\xi - \frac{\eta}{2}|_{\delta_2}^{-1-\nu_2} \, d\eta, \tag{5.27}$$

a genuinely convergent integral (for $\xi \neq 0$) in the case when $-1 < \mathrm{Re}\,\nu_1 < \mathrm{Re}\,\nu_2 < 0$. Using complex continuation, we may indeed assume that

$$-1 - \mathrm{Re}\,\nu_2 < \mathrm{Re}\,\nu_1 < \mathrm{Re}\,\nu_2 < 0. \tag{5.28}$$

Since $h(-x, -\xi) = (-1)^{\delta_1+\delta_2} h(x, \xi)$, one has from (5.7) and (5.5) (which here extends if both sides are lifted to functions of (x, ξ) as indicated in the beginning of the present proof and both sides are tested against some function of x in $\mathcal{S}(\mathbb{R})$)

$$h^b_{\lambda,\delta}(s) = h^b_{\lambda}(s) = \frac{1}{2\pi} \int_0^\infty r^{i\lambda} h(rs, r)\, dr$$

$$= \frac{1}{2}(-i)^{\delta_1} \pi^{\nu_1-\frac{1}{2}} \frac{\Gamma(\frac{-\nu_1+\delta_1}{2})}{\Gamma(\frac{\nu_1+\delta_1+1}{2})} \int_{-\infty}^\infty |\eta|^{\nu_1}_{\delta_1}\, d\eta \int_0^\infty r^{i\lambda} e^{2i\pi rsn} \left|r - \frac{\eta}{2}\right|^{-1-\nu_2}_{\delta_2} dr \tag{5.29}$$

or, after the change of variable $\eta = 2\,rt$,

$$h^b_{\lambda,\delta}(s) = 2^{\nu_1} (-i)^{\delta_1} \pi^{\nu_1-\frac{1}{2}} \frac{\Gamma(\frac{-\nu_1+\delta_1}{2})}{\Gamma(\frac{\nu_1+\delta_1+1}{2})} \int_{-\infty}^\infty |t|^{\nu_1}_{\delta_1} |1-t|^{-1-\nu_2}_{\delta_2} dt$$

$$\int_0^\infty r^{\nu_1-\nu_2+i\lambda} e^{4i\pi r^2 st}\, dr, \tag{5.30}$$

where the dr–integral is an improper one. Using (5.57), one gets

$$h^b_{\lambda,\delta}(s) = 2^{-2+\nu_2-i\lambda} (-i)^{\delta_1} \pi^{\frac{\nu_1+\nu_2-i\lambda-2}{2}} \frac{\Gamma(\frac{-\nu_1+\delta_1}{2}) \Gamma(\frac{1+\nu_1-\nu_2+i\lambda}{2})}{\Gamma(\frac{\nu_1+\delta_1+1}{2})} |s|^{\frac{-1-\nu_1+\nu_2-i\lambda}{2}}$$

$$\int_{-\infty}^\infty |t|^{\frac{-1+\nu_1+\nu_2-i\lambda}{2}}_{\delta_1} |1-t|^{-1-\nu_2}_{\delta_2} e^{\frac{i\pi}{4}(1+\nu_1-\nu_2+i\lambda)\mathrm{sign}(st)}\, dt, \tag{5.31}$$

again a genuinely convergent integral in view of (5.28). Breaking the exponential into its two trigonometric terms and using (5.56), one gets

$$e^{\frac{i\pi}{4}(1+\nu_1-\nu_2+i\lambda)\mathrm{sign}(st)}$$

$$= \frac{\pi}{\Gamma(\frac{1-\nu_1+\nu_2-i\lambda}{4})\Gamma(\frac{3+\nu_1-\nu_2+i\lambda}{4})} + \frac{i\pi\, \mathrm{sign}(st)}{\Gamma(\frac{1+\nu_1-\nu_2+i\lambda}{4})\Gamma(\frac{3-\nu_1+\nu_2-i\lambda}{4})}. \tag{5.32}$$

Next, using (5.55) with $2z = \frac{1+\nu_1-\nu_2+i\lambda}{2}$,

$$\Gamma(\frac{1+\nu_1-\nu_2+i\lambda}{2}) e^{\frac{i\pi}{4}(1+\nu_1-\nu_2+i\lambda)\mathrm{sign}(st)} =$$

$$(\frac{\pi}{2})^{\frac{1}{2}} 2^{\frac{\nu_1-\nu_2+i\lambda}{2}} \left[\frac{\Gamma(\frac{1+\nu_1-\nu_2+i\lambda}{4})}{\Gamma(\frac{1-\nu_1+\nu_2-i\lambda}{4})} + i\,(\mathrm{sign}\,(st)) \frac{\Gamma(\frac{3+\nu_1-\nu_2+i\lambda}{4})}{\Gamma(\frac{3-\nu_1+\nu_2-i\lambda}{4})} \right]. \tag{5.33}$$

Starting from (5.31) again, one finds

$$h^b_{\lambda,\delta}(s) = 2^{\frac{\nu_1+\nu_2-i\lambda-5}{2}} (-i)^{\delta_1} \pi^{\frac{\nu_1+\nu_2-i\lambda-1}{2}} \frac{\Gamma(\frac{-\nu_1+\delta_1}{2})}{\Gamma(\frac{\nu_1+\delta_1+1}{2})} \times$$

$$\left[\frac{\Gamma(\frac{1+\nu_1-\nu_2+i\lambda}{4})}{\Gamma(\frac{1-\nu_1+\nu_2-i\lambda}{4})} |s|^{\frac{-1-\nu_1+\nu_2-i\lambda}{2}} \int_{-\infty}^\infty |t|^{\frac{-1+\nu_1+\nu_2-i\lambda}{2}}_{\delta_1} |1-t|^{-1-\nu_2}_{\delta_2} dt \right.$$

$$\left. + i \frac{\Gamma(\frac{3+\nu_1-\nu_2+i\lambda}{4})}{\Gamma(\frac{3-\nu_1+\nu_2-i\lambda}{4})} \langle s \rangle^{\frac{-1-\nu_1+\nu_2-i\lambda}{2}} \int_{-\infty}^\infty |t|^{\frac{-1+\nu_1+\nu_2-i\lambda}{2}}_{1-\delta_1} |1-t|^{-1-\nu_2}_{\delta_2} dt \right]. \tag{5.34}$$

Using (5.54), we are done. □

Let $\mathbf{L}_{i\lambda_1,i\lambda_2;i\lambda}^{\varepsilon_1,\varepsilon_2;\varepsilon}$ be the bilinear operator with integral kernel $\chi_{i\lambda_1,i\lambda_2;i\lambda}^{\varepsilon_1,\varepsilon_2;\varepsilon}$. If $h_1(x,\xi) = |x|_{\delta_1}^{-1-i\lambda_1}$, $h_2(x,\xi) = |\xi|_{\delta_2}^{-1-i\lambda_2}$, one has $(h_1)_{\lambda_1,\delta_1}^{\flat}(s_1) = |s_1|_{\delta_1}^{-1-i\lambda_1}$ and $(h_2)_{\lambda_2,\delta_2}^{\flat}(s_2) = 1$, so that the sum on the right–hand side of (5.12) reduces, if $K_{i\lambda_1,i\lambda_2;i\lambda}^{\delta_1,\delta_2;\delta}$ is given by (5.19), to

$$\sum_{j=0}^{1} a_{\varepsilon_1,\varepsilon_2;\varepsilon}^{\delta_1,\delta_2;\delta}(\lambda_1,\lambda_2;\lambda)\, \mathbf{L}_{i\lambda_1,i\lambda_2;i\lambda}^{\varepsilon_1,\varepsilon_2;\varepsilon}(|s_1|_{\delta_1}^{-1-i\lambda_1}, 1)(s), \qquad (5.35)$$

where (5.20) expresses $(\varepsilon_1,\varepsilon_2;\varepsilon)$ in terms of $(\delta_1,\delta_2;\delta)$ and j. However, before making the appropriate computation, we substitute for $i\lambda_1$ and $i\lambda_2$, as indicated in the remark which follows (5.5), complex numbers ν_1 and ν_2 with Re $\nu_1 < 0$, Re $\nu_2 < 0$.

LEMMA 5.2. *Assuming that* $\delta_1 + \delta_2 \equiv \delta \bmod 2$, *set, in analogy with* (5.19),

$$K_{\nu_1,\nu_2;i\lambda}^{\delta_1,\delta_2;\delta}(s_1,s_2;s) = \sum_{j=0}^{1} a_{\varepsilon_1,\varepsilon_2;\varepsilon}^{\delta_1,\delta_2;\delta}(\nu_1,\nu_2;i\lambda)\, \chi_{\nu_1,\nu_2;i\lambda}^{\varepsilon_1,\varepsilon_2;\varepsilon}(s_1,s_2;s),$$
$$\qquad (5.36)$$

with

$$a_{\varepsilon_1,\varepsilon_2;\varepsilon}^{\delta_1,\delta_2;\delta}(\nu_1,\nu_2;i\lambda) = (-1)^{\delta_1\delta_2+\delta}\,(-i)^{j}\,2^{-\frac{3}{2}}\,(2\pi)^{\frac{-i\lambda+\nu_1+\nu_2)-2}{2}} \times$$

$$\frac{\Gamma(\frac{1+i\lambda+\nu_1-\nu_2+2\varepsilon_1}{4})\Gamma(\frac{1+i\lambda-\nu_1+\nu_2+2\varepsilon_2}{4})\Gamma(\frac{1-i\lambda-\nu_1-\nu_2+2\varepsilon}{4})}{\Gamma(\frac{1-i\lambda-\nu_1+\nu_2+2\varepsilon_1}{4})\Gamma(\frac{1-i\lambda+\nu_1-\nu_2+2\varepsilon_2}{4})\Gamma(\frac{1+i\lambda+\nu_1+\nu_2+2\varepsilon}{4})}, \quad (5.37)$$

where $\varepsilon_1, \varepsilon_2, \varepsilon$ *are characterized by* (5.20).
   Assume that

$$|\mathrm{Re}\,(\nu_1 \pm \nu_2)| < 1, \quad \mathrm{Re}\,\nu_1 < 0, \quad \mathrm{Re}\,\nu_2 < 0. \qquad (5.38)$$

Then, with $h_1(x,\xi) = |x|_{\delta_1}^{-1-\nu_1}$, $h_2(x,\xi) = |\xi|_{\delta_2}^{-1-\nu_2}$, *one has for all real* λ *and almost all* s

$$(h_1\#h_2)_{\lambda,\delta}^{\flat}(s) = \int_{\mathbf{R}^2} K_{\nu_1,\nu_2;i\lambda}^{\delta_1,\delta_2;\delta}(s_1,s_2;s)\,|s_1|_{\delta_1}^{-1-\nu_1}\,ds_1\,ds_2.$$
$$\qquad (5.39)$$

PROOF. The (convergent) integral

$$\int \chi_{\nu_1,\nu_2;i\lambda}^{\varepsilon_1,\varepsilon_2;\varepsilon}(s_1,s_2;s)\,|s_1|_{\delta_1}^{-1-\nu_1}\,ds_1\,ds_2 =$$

$$\int_{\mathbf{R}^2} |s_1-s_2|_{\varepsilon}^{\frac{-1+i\lambda+\nu_1+\nu_2}{2}}\,|s_2-s|_{\varepsilon_1}^{\frac{-1-i\lambda-\nu_1+\nu_2}{2}}\,|s-s_1|_{\varepsilon_2}^{\frac{-1-i\lambda+\nu_1-\nu_2}{2}}\,|s_1|_{\delta_1}^{-1-\nu_1}\,ds_1\,ds_2$$
$$\qquad (5.40)$$

can be computed with the help of (5.54), reducing it first (since $\varepsilon + \varepsilon_1 \equiv \delta_2 \bmod 2$) to

$$(-1)^{\delta_2} \, i^{\varepsilon+\varepsilon_1-\delta_2} \, \pi^{\frac{1}{2}} \, \frac{\Gamma(\frac{-\nu_2+\delta_2}{2})\Gamma(\frac{1+i\lambda+\nu_1+\nu_2+2\varepsilon}{4})\Gamma(\frac{1-i\lambda-\nu_1+\nu_2+2\varepsilon_1}{4})}{\Gamma(\frac{1+\nu_2+\delta_2}{2})\Gamma(\frac{1-i\lambda-\nu_1-\nu_2+2\varepsilon}{4})\Gamma(\frac{1+i\lambda+\nu_1-\nu_2+2\varepsilon_1}{4})}$$

$$\int_{-\infty}^{\infty} |s-s_1|_j^{\frac{-1-i\lambda+\nu_1+\nu_2}{2}} \, |s_1|_{\delta_1}^{-1-\nu_1} \, ds_1 \,, \quad (5.41)$$

finally (using (5.54) again together with $j + \delta_1 \equiv \varepsilon_1$, and noting the simplification of four Gamma factors), to

$$i^{\varepsilon+j+\delta_1+\delta_2} \, \pi \, \frac{\Gamma(\frac{-\nu_1+\delta_1}{2})\Gamma(\frac{-\nu_2+\delta_2}{2})}{\Gamma(\frac{1+\nu_1+\delta_1}{2})\Gamma(\frac{1+\nu_2+\delta_2}{2})} \times$$

$$\frac{\Gamma(\frac{1+i\lambda+\nu_1+\nu_2+2\varepsilon}{4})\Gamma(\frac{1-i\lambda+\nu_1+\nu_2+2j}{4})}{\Gamma(\frac{1-i\lambda-\nu_1-\nu_2+2\varepsilon}{4})\Gamma(\frac{1+i\lambda-\nu_1-\nu_2+2j}{4})} \, |s|_{\varepsilon_1}^{\frac{-1-i\lambda-\nu_1+\nu_2}{2}} \,. \quad (5.42)$$

Simplifying four Gamma factors again in each case, one sees that the contribution to the right–hand side of (5.39) corresponding to the term with $\varepsilon_1 = 0$ (so that $j = \delta_1, \varepsilon = \delta_2, \varepsilon_2 = \delta$) from the expression (5.36) of $K^{\delta_1,\delta_2;\delta}_{\nu_1,\nu_2;i\lambda}(s_1,s_2;s)$ just agrees with the first term on the right–hand side of (5.22); when $\varepsilon_1 = 1$, one has $j = 1-\delta_1, \varepsilon = 1-\delta_2, \varepsilon_2 = 1-\delta$, and it may be checked that the corresponding term from the right–hand side of (5.39) agrees with the second term on the right–hand side of (5.22). One notes, to finish the proof, that $i^{\varepsilon+j+\delta_1+\delta_2} i^{-j} (-1)^{\delta_1\delta_2+\delta} = i^{\delta_2-\delta}$ if $\varepsilon_1 = 0$, $i^{-\delta_2-\delta+1}$ if $\varepsilon_1 = 1$, as in both cases it suffices to remark that $2\delta_1\delta_2 + 3\delta + \delta_1 + \delta_2 \equiv 0 \bmod 4$ when $\delta \equiv \delta_1 + \delta_2 \bmod 2$ and all three numbers $\delta_1, \delta_2, \delta$ are 0 or 1. $\qquad \square$

THEOREM 5.3. *Let*

$$K^{\delta_1,\delta_2;\delta}_{i\lambda_1,i\lambda_2;i\lambda}(s_1,s_2;s) = 2^{-\frac{3}{2}} (2\pi)^{\frac{i(-\lambda+\lambda_1+\lambda_2)-2}{2}} (-1)^{\delta_1\delta_2+\delta} \times$$

$$\sum_{j=0}^{1}(-i)^j \, \frac{\Gamma(\frac{1+i(\lambda+\lambda_1-\lambda_2)+2\varepsilon_1}{4})\Gamma(\frac{1+i(\lambda-\lambda_1+\lambda_2)+2\varepsilon_2}{4})\Gamma(\frac{1+i(-\lambda-\lambda_1-\lambda_2)+2\varepsilon}{4})}{\Gamma(\frac{1+i(-\lambda-\lambda_1+\lambda_2)+2\varepsilon_1}{4})\Gamma(\frac{1+i(-\lambda+\lambda_1-\lambda_2)+2\varepsilon_2}{4})\Gamma(\frac{1+i(\lambda+\lambda_1+\lambda_2)+2\varepsilon}{4})}$$

$$\times \, \chi^{\varepsilon_1,\varepsilon_2;\varepsilon}_{i\lambda_1,i\lambda_2;i\lambda}(s_1,s_2;s)\,, \quad (5.43)$$

where

$$\varepsilon \equiv j+\delta, \quad \varepsilon_1 \equiv j+\delta_1, \quad \varepsilon_2 \equiv j+\delta_2 \quad\quad\quad (5.44)$$

and

$$\chi^{\varepsilon_1,\varepsilon_2;\varepsilon}_{i\lambda_1,i\lambda_2;i\lambda}(s_1,s_2;s) = |s_1-s_2|_{\varepsilon}^{\frac{-1+i(\lambda+\lambda_1+\lambda_2)}{2}} \, |s_2-s|_{\varepsilon_1}^{\frac{-1+i(-\lambda-\lambda_1+\lambda_2)}{2}} \, |s-s_1|_{\varepsilon_2}^{\frac{-1+i(-\lambda+\lambda_1)}{2}}$$

Then, for every pair (h_1,h_2) of symbols in $L^2(\mathbf{R}^2)$, the formula (5.12) holds in the weak sense indicated in the beginning of the present section, namely after both sides have been integrated against $(\bar{h}_3)^{\flat}_{\lambda,\delta}(s) \, ds \, d\lambda$, where $h_3 \in L^2(\mathbf{R}^2)$.

PROOF. Let $\mathcal{G}$ be the (slightly modified) symplectic Fourier transform on $L^2(\mathbb{R}^2)$ as defined in (4.12): it satisfies $\mathcal{G}^2 = I$, and commutes with the linear transformations in $SL(2, \mathbb{R})$. As a consequence of the first of these two properties, any symbol h_1 in $S(\mathbb{R}^2)$ can be written as

$$h_1(x, \xi) = \int_0^{2\pi} g_1^{\theta_1}(x \sin \theta_1 - \xi \cos \theta_1) \, d\theta_1, \tag{5.45}$$

with

$$g_1^{\theta_1}(x) = 2 \int_{-\infty}^{\infty} (\mathcal{G}h_1)(t \cos \theta_1, t \sin \theta_1) \, e^{4i\pi tx} |t| \, dt. \tag{5.46}$$

Since the integrand in the last integral, as a function of t, lies in the Sobolev space $H^1(\mathbb{R})$, the function $g_1^{\theta_1}$ satisfies $\int_{-\infty}^{\infty} |x| \, |g_1^{\theta_1}(x)|^2 \, dx < \infty$, which permits to write it as

$$g_1^{\theta_1}(x) = \sum_{\delta_1=0}^{1} \int_{-\infty}^{\infty} |x|_{\delta_1}^{-1-i\lambda_1} k_1(\lambda_1) \, d\lambda_1$$

with $k_1 \in L^2(\mathbb{R})$. Using now some easy approximation argument, it is now sufficient, in order to complete the proof of theorem 5.3, to extend lemma 5.2 to the case when the functions

$$h_1(x, \xi) = |x \sin \theta_1 - \xi \cos \theta_1|_{\delta_1}^{-1-\nu_1}, \qquad h_2(x, \xi) = |x \sin \theta_2 - \xi \cos \theta_2|_{\delta_2}^{-1-\nu_2},$$

where we may assume that $\theta_1 - \theta_2 \notin \pi \mathbf{Z}$, are substituted for the functions $|x|_{\delta_1}^{-1-\nu_1}$ and $|\xi|_{\delta_2}^{-1-\nu_2}$ respectively, and ν_1 and ν_2 satisfy the hypotheses in lemma 5.2. Since there exists a linear transformation in $SL(2, \mathbb{R})$ which transforms the function x into $x \sin \theta_1 - \xi \cos \theta_1$ and the function ξ into $(\sin(\theta_2 - \theta_1))^{-1} (x \sin \theta_2 - \xi \cos \theta_2)$, we are done, thanks to the invariance formula (5.13). $\qquad\square$

Remark. Note that $j = 0$ or 1, as associated with the set $(\delta_1, \delta_2; \delta; \varepsilon_1, \varepsilon_2; \varepsilon)$, is the same as the one associated with the set $(\delta_2, \delta_1; \delta; \varepsilon_2, \varepsilon_1; \varepsilon)$. As a consequence,

$$a_{\varepsilon_2, \varepsilon_1; \varepsilon}^{\delta_2, \delta_1; \delta}(\lambda_2, \lambda_1; \lambda) = a_{\varepsilon_1, \varepsilon_2; \varepsilon}^{\delta_1, \delta_2; \delta}(\lambda_1, \lambda_2; \lambda).$$

On the other hand, as is easily seen from (5.16) together with $\varepsilon + \varepsilon_1 + \varepsilon \equiv j$ mod 2, one has

$$\chi_{i\lambda_2, i\lambda_1; i\lambda}^{\varepsilon_2, \varepsilon_1; \varepsilon}(s_2, s_1; s) = (-1)^j \, \chi_{i\lambda_1, i\lambda_2; i\lambda}^{\varepsilon_1, \varepsilon_2; \varepsilon}(s_1, s_2; s). \tag{5.47}$$

It follows that the terms with $j = 0$ and $j = 1$ respectively, taken from (5.43) and substituted for $K_{i\lambda_1, i\lambda_2; i\lambda}^{\delta_1, \delta_2; \delta}(s_1, s_2; s)$ in (5.12), would yield instead of $(h_1 \# h_2)_{\lambda, \delta}^{\flat}(s)$ the half-anticommutator $\frac{1}{2}[(h_1 \# h_2)_{\lambda, \delta}^{\flat}(s) + (h_2 \# h_1)_{\lambda, \delta}^{\flat}(s)]$ and the half-commutator $\frac{1}{2}[(h_1 \# h_2)_{\lambda, \delta}^{\flat}(s) - (h_2 \# h_1)_{\lambda, \delta}^{\flat}(s)]$ respectively.

For even symbols, $h^\flat_{\lambda,\delta}$ can be nonzero only when $\delta = 0$ so that the composition formula (5.12) reduces in that case to

$$(h_1 \# h_2)^\flat_\lambda(s) = \int_{-\infty}^{\infty} \int_{-\infty}^{\infty} d\lambda_1 \, d\lambda_2$$

$$\int_{\mathbf{R}^2} K^{0,0;0}_{i\lambda_1, i\lambda_2; i\lambda}(s_1, s_2; s) \, (h_1)^\flat_{\lambda_1}(s_1) \, (h_2)^\flat_{\lambda_2}(s_2) \, ds_1 \, ds_2 \quad (5.48)$$

with

$$K^{0,0;0}_{i\lambda_1, i\lambda_2; i\lambda}(s_1, s_2; s) = 2^{-\frac{3}{2}} \, (2\pi)^{\frac{i(-\lambda + \lambda_1 + \lambda_2) - 2}{2}} \times$$

$$\sum_{j=0}^{1} (-i)^j \frac{\Gamma(\frac{1 + i(\lambda + \lambda_1 - \lambda_2) + 2j}{4}) \Gamma(\frac{1 + i(\lambda - \lambda_1 + \lambda_2) + 2j}{4}) \Gamma(\frac{1 + i(-\lambda - \lambda_1 - \lambda_2) + 2j}{4})}{\Gamma(\frac{1 + i(-\lambda - \lambda_1 + \lambda_2) + 2j}{4}) \Gamma(\frac{1 + i(-\lambda + \lambda_1 - \lambda_2) + 2j}{4}) \Gamma(\frac{1 + i(\lambda + \lambda_1 + \lambda_2) + 2j}{4})}$$

$$\times \chi^{j,j;j}_{i\lambda_1, i\lambda_2; i\lambda}(s_1, s_2; s). \quad (5.49)$$

We abbreviate $\chi^{0,0;0}_{i\lambda_1, i\lambda_2; i\lambda}$ as $\chi_{i\lambda_1, i\lambda_2; i\lambda}$, which is consistent with (1.5), and set

$$\chi^-_{i\lambda_1, i\lambda_2; i\lambda} := \chi^{1,1;1}_{i\lambda_1, i\lambda_2; i\lambda}, \quad (5.50)$$

i.e.,

$$\chi^-_{\nu_1, \nu_2; \nu}(s_1, s_2; s) =$$

$$|s_1 - s_2|^{\frac{1}{2}(-1 + \nu + \nu_1 + \nu_2)} |s_1 - s|^{\frac{1}{2}(-1 - \nu + \nu_1 - \nu_2)} |s_2 - s|^{\frac{1}{2}(-1 - \nu - \nu_1 + \nu_2)}$$

$$\times \operatorname{sign}\left(\frac{s_1 - s_2}{(s - s_1)(s_2 - s)}\right). \quad (5.51)$$

Appendix to section 5: some elementary formulas

We here display a few formulas we had to rely on. Recall that, if $\varepsilon = 0$ or 1 and $\mu \in \mathbf{C}$, we set

$$|s|^\mu_\varepsilon = |s|^\mu (\operatorname{sign} s)^\varepsilon. \quad (5.52)$$

This function of s is well–defined as a tempered distribution when $\operatorname{Re} \mu > -1$ and can also be defined (cf. [32, p. 43]), by analytic continuation with respect to μ, in the case (sufficient for our purpose) when $\operatorname{Re} \mu > -2, \mu \neq -1$: we shall dispense with the sign "finite part" sometimes used in this context. One has [27]:

$$(\mathcal{F}(|s|^\mu_\varepsilon))(\sigma) = (-i)^\varepsilon \, \pi^{-\frac{1}{2} - \mu} \frac{\Gamma(\frac{\mu + 1 + \varepsilon}{2})}{\Gamma(\frac{-\mu + \varepsilon}{2})} |\sigma|^{-\mu - 1}_\varepsilon, \qquad -2 < \operatorname{Re} \mu < 1, \mu \neq 0, -1. \quad (5.53)$$

As a consequence, one has the convolution formula

$$|s|^\alpha_{\varepsilon_1} * |s|^\beta_{\varepsilon_2} = i^{\varepsilon_1 + \varepsilon_2 - \varepsilon} \, \pi^{\frac{1}{2}} \frac{\Gamma(\frac{\alpha + 1 + \varepsilon_1}{2}) \Gamma(\frac{\beta + 1 + \varepsilon_2}{2}) \Gamma(\frac{-\alpha - \beta - 1 + \varepsilon}{2})}{\Gamma(\frac{-\alpha + \varepsilon_1}{2}) \Gamma(\frac{-\beta + \varepsilon_2}{2}) \Gamma(\frac{\alpha + \beta + 2 + \varepsilon}{2})} |s|^{\alpha + \beta + 1}_\varepsilon \quad (5.54)$$

if $\varepsilon = 0$ or 1 satisfies $\varepsilon \equiv \varepsilon_1 + \varepsilon_2 \bmod 2$: this formula is valid in the usual sense (a convergent integral) in the case when $\operatorname{Re} \alpha > -1, \operatorname{Re} \beta >$

-1, Re $(\alpha + \beta) < -1$.

Note for easy reference the duplication formula

$$\Gamma(2z) = (2\pi)^{-\frac{1}{2}} 2^{2z-\frac{1}{2}} \Gamma(z)\Gamma(z + \frac{1}{2}), \qquad (5.55)$$

as well as

$$\sin \pi z = \frac{\pi}{\Gamma(z)\Gamma(1-z)}, \qquad \cos \pi z = \frac{\pi}{\Gamma(\frac{1}{2}-z)\Gamma(\frac{1}{2}+z)}. \qquad (5.56)$$

Let us finally mention the following immediate consequence of the definition of Eulerian integrals of the first species, involving an improper integral:

$$\int_0^\infty r^\alpha e^{2i\pi t r^2} dr = \frac{1}{2} (2\pi|t|)^{\frac{-\alpha-1}{2}} \Gamma(\frac{\alpha+1}{2}) e^{\frac{i\pi}{4}(\alpha+1)\,\text{sign}\,t}, \qquad -1 < \text{Re}\,\alpha < 1.$$

$$(5.57)$$

Remarks on the composition of Weyl symbols

The reader may be familiar with the formula

$$(h_1 \# h_2)(X) = 4 \int_{\mathbb{R}^4} h_1(Y) h_2(Z) e^{-4i\pi[Y-X,Z-X]} dY\, dZ \qquad (5.58)$$

where $X = (x, \xi)$, etc... and

$$[(y, \eta), (z, \zeta)] = -y\zeta + z\eta. \qquad (5.59)$$

This formula is nothing but Weyl's exponential form of Heisenberg's uncertainty relations, and may also be stated as follows: if $h_j(x, \xi) = e^{2i\pi(a_j x + \alpha_j \xi)}$, $j = 1, 2$, then

$$(h_1 \# h_2)(x, \xi) = e^{2i\pi(\alpha_1 a_2 - a_1 \alpha_2)} h_1(x, \xi) h_2(x, \xi). \qquad (5.60)$$

Despite appearances, formula (5.12), the analysis of which was the object of the present section, is, on a deep level, similar to (5.60), once one has accepted the following rule of thumb: to properly state the composition rule in some *good* (*i.e.*, with a large covariance group) symbolic calculus, expand symbols (when possible!) as sums of joint eigenfunctions of the set of differential operators on the phase space which commute with the action of the covariance group G. Now, the Weyl calculus benefits from two covariance properties: the one under the Heisenberg representation, for which the covariance group is $\mathbb{R}^2$, acting on $\mathbb{R}^2$ by translations; and the one under the metaplectic representation, for which the covariance group is $SL(2, \mathbb{R})$, acting on $\mathbb{R}^2$ in a linear way. In the two cases under consideration, our rule calls for the consideration of joint eigenfunctions of the set of differential operators on $\mathbb{R}^2$ with constant coefficients, *resp.* that of eigenfunctions of the Euler operator. Thus, we are led to decomposing symbols into (continuous) sums of exponentials of the type $(x, \xi) \mapsto e^{2i\pi(ax+\alpha\xi)}$, *resp.* into homogeneous functions. This leads to (5.60) and (5.12) respectively.

Another version of (5.58) is

$$(h_1 \# h_2)(X) = \{e^{i\pi L}(h_1(X + Y) \, h_2(X + Z))\} \, (Y = Z = 0),$$

$$(5.61)$$

with

$$i\pi L = \frac{1}{4i\pi} [-\frac{\partial^2}{\partial y \, \partial \zeta} + \frac{\partial^2}{\partial z \, \partial \eta}].$$

$$(5.62)$$

As will be recalled in section 17, this formula admits an extension to a certain situation (this is far from being the only one) in which the Poincaré upper half–plane rather than $\mathbf{R}^2$ has the role of a phase space: of course, the X–dependent chart $Y \mapsto X + Y$ has to be replaced by some chart Φ_X (17.33) dependent on hyperbolic geometry. But, most important (cf. (17.36)–(17.35)) the exponential $e^{i\pi L}$ must be replaced by some function $E(i\pi L)$ with an essential singularity (of Fuchs type: cf. [42], p.152) at zero. As a consequence, the quite familiar (and often useful as well) Moyal–type expansion which one derives from (5.61) by expanding the exponential into a powers series in L cannot be generalized to situations involving the quantization of Riemannian symmetric spaces. However, our "rule of thumb" still applies to this case, yielding (17.32).

6. The Roelcke–Selberg decomposition and the Radon transform

Since the quasi–regular action of G in $L^2_{even}(\mathbb{R}^2\backslash\{0\})$ decomposes, as indicated in (4.24), as the continuous sum of the representations $\pi_{i\lambda}$, no smooth dependence with respect to λ is involved in the space C^∞_G of $C^\infty-$vectors for this action. However, consider the representation $\tilde{\pi}$ of $\mathbb{R}$ into C^∞_G defined by

$$(\tilde{\pi}(\mu)h)(\xi) = |\xi|^{-i\mu}h(\xi) \tag{6.1}$$

and the space $C^\infty(\mathbb{R}, C^\infty_G)$ of $C^\infty-$vectors of this action: as it follows from $(4.20),(4.21)$ and (6.1),

$$(\tilde{\pi}(\mu)h)^\flat_\lambda(s) = \frac{1}{2\pi}\int_0^\infty t^{i(\lambda-\mu)}(1+s^2)^{-\frac{i\mu}{2}}h(ts,t)\,dt\,, \tag{6.2}$$

so that

$$\left(\frac{\partial}{\partial\lambda} + \frac{\partial}{\partial\mu}\right)\left((1+s^2)^{\frac{i\lambda}{2}}(\tilde{\pi}(\mu)h)^\flat_\lambda\right) = 0\,. \tag{6.3}$$

From the definition, given just before definition 2.3, of the notion of C^∞ map associating with λ an element of the space $C^\infty_{i\lambda}$, it then follows that if $h \in C^\infty(\mathbb{R}, C^\infty_G)$, the function $h^\flat_\lambda$, as an element of $C^\infty_{i\lambda}$, is a C^∞ function of λ: recall that this means that $s \mapsto (1+s^2)^{\frac{i\lambda}{2}}h^\flat_\lambda(s)$, as an element of the space of $C^\infty-$vectors of the representation π_ν taken at $\nu = 0$, depends smoothly on λ. In particular, one can then define $h^\flat_\lambda$ for *all* λ. We shall denote as $C^\infty_0(\mathbb{R}, C^\infty_G)$ the subspace of $h \in C^\infty(\mathbb{R}, C^\infty_G)$ for which h_λ is zero for large $|\lambda|$, or λ near 0.

One should note that if $h \in C^\infty(\mathbb{R}, C^\infty_G)$, then $\hat{h}^\flat_\lambda(n)$, $n \neq 0$, as well as $\mathrm{res}_0(\hat{h}^\flat_\lambda)$ and $\mathrm{res}_{i\lambda}(\hat{h}^\flat_\lambda)$ (*cf.* definition 3.5), are C^∞ functions of λ: indeed, $h^\flat_\lambda$ depends smoothly on λ as an element of $L^1(\mathbb{R})$, which proves the first point; (3.15) then connects $\mathrm{res}_{i\lambda}(\hat{h}^\flat_\lambda)$ to the value at 0 of $\pi_{i\lambda}(\left(\begin{smallmatrix}0 & -1\\ 1 & 0\end{smallmatrix}\right))h_\lambda$, and the other residue at zero of $\hat{h}^\flat_\lambda$ is taken care of in the same way.

Our plan is to show that if $f \in L^2(\Gamma\backslash\Pi)$ is orthogonal to the constants, the formula $f = V^*T^*TVf$, or (*cf.* (4.29))

$$f = \int_{-\infty}^\infty V^*T^*(TVf)_\lambda\,d\lambda\,, \tag{6.4}$$

when properly understood, is still valid. Since now the spectrum of Δ has a discrete part, we cannot expect the (distribution–valued) "map" $\lambda \mapsto (TVf)_\lambda$ to be a genuine one, but we must interpret it as a distribution (actually a measure), or even (this will lie deeper) as a hyperfunction with respect to λ too, somewhat in the spirit of the following lemma.

LEMMA 6.1. *Let λ_1 and σ be fixed non–zero real numbers. As a distribution in λ, one has*

$$[\Gamma(\frac{i\lambda}{2})\Gamma(\frac{-i\lambda}{2})]^{-1} \int_0^\infty K_{\frac{i\lambda_1}{2}}(2\pi|\sigma|y) \, K_{\frac{i\lambda}{2}}(2\pi|\sigma|y) \, \frac{dy}{y} =$$
$$\pi \, [\delta(\lambda - \lambda_1) + \delta(\lambda + \lambda_1)]. \quad (6.5)$$

Writing the right–hand side of this equation (in the sense of hyperfunction theory) as the difference of the values on the real line of the two holomorphic functions both given as $\lambda \mapsto \frac{i}{2}[\frac{1}{\lambda-\lambda_1} + \frac{1}{\lambda+\lambda_1}]$, the first one in the upper half-plane, the second in the lower half–plane, and recalling that

$$[\Gamma(\frac{i\lambda}{2})\Gamma(\frac{-i\lambda}{2})]^{-1} K_{\frac{i\lambda}{2}} = \frac{i\lambda}{4} [I_{\frac{i\lambda}{2}} - I_{-\frac{i\lambda}{2}}], \quad (6.6)$$

one should also note that, when $\text{Im } \lambda > 0$, one has the formula

$$-\frac{i\lambda}{4} \int_0^\infty K_{\frac{i\lambda_1}{2}}(2\pi|\sigma|y) \, I_{-\frac{i\lambda}{2}}(2\pi|\sigma|y) \, \frac{dy}{y} = \frac{i\lambda}{\lambda^2 - \lambda_1^2}, \quad (6.7)$$

and that, when $\text{Im } \lambda < 0$, the same equation is true after one has substituted $I_{\frac{i\lambda}{2}}$ for $I_{-\frac{i\lambda}{2}}$.

PROOF. What we mean in the first part is that, given $\chi \in C_0^\infty(\mathbb{R})$, one has

$$\int_0^\infty K_{\frac{i\lambda_1}{2}}(2\pi|\sigma|y) \left[\int_{-\infty}^\infty K_{\frac{i\lambda}{2}}(2\pi|\sigma|y)\chi(\lambda)\,d\lambda\right] \frac{dy}{y} =$$
$$\pi \, \Gamma(\frac{i\lambda_1}{2})\Gamma(\frac{-i\lambda_1}{2}) \, [\chi(\lambda_1) + \chi(-\lambda_1)]. \quad (6.8)$$

Note that, as shown by an integration by parts corresponding to the formula

$$y^{\pm\frac{i\lambda}{2}} = (-2i\frac{d}{d\lambda})^2 \, [(\log y)^{-2} \, y^{\pm\frac{i\lambda}{2}}], \quad (6.9)$$

performed only, say, when $y < \frac{1}{2}$, the left–hand side of (6.8) makes sense, and is also the limit, as ε goes to zero, of the same integral regularized by means of the extra factor $y^{\frac{\varepsilon}{2}}$. Starting from the Weber–Schafheitlin integral ([27], p.101)

$$I_\varepsilon(\lambda) := \int_0^\infty K_{\frac{i\lambda_1}{2}}(2\pi|\sigma|y) \, K_{\frac{i\lambda}{2}}(2\pi|\sigma|y) \, y^{\frac{\varepsilon}{2}-1} \, dy = \frac{2^{\frac{\varepsilon}{2}-3}(2\pi|\sigma|)^{-\frac{\varepsilon}{2}}}{\Gamma(\frac{\varepsilon}{2})} \times$$
$$\Gamma(\frac{\varepsilon + i(\lambda + \lambda_1)}{4})\Gamma(\frac{\varepsilon + i(\lambda - \lambda_1)}{4})\Gamma(\frac{\varepsilon + i(-\lambda + \lambda_1)}{4})\Gamma(\frac{\varepsilon - i(\lambda + \lambda_1)}{4}), \quad (6.10)$$

let us first observe that, as $\varepsilon \to 0$, this converges uniformly to zero (because of the $\Gamma(\frac{\varepsilon}{2})$ factor) when λ stays away from $\pm\lambda_1$ in a fixed way. Assuming that $\lambda_1 > 0$ and that χ is supported in $]0,\infty[$, we get, up to error terms tending to zero as $\varepsilon \to 0$,

$$\int_{-\infty}^{\infty} I_\varepsilon(\lambda)\,\chi(\lambda)\,d\lambda \sim \frac{1}{8}\Gamma(\frac{i\lambda_1}{2})\,\Gamma(-\frac{i\lambda_1}{2})\int_{-\infty}^{\infty} \frac{\Gamma(\frac{\varepsilon+i(\lambda-\lambda_1)}{4})\,\Gamma(\frac{\varepsilon-i(\lambda-\lambda_1)}{4})}{\Gamma(\frac{\varepsilon}{2})}\chi(\lambda)\,d\lambda$$

$$\sim \pi\,\Gamma(\frac{i\lambda_1}{2})\,\Gamma(-\frac{i\lambda_1}{2})\times\frac{1}{\pi}\int_{-\infty}^{\infty}\frac{\varepsilon}{(\lambda-\lambda_1)^2+\varepsilon^2}\chi(\lambda)\,d\lambda$$

$$\sim \pi\,\Gamma(\frac{i\lambda_1}{2})\,\Gamma(-\frac{i\lambda_1}{2})\,\chi(\lambda_1). \tag{6.11}$$

For the second part, one needs help too: [13], p.80, gives

$$\int_0^\infty y^{-\lambda}K_\mu(ay)I_\nu(ay)\,dy = a^{\lambda-1}\frac{\Gamma(\lambda)}{2^{\lambda+1}}\frac{\Gamma(\frac{1-\lambda+\mu+\nu}{2})\Gamma(\frac{1-\lambda-\mu+\nu}{2})}{\Gamma(\frac{1+\lambda+\mu+\nu}{2})\Gamma(\frac{1+\lambda-\mu+\nu}{2})} \tag{6.12}$$

if $a > 0$, $0 < \operatorname{Re}\lambda < 1 + \operatorname{Re}(\nu\pm\mu)$; in particular,

$$\int_0^\infty K_\mu(ay)\,I_\nu(ay)\,\frac{dy}{y} = \frac{1}{\nu^2-\mu^2} \tag{6.13}$$

if $\operatorname{Re}\nu > |\operatorname{Re}\mu|$. $\qquad\qquad\square$

LEMMA 6.2. *Let $h \in C_0^\infty(\mathbf{R}, C_G^\infty)$. For every $y > 0$, the function*

$$\sigma \mapsto (Lh)(\sigma, y) = \int_{-\infty}^{\infty}[\Gamma(\frac{i\lambda}{2})\Gamma(\frac{-i\lambda}{2})]^{-1}|\sigma|^{\frac{i\lambda}{2}}\,K_{\frac{i\lambda}{2}}(2\pi|\sigma|y)\,\hat{h}_{-\lambda}^b(\sigma)\,d\lambda \tag{6.14}$$

is continuous on the real line. One has

$$(Lh)(0, y) = \frac{1}{2}\int_{-\infty}^{\infty}\left[\frac{(\pi y)^{-\frac{i\lambda}{2}}}{\Gamma(-\frac{i\lambda}{2})}\operatorname{res}_0(\hat{h}_{-\lambda}^b) + \frac{(\pi y)^{\frac{i\lambda}{2}}}{\Gamma(\frac{i\lambda}{2})}\operatorname{res}_{-i\lambda}(\hat{h}_{-\lambda}^b)\right]d\lambda\,; \tag{6.15}$$

this latter function is $O(|\log y|^{-\infty})$ near $y = 0$, and so is $(Lh)(\sigma, y)$ for any fixed $\sigma \neq 0$.

PROOF. Obviously, the function under consideration is C^∞ outside $\sigma = 0$, and rapidly decreasing at infinity. Near $\sigma = 0$, proposition 2.1 provides us with the expansion

$$\hat{h}_{-\lambda}^b(\sigma) = \operatorname{res}_0(\hat{h}_{-\lambda}^b) + \operatorname{res}_{-i\lambda}(\hat{h}_{-\lambda}^b)|\sigma|^{-i\lambda} + O(|\sigma|) \tag{6.16}$$

and, classically,

$$2|\sigma|^{\frac{i\lambda}{2}}K_{\frac{i\lambda}{2}}(2\pi|\sigma|y) = \Gamma(\frac{i\lambda}{2})(\pi y)^{-\frac{i\lambda}{2}} + \Gamma(-\frac{i\lambda}{2})(\pi y)^{\frac{i\lambda}{2}}|\sigma|^{i\lambda} + O(|\sigma|). \tag{6.17}$$

If one groups the first term in the expansion of $\hat{h}_{-\lambda}^b(\sigma)$ with the second term in that of $|\sigma|^{\frac{i\lambda}{2}}K_{\frac{i\lambda}{2}}(2\pi|\sigma|y)$, one gets as a contribution to $(Lh)(\sigma, y)$ the integral

$$\frac{1}{2} \int_{-\infty}^{\infty} (\Gamma(\frac{i\lambda}{2}))^{-1} (\pi y)^{\frac{i\lambda}{2}} \operatorname{res}_0(\hat{h}^{\flat}_{-\lambda}) |\sigma|^{i\lambda} d\lambda =$$

$$\frac{1}{2} \int_{-\infty}^{\infty} i \frac{d}{d\lambda} \left[\frac{(\pi y)^{\frac{i\lambda}{2}}}{\Gamma(\frac{i\lambda}{2})} \operatorname{res}_0(\hat{h}^{\flat}_{-\lambda}) \right] (\log|\sigma|)^{-1} |\sigma|^{i\lambda} d\lambda, \quad (6.18)$$

where the right–hand side has been obtained after an integration by parts. This shows that the corresponding term in the integral which defines $L(\sigma, y)$ goes to zero as $\sigma \to 0$. We are thus left with a function which has the same limit, as σ goes to 0, as that obtained from pairing only the terms, in the expansions (6.16) and (6.17), which associate so as to produce a result independant of σ. This yields the first part of lemma 6.2 together with the value $(Lh)(0, y)$, and an integration by parts in this latter integral or the one defining $(Lh)(\sigma, y)$ when $\sigma \neq 0$, based on

$$y^{\pm \frac{i\lambda}{2}} = (\mp 2i \frac{d}{d\lambda})^k ((\log y)^{-k} y^{\pm \frac{i\lambda}{2}}),$$

permits to conclude.   □

Before introducing the definition of TVf in the case when $f \in L^2(\Gamma \backslash \Pi)$, the Hilbert space consisting of all Γ–invariant functions f on Π such that

$$\int_D |f(z)|^2 d\mu(z) < \infty \quad (6.19)$$

if D is a fundamental domain of Γ in Π — recall that the most usual one is characterized by the inequalities $-\frac{1}{2} < \operatorname{Re} z < \frac{1}{2}$, $|z| > 1$ — we substitute for T (which was first defined in (4.10)) the operator

$$S = 2^{-\frac{1}{2}} \pi^{i\pi\mathcal{E}} (\Gamma(i\pi\mathcal{E}))^{-1} T$$

introduced in (4.46). This will only change $(TVf)^{\flat}_{\lambda}$ by a numerical factor, and one should notice that the former choice had been made so that TV should be an isometry from $L^2(\Pi)$ onto some subspace of $L^2(\mathbb{R}^2 \backslash \{0\})$: now, this has become irrelevant since, on Π, the Hilbert space we are interested in is no longer $L^2(\Pi)$ but $L^2(\Gamma \backslash \Pi)$ as defined by the Petersson scalar product. On the other hand, besides leading to the simple equation (4.45) for the inverse of $\Theta_{i\lambda}$, our new choice will permit us to canonically define the distribution $\lambda \mapsto (SVf)^{\flat}_{\lambda}$ as the difference of boundary values of two holomorphic functions on the upper and lower half–planes.

In view of (4.36) and (4.46), the formula (6.4) takes the form

$$f = \int_{-\infty}^{\infty} \Theta_{i\lambda}(SVf)^{\flat}_{\lambda} d\lambda, \quad (6.20)$$

and we shall show that, indeed, it is valid in the Γ-invariant setting.

DEFINITION 6.3. Let $f \in L^2_{[\frac{1}{4},\infty[}(\Gamma\backslash\Pi)$, the hyperplane in the Hilbert space $L^2(\Gamma\backslash\Pi)$ orthogonal to the constants; assume, moreover, that f lies in the image of the spectral projection corresponding to some interval $[a,b] \subset]\frac{1}{4},\infty[$ in the spectrum of Δ. We then define SVf, as a linear form on $C_0^\infty(\mathbb{R}, C_G^\infty)$, by the equation

$$< SVf, h >= \sum_{n\in\mathbf{Z}} \int_0^\infty a_n(y)\, y^{-\frac{3}{2}}\, (Lh)(n,y)\, dy, \qquad (6.21)$$

where Lh has been defined in the preceding lemma and

$$f(x+iy) = \sum_{n\in\mathbf{Z}} a_n(y)\, e^{-2i\pi nx} \qquad (6.22)$$

is the Fourier expansion of f.

Remarks. 1. The notation $L^2_{[\frac{1}{4},\infty[}(\Gamma\backslash\Pi)$ refers to the fact that we omit the isolated point 0 of the spectrum of Δ from our considerations.

2. The definition generalizes the one that would follow from the considerations in section 4 in the case when $f \in C_0^\infty(\Pi)$. Indeed, if h_1 and h lie in $L^2_{\text{even}}(\mathbb{R}^2\backslash\{0\})$, the polarized version of the identity

$$\|h\|^2_{L^2(\mathbf{R}^2\backslash\{0\})} = 4\pi \int_{-\infty}^\infty \|h^\flat_\lambda\|^2_{L^2(\mathbf{R})}\, d\lambda \qquad (6.23)$$

(recalling (4.22)), together with

$$\overline{(h^\flat_\lambda)}(s) = (\bar{h})^\flat_{-\lambda}(s), \qquad (6.24)$$

yields

$$\int_{-\infty}^\infty h_1(\xi)\, h(\xi)\, d\xi = 4\pi \int_{-\infty}^\infty d\lambda \int_{-\infty}^\infty (h_1)^\flat_\lambda(s)\, h^\flat_{-\lambda}(s)\, ds. \qquad (6.25)$$

In particular, if $f \in C_0^\infty(\Pi)$ and $h \in C_0^\infty(\mathbb{R}, C_G^\infty)$, one may write

$$< TVf, h >= 4\pi \int_{-\infty}^\infty < (TVf)^\flat_\lambda, h^\flat_{-\lambda} > d\lambda \qquad (6.26)$$

or

$$< SVf, h >= 4\pi \int_{-\infty}^\infty < (SVf)^\flat_\lambda, h^\flat_{-\lambda} > d\lambda, \qquad (6.27)$$

where, from the definition (4.46) of S,

$$(SVf)^\flat_\lambda = \frac{2^{-\frac{1}{2}}\pi^{-\frac{i\lambda}{2}}}{\Gamma(-\frac{i\lambda}{2})}(TVf)^\flat_\lambda. \qquad (6.28)$$

From (6.28) and (4.34), $(SVf)^\flat_\lambda \in \mathcal{D}'_{i\lambda}$ should be defined as

$$< (SVf)^\flat_\lambda, u >= \frac{1}{8}\, \pi^{-\frac{3}{2}-\frac{i\lambda}{2}} \frac{\Gamma(\frac{1+i\lambda}{2})}{\Gamma(\frac{i\lambda}{2})\Gamma(-\frac{i\lambda}{2})} \int_\Pi f(z)\, d\mu(z)$$

$$\int_{-\infty}^\infty \left(\frac{|z-s|^2}{y}\right)^{-\frac{1}{2}-\frac{i\lambda}{2}} u(s)\, ds\,, \qquad u \in C^\infty_{-i\lambda}. \quad (6.29)$$

Next, as it follows from (3.25),

$$\int_{-\infty}^\infty \left(\frac{|z-s|^2}{y}\right)^{-\frac{1}{2}-\frac{i\lambda}{2}} u(s)\, ds = \frac{2\pi^{\frac{1+i\lambda}{2}}}{\Gamma(\frac{1+i\lambda}{2})}\, y^{\frac{1}{2}}$$

$$\int_{-\infty}^\infty |\sigma|^{\frac{i\lambda}{2}} K_{\frac{i\lambda}{2}}(2\pi|\sigma|y)\, e^{2i\pi\sigma x}\, \hat{u}(\sigma)\, d\sigma. \quad (6.30)$$

From (6.27), (6.29) and (6.30), we thus get

$$< SVf, h >= \int_0^\infty \int_{-\infty}^\infty f(x+iy)\, y^{-\frac{3}{2}}\, dy\, dx \int_{-\infty}^\infty \frac{d\lambda}{\Gamma(\frac{1+i\lambda}{2})\Gamma(\frac{1-i\lambda}{2})}$$

$$\int_{-\infty}^\infty |\sigma|^{\frac{i\lambda}{2}}\, K_{\frac{i\lambda}{2}}(2\pi|\sigma|y)\, e^{2i\pi\sigma x}\, \hat{h}^\flat_{-\lambda}(\sigma)\, d\sigma \quad (6.31)$$

or, in terms of Lh as defined in lemma 6.2,

$$< SVf, h >= \int_0^\infty \int_{-\infty}^\infty f(x+iy)\, y^{-\frac{3}{2}}\, dy\, dx$$

$$\int_{-\infty}^\infty (Lh)(\sigma, y)\, e^{2i\pi\sigma x}\, d\sigma. \quad (6.32)$$

If, instead of lying in $C_0^\infty(\Pi)$, f is periodic with the Fourier expansion (6.22), this is no longer convergent since the Fourier transform that occurs on the second line of the preceding equation is not integrable as a function of x (it is only L^p for $p > 1$): however, *formally*, we would get just (6.21), which we shall now take as a definition.

Our program in the rest of this section is to first show that the definition makes sense and to compute $< SVf, h >$ (or, in view of (6.27), $(SVf)^\flat_\lambda$) in terms of the Roelcke–Selberg decomposition of f. Next, to show that the "map" $\lambda \mapsto (SVf)^\flat_\lambda$, actually a distribution, can be written in a canonical way as the difference of two (distribution–valued) holomorphic functions in the upper and lower half-planes respectively.

The Roelcke–Selberg theorem ([36], p.254 or [23], chapter 5) expresses $f \in L^2_{[\frac{1}{4},\infty[}(\Gamma\backslash\Pi)$ as

$$f(z) = \sum_{k\geq 1} \Phi^k\, \mathcal{M}_k(z) + \frac{1}{8\pi} \int_{-\infty}^\infty \Phi(\rho)\, E_{\frac{1-i\rho}{2}}(z)\, d\rho\,, \quad (6.33)$$

in which $(\mathcal{M}_k)$ is a complete orthonormal set of Maass cusp forms, $\Phi^k = (\mathcal{M}_k|f)$ for all k and, for almost all ρ, $\Phi(\rho) = (E_{\frac{1-i\rho}{2}}|f)$ (we define the inner product $(\ |\ )$ on $L^2(\Gamma\backslash\Pi)$ to be linear with respect to the second variable).

Let

$$\mathcal{M}_k(x + iy) = y^{\frac{1}{2}} \sum_{n\neq 0} b_n^k\, K_{\frac{i\lambda_k}{2}}(2\pi|n|y)\, e^{-2i\pi nx} \tag{6.34}$$

be the Fourier expansion ([36], p.208) of the cusp–form $\mathcal{M}_k$, corresponding to the eigenvalue $\frac{1+\lambda_k^2}{4}$ of Δ: we may assume that the sequence (λ_k) is not decreasing and consists of positive numbers. Also, let

$$E_{\frac{1-i\rho}{2}}(x + iy) = y^{\frac{1-i\rho}{2}} + b_0^\rho\, y^{\frac{1+i\rho}{2}} + \sum_{n\neq 0} b_n^\rho\, y^{\frac{1}{2}}\, K_{\frac{i\rho}{2}}(2\pi|n|y)\, e^{-2i\pi nx} \tag{6.35}$$

be the Fourier expansion (recalled in (3.29)) of the Eisenstein series on the left–hand side. Then the coefficients $a_n(y)$ of the expansion (6.22) are given by

$$a_0(y) = \frac{1}{8\pi} \int_{-\infty}^{\infty} \Phi(\rho)\, [y^{\frac{1-i\rho}{2}} + b_0^\rho\, y^{\frac{1+i\rho}{2}}]\, d\rho \tag{6.36}$$

and, if $n \neq 0$,

$$a_n(y) = y^{\frac{1}{2}} \sum_{k\geq 1} b_n^k\, \Phi^k\, K_{\frac{i\lambda_k}{2}}(2\pi|n|y)$$

$$+ \frac{1}{8\pi}\, y^{\frac{1}{2}} \int_{-\infty}^{\infty} \Phi(\rho)\, b_n^\rho\, K_{\frac{i\rho}{2}}(2\pi|n|y)\, d\rho. \tag{6.37}$$

In both cases the series in k has finitely many non–zero terms, and Φ has a compact support disjoint from $\{0\}$.

From (3.29) and (6.35), we get

$$b_0^\rho = \pi^{-i\rho}\frac{\Gamma(\frac{1+i\rho}{2})\,\zeta(1 + i\rho)}{\Gamma(\frac{1-i\rho}{2})\,\zeta(1 - i\rho)} \tag{6.38}$$

and since, as a consequence of its definition right after (6.33), $\Phi(\rho)$ satisfies the property that

$$\pi^{-\frac{i\rho}{2}}\Gamma(\frac{1+i\rho}{2})\zeta(1 + i\rho)\, \Phi(\rho) \qquad \text{is even}, \tag{6.39}$$

(6.36) yields

$$a_0(y) = \frac{1}{8\pi} \int_{-\infty}^{\infty} [\,\Phi(\rho)\, y^{\frac{1-i\rho}{2}} + \Phi(-\rho)\, y^{\frac{1+i\rho}{2}}\,]\, d\rho$$

$$= \frac{1}{4\pi} \int_{-\infty}^{\infty} \Phi(\rho)\, y^{\frac{1-i\rho}{2}}\, d\rho. \tag{6.40}$$

Also, still from (3.29) and (6.35),

$$b_n^\rho = \frac{2\pi^{\frac{1-i\rho}{2}}}{\Gamma(\frac{1-i\rho}{2})\zeta(1-i\rho)} |n|^{-\frac{i\rho}{2}} \sigma_{i\rho}(|n|), \qquad (6.41)$$

and from (6.39) again it follows that

$$\Phi(\rho) b_n^\rho = \Phi(-\rho) b_n^{-\rho}. \qquad (6.42)$$

THEOREM 6.4. Let $f \in L^2(\Gamma\backslash\Pi)$ be a cusp-form corresponding to the eigenvalue $\frac{1+\lambda_1^2}{2}$ of Δ, with Fourier expansion

$$f(z) = \sum_{n\neq 0} b_n \, y^{\frac{1}{2}} \, K_{\frac{i\lambda_1}{2}}(2\pi|n|y) \, e^{-2i\pi nx}. \qquad (6.43)$$

Define $\mathfrak{T}^+ \in \mathcal{D}'_{i\lambda_1}$ by the equation

$$< \mathfrak{T}^+, u > = \sum_{n\neq 0} c_n^+ \, \hat{u}(n), \qquad u \in C^\infty_{-i\lambda_1}, \qquad (6.44)$$

where

$$c_n^+ = \frac{1}{2} |n|^{\frac{i\lambda_1}{2}} b_n. \qquad (6.45)$$

Define $\mathfrak{T}^-$ in a similar way, changing λ_1 into its negative. Then

$$\mathfrak{T}^- = \theta_{i\lambda_1} \mathfrak{T}^+, \qquad (6.46)$$

where the intertwining operator $\theta_{i\lambda_1}$ was defined on distributions in (3.32). Finally, for every $h \in C_0^\infty(\mathbb{R}, C_G^\infty)$, one has

$$\frac{1}{2\pi} < SVf, h > = < \mathfrak{T}^+, h^\flat_{-\lambda_1} > + < \mathfrak{T}^-, h^\flat_{\lambda_1} >. \qquad (6.47)$$

PROOF. That $\mathfrak{T}^+$ belongs to $\mathcal{D}'_{i\lambda_1}$ comes from the fact that the sequence (c_n^+) has polynomial growth at most: then it is an $(i\lambda_1)$-modular distribution since, as follows from (3.27) and (6.43), one has $< \mathfrak{T}^+, \phi_z^{-i\lambda_1} > = f(z)$ for all $z \in \Pi$ and the linear span of the set of functions $\phi_z^{-i\lambda_1}$, $z \in \Pi$, is dense in $C^\infty_{-i\lambda_1}$.

To get the relation between $\mathfrak{T}^+$ and $\mathfrak{T}^-$, start from

$$< \mathfrak{T}^+, \phi_z^{-i\lambda_1} > = f(z) = < \mathfrak{T}^-, \phi_z^{i\lambda_1} > \qquad (6.48)$$

since (3.27) can be applied with $\nu = \pm i\lambda_1$; also remember that $\theta_{-\nu}\phi^{-\nu} = \phi^\nu$ as a consequence of (2.12) and (2.15). Since $\theta_{i\lambda_1}$ is an intertwining operator, one thus gets, with $z = g.i$,

$$
\begin{aligned}
< \theta_{i\lambda_1} \mathfrak{T}^+, \phi_z^{i\lambda_1} > &= < \mathfrak{T}^+, \theta_{i\lambda_1} \phi_z^{i\lambda_1} > \\
&= < \mathfrak{T}^+, \theta_{i\lambda_1} \pi_{i\lambda_1}(g) \phi^{i\lambda_1} > \\
&= < \mathfrak{T}^+, \pi_{-i\lambda_1}(g) \theta_{i\lambda_1} \phi^{i\lambda_1} > \\
&= < \mathfrak{T}^+, \pi_{-i\lambda_1}(g) \phi^{-i\lambda_1} > \\
&= < \mathfrak{T}^+, \phi_z^{-i\lambda_1} > ,
\end{aligned}
\tag{6.49}
$$

which yields the result in view of (6.48).

In this case, $a_0(y) = 0$ and, for $n \neq 0$,

$$
a_n(y) = b_n \, y^{\frac{1}{2}} \, K_{\frac{i\lambda_1}{2}}(2\pi|n|y), \tag{6.50}
$$

so that definition 6.3 and lemma 6.2 permit to write

$$
< SVf, h > = \sum_{n \neq 0} b_n \int_0^\infty K_{\frac{i\lambda_1}{2}}(2\pi|n|y) \frac{dy}{y}
$$
$$
\int_{-\infty}^\infty \frac{1}{\Gamma(\frac{i\lambda}{2})\Gamma(\frac{-i\lambda}{2})} |n|^{\frac{i\lambda}{2}} K_{\frac{i\lambda}{2}}(2\pi|n|y) \hat{h}^b_{-\lambda}(n) \, d\lambda \tag{6.51}
$$

or, using lemma 6.1,

$$
\frac{1}{\pi} < SVf, h > = \sum_{n \neq 0} b_n \left[|n|^{\frac{i\lambda_1}{2}} \hat{h}^b_{-\lambda_1}(n) + |n|^{-\frac{i\lambda_1}{2}} \hat{h}^b_{\lambda_1}(n) \right]. \tag{6.52}
$$

This is just the same as the right–hand side of (6.47), as seen from (6.44) and (6.45). $\qquad\qquad\square$

Remark. In view of (6.27), one may also state (6.47) in the possibly more expressive form

$$
(SVf)^b_\lambda = \frac{1}{2} \left[\delta(\lambda - \lambda_1) \mathfrak{T}^+ + \delta(\lambda + \lambda_1) \mathfrak{T}^- \right]. \tag{6.53}
$$

The formula (6.20), which expresses the spectral decomposition of f in terms of that of its Radon transform, immediately follows since (6.48) can also be written as

$$
\Theta_{i\lambda} \mathfrak{T}^+ = f = \Theta_{-i\lambda} \mathfrak{T}^-. \tag{6.54}
$$

THEOREM 6.5. *Let f, satisfying the hypotheses of definition 6.3, reduce to its continuous part*

$$
f(z) = \frac{1}{8\pi} \int_{-\infty}^\infty \Phi(\rho) \, E_{\frac{1-i\rho}{2}}(z) \, d\rho, \tag{6.55}
$$

taken from (6.33). Then

$$
< SVf, h > = \frac{1}{2} \int_{-\infty}^\infty (\zeta^*(1 - i\lambda))^{-1} \Phi(\lambda) < \mathfrak{E}_{i\lambda}, h^b_{-\lambda} > d\lambda, \tag{6.56}
$$

with

$$\zeta^*(s): \ = \pi^{-\frac{s}{2}} \Gamma(\frac{s}{2}) \zeta(s) = \zeta^*(1-s). \tag{6.57}$$

Remark. Using (6.27) again as a definition of $(SVf)^\flat_\lambda$, one can write this as

$$(SVf)^\flat_\lambda = \frac{1}{8\pi} (\zeta^*(1-i\lambda))^{-1} \Phi(\lambda) \, \mathfrak{E}_{i\lambda}. \tag{6.58}$$

As a consequence of (6.58) and proposition 3.4,

$$\Theta_{i\lambda}(SVf)^\flat_\lambda = \frac{1}{8\pi} \Phi(\lambda) \, E_{\frac{1-i\lambda}{2}}, \tag{6.59}$$

so that the decomposition formula (6.20) is again verified.

PROOF. Let W be the linear form on $C^\infty(\mathbb{R}, C_G^\infty)$ such that $< W, h >$ is defined as the right–hand side of (6.56). As a consequence of proposition 3.8, one has

$$< W, h > = \frac{1}{2} \int_{-\infty}^{\infty} (\zeta^*(1-i\lambda))^{-1} \Phi(\lambda)$$

$$[\zeta(-i\lambda) \operatorname{res}_0(\hat{h}^\flat_{-\lambda}) + \zeta(i\lambda) \operatorname{res}_{-i\lambda}(\hat{h}^\flat_{-\lambda}) + \sum_{n \neq 0} \sigma_{i\lambda}(|n|) \, \hat{h}^\flat_{-\lambda}(n)] \, d\lambda. \tag{6.60}$$

On the other hand, let us split $< SVf, h >$, as expressed by definition 6.3, as

$$< SVf, h > = < SVf, h >_0 + < SVf, h >_1, \tag{6.61}$$

where the first term is the one with $n = 0$ taken from the right–hand side of (6.21). Then, from (6.40) and (6.15),

$$< SVf, h >_0 = \frac{1}{8\pi} \int_{-\infty}^{\infty} \Phi(\rho) \, d\rho \int_{0}^{\infty} y^{-\frac{i\rho}{2}} \frac{dy}{y}$$

$$\int_{-\infty}^{\infty} [\frac{(\pi y)^{-\frac{i\lambda}{2}}}{\Gamma(-\frac{i\lambda}{2})} \operatorname{res}_0(\hat{h}^\flat_{-\lambda}) + \frac{(\pi y)^{\frac{i\lambda}{2}}}{\Gamma(\frac{i\lambda}{2})} \operatorname{res}_{-i\lambda}(\hat{h}^\flat_{-\lambda})] \, d\lambda. \tag{6.62}$$

Using

$$\int_{0}^{\infty} y^{-\frac{i(\rho+\lambda)}{2}} \frac{dy}{y} = 4\pi \, \delta(\rho + \lambda) \tag{6.63}$$

with the same meaning as in lemma 6.1 (i.e., first integrate against a smooth function of λ with compact support), we get

$$< SVf, h >_0 = \frac{1}{2} \int_{-\infty}^{\infty} \frac{\pi^{\frac{i\rho}{2}}}{\Gamma(\frac{i\rho}{2})} \Phi(\rho) [\operatorname{res}_0(\hat{h}^\flat_\rho) + \operatorname{res}_{-i\rho}(\hat{h}^\flat_{-\rho})] \, d\rho, \tag{6.64}$$

an integral that we need to compare to the sum of the two cuspidal terms on the right–hand side of (6.60), in which we let $\lambda \mapsto -\rho$ in the first term,

$\lambda \mapsto \rho$ in the second one. The coefficients which accompany $\text{res}_{-i\rho}(\hat{h}_\rho^b)$ just fit as follows from the equation

$$\zeta^*(1 - i\rho) = \pi^{-\frac{i\rho}{2}} \Gamma(\frac{i\rho}{2}) \zeta(i\rho). \tag{6.65}$$

To compare the coefficients of $\text{res}_0(\hat{h}_\rho^b)$, we also use (6.39).

Finally, from definition 6.3, (6.37) and (6.14),

$$< SVf, h >_1 = \frac{1}{8\pi} \sum_{n \neq 0} \int_0^\infty \frac{dy}{y} \int_{-\infty}^\infty \Phi(\rho)\, b_n^\rho\, K_{\frac{i\rho}{2}}(2\pi|n|y)\, d\rho$$
$$\int_{-\infty}^\infty [\Gamma(\frac{i\lambda}{2})\Gamma(-\frac{i\lambda}{2})]^{-1}\, |n|^{\frac{i\lambda}{2}}\, K_{\frac{i\lambda}{2}}(2\pi|n|y)\, \hat{h}_{-\lambda}^b(n)\, d\lambda. \tag{6.66}$$

Since Φ is square–integrable (as a consequence of the Roelcke–Selberg theorem) and has compact support, we may integrate with respect to y before ρ, then use lemma 6.1, getting

$$< SVf, h >_1 = \frac{1}{8} \sum_{n \neq 0} \int_{-\infty}^\infty \Phi(\rho)\, b_n^\rho [\, |n|^{\frac{i\rho}{2}}\, \hat{h}_{-\rho}^b(n) + |n|^{-\frac{i\rho}{2}}\, \hat{h}_\rho^b(n)\,]\, d\rho \tag{6.67}$$

as a result. From the symmetry (6.42), it follows that

$$< SVf, h >_1 = \frac{1}{4} \sum_{n \neq 0} \int_{-\infty}^\infty \Phi(\rho)\, b_n^\rho\, |n|^{\frac{i\rho}{2}}\, \hat{h}_{-\rho}^b(n)\, d\rho, \tag{6.68}$$

a series readily compared, with the help of (6.41), to that which is contributed, on the right–hand side of (6.60), by the terms with $n \neq 0$.   □

Remark. Let us now bring the results of theorems 6.4 and 6.5 together. Let f, satisfying the hypotheses of definition 6.3, be given by the expansion (6.33). Set

$$\Psi(\lambda) = (\zeta^*(1 - i\lambda))^{-1}\, \Phi(\lambda), \tag{6.69}$$

so that (6.39) expresses that Ψ is an even function. Recall that the coefficients b_n^k in the Fourier expansions of the cusp–form $\mathcal{M}_k$ were defined in (6.34).

Using (6.53) and (6.59), one gets

$$< (SVf)_\lambda^b, u > = C_0(\lambda)\, \text{res}_0(\hat{u}) + C_\infty(\lambda)\, \text{res}_{-i\lambda}(\hat{u}) + \sum_{n \neq 0} c_n(\lambda)\hat{u}(n) \tag{6.70}$$

for every $u \in C^{\infty}_{-i\lambda}$, with

$$C_0(\lambda) = \frac{1}{8\pi} \zeta(-i\lambda) \Psi(\lambda),$$

$$C_{\infty}(\lambda) = \frac{1}{8\pi} \zeta(i\lambda) \Psi(\lambda),$$

$$c_n(\lambda) = \frac{1}{4} |n|^{\frac{i\lambda}{2}} \sum_{k \geq 0} \Phi^k b_n^k [\delta(\lambda - \lambda_k) + \delta(\lambda + \lambda_k)]$$

$$+ \frac{1}{8\pi} \sigma_{i\lambda}(|n|) \Psi(\lambda). \tag{6.71}$$

We now reformulate (6.71) in terms of hyperfunction theory. Let us also comment at once that part of the theorem which follows will be connected (under certain conditions) to the Rankin–Selberg method (*cf.* remark right after the proof of theorem 7.3). We adopt the notations of [4] and set

$$\varepsilon(\mu) = \begin{cases} 1 \text{ if } \mathrm{Im}\,\mu > 0 \\ 0 \text{ if } \mathrm{Im}\,\mu < 0 \end{cases}, \qquad \bar{\varepsilon}(\mu) = \begin{cases} 0 \quad \text{if } \mathrm{Im}\,\mu > 0 \\ -1 \text{ if } \mathrm{Im}\,\mu < 0 \end{cases}. \tag{6.72}$$

If $c = c(\mu)$ is a holomorphic function in a complex neighbourhood of $\mathbb{R}$ with $\mathbb{R}$ deleted, we also denote as $\{c(\mu)\}_{\mu=\lambda}$ the hyperfunction associated with c, i.e., the difference of the two traces of c on $\mathbb{R}$ from above and from below. In particular, if $\chi \in L^2(\mathbb{R})$, χ is associated, if so wished, with the analytic function

$$\mu \mapsto \frac{1}{2i\pi} \int_{-\infty}^{\infty} \chi(\rho) \frac{d\rho}{\rho - \mu}, \qquad \mu \in \mathbb{C} \backslash \mathbb{R}. \tag{6.73}$$

THEOREM 6.6. *Under the assumptions of definition 6.3, the distributions C_0, C_{∞}, c_n that make (6.70) valid are given as*

$$C_0(\lambda) = \left\{ \frac{1}{8\pi} \int_0^1 a_0(y) \, y^{-\frac{3}{2}} \frac{(\pi y)^{-\frac{i\mu}{2}}}{\Gamma(-\frac{i\mu}{2})} \, dy \, . \, \varepsilon(\mu) \right\}_{\mu=\lambda}$$

$$+ \left\{ \frac{1}{8\pi} \int_1^{\infty} a_0(y) \, y^{-\frac{3}{2}} \frac{(\pi y)^{-\frac{i\mu}{2}}}{\Gamma(-\frac{i\mu}{2})} \, dy \, . \, \bar{\varepsilon}(\mu) \right\}_{\mu=\lambda}, \tag{6.74}$$

$$C_{\infty}(\lambda) = C_0(-\lambda), \tag{6.75}$$

$$c_n(\lambda) = \left\{ -\frac{i\mu}{16\pi} \int_0^{\infty} a_n(y) \, y^{-\frac{3}{2}} I_{-\frac{i\mu}{2}}(2\pi|n|y) \, |n|^{\frac{i\mu}{2}} \, dy \, . \, \varepsilon(\mu) \right\}_{\mu=\lambda}$$

$$+ \left\{ \frac{i\mu}{16\pi} \int_0^{\infty} a_n(y) \, y^{-\frac{3}{2}} I_{\frac{i\mu}{2}}(2\pi|n|y) \, |n|^{\frac{i\mu}{2}} \, dy \, . \, \bar{\varepsilon}(\mu) \right\}_{\mu=\lambda}. \tag{6.76}$$

PROOF. Define $C_0^\pm$, $c_n^\pm$ in such a way that the formulas above should read

$$
C_0(\lambda) = \{C_0^+(\mu) . \varepsilon(\mu)\}_{\mu=\lambda} - \{C_0^-(\mu) . \bar\varepsilon(\mu)\}_{\mu=\lambda} ,
$$
$$
c_n(\lambda) = \{c_n^+(\mu) . \varepsilon(\mu)\}_{\mu=\lambda} - \{c_n^-(\mu) . \bar\varepsilon(\mu)\}_{\mu=\lambda} . \tag{6.77}
$$

In the case when $f \in L^2(\Gamma\backslash\Pi)$ is the cusp–form introduced in theorem 6.4, one has, as a consequence of (6.43) and (6.13),

$$
c_n^+(\mu) = - b_n \frac{i\mu}{16\,\pi} |n|^{\frac{i\mu}{2}} \int_0^\infty K_{\frac{i\lambda_1}{2}}(2\pi|n|y)\, I_{-\frac{i\mu}{2}}(2\pi|n|y)\, \frac{dy}{y}
$$
$$
= b_n \frac{i\mu}{4\pi} |n|^{\frac{i\mu}{4}} (\mu^2 - \lambda_1^2)^{-1} , \tag{6.78}
$$

and $c_n^-(\mu)$ is given by the same formula in the lower half-plane, so that our expression of $c_n(\lambda)$ as a hyperfunction gives (6.71) as a result in view of the equation

$$
\delta(\lambda - \lambda_1) + \delta(\lambda + \lambda_1) = \left\{ \frac{i\mu}{\pi} (\mu^2 - \lambda_1^2)^{-1} \right\}_{\mu=\lambda} . \tag{6.79}
$$

Now let f be the continuous superposition of Eisenstein series given in (6.55), so that $a_n(y)$ is given as the second term on the right–hand side of (6.37). Thus, using (6.13) again, then (6.41), finally (6.69), one gets

$$
c_n^+(\mu) = -\frac{i\mu}{2^7\pi^2} \int_{-\infty}^\infty \Phi(\rho)\, b_n^\rho\, d\rho \int_0^\infty K_{\frac{i\rho}{2}}(2\pi|n|y)\, I_{-\frac{i\mu}{2}}(2\pi|n|y)\, |n|^{\frac{i\mu}{2}} \frac{dy}{y}
$$
$$
= -\frac{i\mu}{2^5\pi^2} |n|^{\frac{i\mu}{2}} \int_\infty^\infty \Phi(\rho)\, b_n^\rho \frac{d\rho}{\rho^2 - \mu^2}
$$
$$
= -\frac{i\mu}{16\,\pi^2} |n|^{\frac{i\mu}{2}} \int_{-\infty}^\infty (\zeta^*(1-\rho))^{-1}\, \Phi(\rho)\, |n|^{-\frac{i\rho}{2}}\, \sigma_{i\rho}(|n|) \frac{d\rho}{\rho^2 - \mu^2}
$$
$$
= \frac{i}{2^5\,\pi^2} |n|^{\frac{i\mu}{2}} \int_{-\infty}^\infty \Psi(\rho)\, |n|^{-\frac{i\rho}{2}}\, \sigma_{i\rho}(|n|) \left[\frac{1}{\mu-\rho} + \frac{1}{\mu+\rho}\right] d\rho , \tag{6.80}
$$

and the same formula goes for c_n^- in the lower half–plane. By (6.73), the distribution associated with $c_n^+.\varepsilon - c_n^-.\bar\varepsilon$ is

$$
\lambda \mapsto \frac{1}{16\,\pi} |n|^{\frac{i\lambda}{2}} [\Psi(\lambda)\, |n|^{-\frac{i\lambda}{2}}\, \sigma_{i\lambda}(|n|) + \Psi(-\lambda)\, |n|^{\frac{i\lambda}{2}}\, \sigma_{-i\lambda}(|n|)] , \tag{6.81}
$$

which is just the same as the second term on the right–hand side of the last equation (6.71) since Ψ is an even function as mentioned just after (6.69) and $|n|^{i\lambda}\sigma_{-i\lambda}(|n|) = \sigma_{i\lambda}(|n|)$.

Finally, the cuspidal coefficients, still under the same assumptions about f: getting a_0 from (6.40), one gets for the two functions $C_0^\pm$, each one on its half-plane, the equations

$$C_0^+(\mu) = \frac{1}{2^5\,\pi^2}\,\frac{\pi^{-\frac{i\mu}{2}}}{\Gamma(-\frac{i\mu}{2})} \int_{-\infty}^{\infty} \Phi(\rho)\,d\rho \int_0^1 y^{-\frac{i(\rho+\mu)}{2}}\,\frac{dy}{y}$$

$$= \frac{i}{16\,\pi^2}\,\frac{\pi^{-\frac{i\mu}{2}}}{\Gamma(-\frac{i\mu}{2})} \int_{-\infty}^{\infty} \Phi(\rho)\,\frac{d\rho}{\rho+\mu}. \tag{6.82}$$

and

$$C_0^-(\mu) = -\frac{1}{2^5\,\pi^2}\,\frac{\pi^{-\frac{i\mu}{2}}}{\Gamma(-\frac{i\mu}{2})} \int_{-\infty}^{\infty} \Phi(\rho)\,d\rho \int_1^{\infty} y^{-\frac{i(\rho+\mu)}{2}}\,\frac{dy}{y}$$

$$= \frac{i}{16\,\pi^2}\,\frac{\pi^{-\frac{i\mu}{2}}}{\Gamma(-\frac{i\mu}{2})} \int_{-\infty}^{\infty} \Phi(\rho)\,\frac{d\rho}{\rho+\mu}. \tag{6.83}$$

From (6.73) again, the associated distribution is

$$\lambda \mapsto \frac{1}{8\,\pi}\,\frac{\pi^{-\frac{i\lambda}{2}}}{\Gamma(-\frac{i\lambda}{2})}\,\Phi(-\lambda)$$

$$= \frac{1}{8\pi}\,\frac{\pi^{-\frac{i\lambda}{2}}}{\Gamma(-\frac{i\lambda}{2})}\,\zeta^*(1+i\lambda)\,\Psi(\lambda)$$

$$= \frac{1}{8\,\pi}\,\zeta(-i\lambda)\,\Psi(\lambda) \tag{6.84}$$

where we have used (6.69), the parity of Ψ and (6.65).   □

Let us sum up the results of this section. As a result of our attempt to generalize the Radon transformation inversion formula (6.4) to an arithmetic environment ($L^2(\Pi)$ giving way to $L^2_{\text{even}}(\mathbb{R}^2\backslash\{0\})$), we were led to substituting for the operator TV (an isometry from $L^2(\Pi)$ to some subspace of $L^2_{\text{even}}(\mathbb{R}^2\backslash\{0\})$) the operator SV, where S was introduced in (4.46). Given a function $f \in L^2(\Gamma\backslash\Pi)$, orthogonal to the constants, with the Roelcke–Selberg expansion (i.e., the spectral decomposition with respect to the automorphic Laplacian) (6.33), satisfying some technically more demanding spectral–theoretic assumption (cf. definition 6.3), we constructed distributions C_0, C_∞, c_n with the property that

$$< SVf, h >=$$

$$4\pi \int_{-\infty}^{\infty} [\,C_0(\lambda)\,\mathrm{res}_0(\hat{h}^b_{-\lambda}) + C_\infty(\lambda)\,\mathrm{res}_{-i\lambda}(\hat{h}^b_{-\lambda}) + \sum_{n\neq 0} c_n(\lambda)\,\hat{h}^b_{-\lambda}(n)\,]\,d\lambda \tag{6.85}$$

for every $h \in C_0^\infty(\mathbb{R}, C_G^\infty)$.

As shown by the formulas (6.71), knowing the coefficients C_0, C_∞, c_n is equivalent to knowing the coefficients of the Roelcke–Selberg decomposition (6.33) of f: not the coefficients Φ^k individually, of course (since we have not, as yet, singled out any orthonormal basis of the eigenspace of Δ corresponding to any given eigenvalue $\frac{1+\lambda_k^2}{4}$), but the orthogonal projection of f on each such eigenspace. What makes (6.85) interesting, however, is that these distributions admit definitions (*cf.* theorem 6.6) that only depend on the Fourier coefficients of f itself.

In the next (self–contained) section, we shall extend theorem 6.6 (at the price of some change in the coefficients $c_n^{\pm}$, more about it to come) so as to get rid of the spectral assumption introduced in definition 6.3: we shall also dispense with the assumption that f is orthogonal to the constants. Let us emphasize that it is not our aim to give a new proof of the Roelcke–Selberg theorem — we rely on it — but to obtain a new way to compute the coefficients of such a decomposition.

Even though the functions $c_n^{\pm}(\mu)$, as defined in (6.76), occurred in a very natural way, we found it necessary to replace them, in the next section, by some modified version $\tilde{c}_n^{\pm}(\mu)$, keeping only, in the defining integrals (6.76), the first term in the power series expansion of the function $I_{\pm\frac{i\mu}{2}}$ involved: the reason is simply that this modification made the evaluation of some integrals, in sections 11 and 12, possible. On the other hand, due to the presence of Gamma factors, the functions $\tilde{c}_n^{\pm}$, contrary to $c_n^{\pm}$, are burdened with new poles inessential to our story, none of which, however, is real. Actually, we shall not bother at all with $\tilde{c}_n^{-}$, since the consideration of $\tilde{c}_n^{+}$ will suffice.

7. Recovering the Roelcke–Selberg coefficients of a function in $L^2(\Gamma\backslash\Pi)$

As some readers, interested in modular form theory but possibly not in quantization, may have followed our advice in the foreword suggesting to start with the present section, we recall a few facts and notations: some repetition has been found unavoidable as going from section 6 to section 7.

The Eisenstein series $E_{\frac{1-\nu}{2}}$ is classically defined, for $\mathrm{Re}\,\nu < -1$, as the function on the upper half–plane Π such that $(cf.\,(3.14))$

$$E_{\frac{1-\nu}{2}}(z) = \frac{1}{2}(\mathrm{Im}\,z)^{\frac{1-\nu}{2}} \sum_{\substack{m,n\in\mathbf{Z} \\ (n,m)=1}} |mz - n|^{\nu-1}. \tag{7.1}$$

The function $E^*_{\frac{1-\nu}{2}} := \zeta^*(1-\nu)E_{\frac{1-\nu}{2}}$, where $\zeta^*(s) := \pi^{-\frac{s}{2}}\Gamma(\frac{s}{2})\zeta(s)$, extends as a holomorphic function of ν for $\nu \neq 0,1$, and $E^*_{\frac{1-\nu}{2}} = E^*_{\frac{1+\nu}{2}}$. The well–known Fourier expansion of $E_{\frac{1-\nu}{2}}$ has been recalled in (3.29).

On $L^2(\Gamma\backslash\Pi)$, where $\Gamma = SL(2,\mathbf{Z})$, the Eisenstein series $E_{\frac{1-i\lambda}{2}}$, $\lambda \in \mathbf{R}$, serve as a (redundant, because $E_{\frac{1\pm\nu}{2}}$ are linearly dependent) family of generalized eigenfunctions of the Laplacian Δ. This operator also has a discrete spectrum, and the eigenfunctions (which lie in $L^2(\Gamma\backslash\Pi)$) are the constants and the so–called Maass cusp–forms. In this section, $\{\mathcal{M}_k\}$ shall denote a complete orthonormal set of Maass cusp forms, corresponding to the sequence of eigenvalues $(\frac{1+\lambda_k^2}{4})$: we can assume that $\lambda_k > 0$ and that the sequence (λ_k) is not decreasing: we shall also assume that each $\mathcal{M}_k$ is a common eigenfunction of all the Hecke operators T_n (there is more about T_n in the sections to come). Let us note that in some sections, especially the ones dealing with Hecke's theory, we shall, on the contrary, denote as $(\frac{1+\lambda_k^2}{4})$ the strictly increasing sequence of non–zero eigenvalues of Δ, and denote as $(\mathcal{M}_{k,\ell})_{1\leq\ell\leq n_k}$ a (Hecke) orthonormal basis of the corresponding eigenspace.

The Roelcke–Selberg theorem is the statement $(cf.\,[\mathbf{36}, \mathrm{p.\,254}])$ that the space $L^2(\Gamma\backslash\Pi)$ is identical to the space of all functions f with

$$f(z) = \Phi^0 + \sum_{k\geq1} \Phi^k \mathcal{M}_k(z) + \frac{1}{8\pi}\int_{-\infty}^{\infty} \Phi(\rho)\,E_{\frac{1-i\rho}{2}}(z)\,d\rho, \tag{7.2}$$

where $\Phi^0 \in \mathbb{C}$, $\Phi^k \in \mathbb{C}$ for every k and $\sum_k |\Phi^k|^2 < \infty$, finally $\Phi \in L^2(\mathbb{R})$ and

$$\frac{\Phi(\rho)}{\zeta^*(1 - i\rho)} = \frac{\Phi(-\rho)}{\zeta^*(1 + i\rho)}. \tag{7.3}$$

If such is the case, noting that all scalar products in this work are *antilinear with respect to the variable on the left*, one has $\Phi^k = (\mathcal{M}_k|f)$ for every k and, if $f(x + iy)$ goes to zero fairly rapidly as $y \to \infty$, $\Phi(\rho) = (E_{\frac{1-i\rho}{2}}|f)$ for almost all ρ.

Our aim in the present section is to introduce an efficient way to recover the Roelcke–Selberg coefficients of any given $f \in L^2(\Gamma\backslash\Pi)$. Set

$$\mathcal{M}_k(x + iy) = y^{\frac{1}{2}} \sum_{n \neq 0} b_n^k \, K_{i\lambda_k}(2\pi|n|y) \, e^{-2i\pi nx} \tag{7.4}$$

By Hecke's theory [**36**, p. 242], the ratios $\frac{b_n^k}{b_{n/|n|}^k}$ can be interpreted, for every n, as eigenvalues of the Hecke operator T_n, thus are bounded by a constant depending only on n. As a consequence of a result of R.A.Smith ([**35**], quoted in [**36**], p.247), and of asymptotics ([**27**], p.13) of the Gamma function on vertical lines, one thus has the estimates

$$|b_n^k| \leq A(n) \, \lambda_k^{\frac{1}{2}} \exp\frac{\pi \lambda_k}{4} : \tag{7.5}$$

we shall have no need for the precise (and hard) estimates of $A(n)$ available.

We denote as $L^2_{[\frac{1}{4},\infty[}(\Gamma\backslash\Pi)$ the subspace of $L^2(\Gamma\backslash\Pi)$ consisting of functions f whose constant coefficient Φ^0 (from the expansion (7.2)) is zero: the notation stems from the fact that the isolated eigenvalue 0, the associated eigenfunctions of which are the constants, is excluded.

PROPOSITION 7.1. *If $f \in L^2_{[\frac{1}{4},\infty[}(\Gamma\backslash\Pi)$ is such that $\Delta f \in L^2(\Gamma\backslash\Pi)$ too, then the Fourier coefficients $(a_n)_{n \in \mathbb{Z}}$ of f, defined by*

$$f(x + iy) = \sum_{n \in \mathbb{Z}} a_n(y) \, e^{-2i\pi nx}, \tag{7.6}$$

satisfy the estimates

$$|a_0(y)| \leq A \, y^{\frac{1}{2}}, \qquad 0 < y < \infty, \tag{7.7}$$

and, for $n \neq 0$,

$$|a_n(y)| \leq A(n) \, y^{\frac{1}{2}} \, |\log y|^{\frac{1}{2}}, \qquad 0 < y < 1, \tag{7.8}$$

where A, $A(n)$ do not depend on y.

We first need some estimates about Bessel functions.

LEMMA 7.2. *For $a > 0$, the integral*

$$\int_0^\infty e^{\frac{\pi\rho}{2}} (K_{\frac{i\rho}{2}}(a))^2 \frac{d\rho}{1+\rho^2} \tag{7.9}$$

converges and is $\leq C |\log a|$ as $a \to 0$.
On the other hand,

$$|K_{\frac{i\lambda_k}{2}}(a)| \leq C \lambda_k^{-\frac{1}{2}} \exp -\frac{\pi\lambda_k}{4}, \qquad 0 < a \leq a_0. \tag{7.10}$$

PROOF. Set, for $0 < s < 1$,

$$\psi(s) = \int_0^\infty (K_{\frac{i\rho}{2}}(a))^2 \cosh(\frac{\pi\rho}{2}(1-s)) \, d\rho. \tag{7.11}$$

By ([27], p.106),

$$\psi(s) = \pi K_0(2 a \sin \frac{\pi s}{2}). \tag{7.12}$$

Let $\Psi(s)$ be the second antiderivative on $(0, 1)$ of $\psi(s)$ with vanishing Cauchy data at $s = 1$, so that

$$\Psi(0) = \frac{4}{\pi^2} \int_0^\infty (K_{\frac{i\rho}{2}}(a))^2 (\cosh \frac{\pi\rho}{2} - 1) \frac{d\rho}{\rho^2}$$
$$= \pi \int_0^1 s \, K_0(2 a \sin \frac{\pi s}{2}) \, ds. \tag{7.13}$$

Comparing the two expressions, we are done for the first part.
Next, the power series expansion

$$I_{\frac{i\lambda_k}{2}}(a) = (\frac{a}{2})^{\frac{i\lambda_k}{2}} \sum_{m \geq 0} \frac{(\frac{a}{2})^{2m}}{m! \, \Gamma(\frac{i\lambda_k}{2} + m + 1)} \tag{7.14}$$

yields

$$|I_{\frac{i\lambda_k}{2}}(a)| \leq \frac{1}{|\Gamma(\frac{i\lambda_k}{2} + 1)|} \exp \frac{a^2}{4\sqrt{1+\frac{\lambda_k^2}{4}}}. \tag{7.15}$$

Using [27, p. 13], one gets the second part of the lemma.   □

PROOF OF PROPOSITION 7.1. Since f and Δf lie in $L^2(\Gamma\backslash\Pi)$, one has

$$\int_{-\infty}^\infty |\Phi(\rho)|^2 (1+\rho^2)^2 \, d\rho < \infty \tag{7.16}$$

and

$$\sum_{k \geq 1} (1+\lambda_k^2)^2 |\Phi^k|^2 < \infty. \tag{7.17}$$

Thus $\Phi(\rho)$ is integrable over $\mathbb{R}$ and, in view of the estimate ([36], p.290) $|\lambda_k| \sim 12\, k$, a consequence of the Selberg trace formula, the series $\sum |\Phi^k|$ is convergent. This makes it possible to write, as a consequence of (3.29),

$$a_0(y) = \frac{1}{8\pi} \int_{-\infty}^{\infty} \Phi(\rho) \, [y^{\frac{1-i\rho}{2}} + \frac{\zeta^*(1+i\rho)}{\zeta^*(1-i\rho)} \, y^{\frac{1+i\rho}{2}}] \, d\rho \,, \qquad (7.18)$$

or, using (7.3),

$$a_0(y) = \frac{1}{4\pi} \int_{-\infty}^{\infty} \Phi(\rho) \, y^{\frac{1-i\rho}{2}} \, d\rho \,, \qquad (7.19)$$

thus proving (7.7). Also, for $n \neq 0$,

$$a_n(y) = (a_n)_{\text{cont}}(y) + (a_n)_{\text{disc}}(y) \,, \qquad (7.20)$$

with

$(a_n)_{\text{cont}}(y) =$

$$\frac{1}{8\pi} \, y^{\frac{1}{2}} \int_{-\infty}^{\infty} \Phi(\rho) \frac{2\pi^{\frac{1-i\rho}{2}}}{\Gamma(\frac{1-i\rho}{2}) \zeta(1-i\rho)} \, |n|^{-\frac{i\rho}{2}} \, \sigma_{i\rho}(|n|) \, K_{\frac{i\rho}{2}}(2\pi|n|y) \, d\rho \quad (7.21)$$

and

$$(a_n)_{\text{disc}}(y) = y^{\frac{1}{2}} \sum_{k \geq 1} \Phi^k \, b_n^k \, K_{\frac{i\lambda_k}{2}}(2\pi|n|y) \,. \qquad (7.22)$$

The proof of (7.8) is then a consequence of (7.10), (7.17) and of the estimate (7.5) for the contribution of the discrete part; for that of the continuous part, use the Cauchy–Schwarz inequality and (7.16), (7.9) together with the estimate ([27], p.13)

$$|\Gamma(\frac{1-i\rho}{2})|^{-2} \leq C \cosh \frac{\pi\rho}{2} \,, \qquad (7.23)$$

and the fact ([37], p.161 or [29], p.100) that $(\zeta(1-i\rho))^{-1}$ is majorized by some power of $\log|\rho|$, $|\rho| \to \infty$.

$\square$

Our formulation of the next theorem depends on hyperfunction theory: this permits a great generality, eliminating the need for any assumption to the effect that f should be sufficiently rapidly decreasing at infinity in the fundamental domain.

Recall that hyperfunctions are classes of holomorphic functions $c = c(\mu)$ defined in complex neighborhoods (depending on c) of $\mathbb{R}$ with $\mathbb{R}$ deleted, under the equivalence relation which identifies such a function to zero if it agrees with the restriction of some function defined and holomorphic in some neighborhhood of $\mathbb{R}$ (with $\mathbb{R}$ not deleted). We denote as $\{c(\mu)\}_{\mu=\lambda}$ the hyperfunction associated with c. It may happen that the function $\lambda \mapsto c(\lambda + i0) - c(\lambda - i0) := \lim_{\varepsilon \to 0}(c(\lambda + i\varepsilon) - c(\lambda - i\varepsilon))$ on the real line exist as a distribution, in which case the hyperfunction $\{c(\mu)\}_{\mu=\lambda}$ is to be identified with the distribution $c(\lambda + i0) - c(\lambda - i0)$, i.e., the difference of the two traces of c on $\mathbb{R}$ from above and from below. This will always be the case in our present investigations: still, the hyperfunction point of view, even in a distribution–only setting, seems to be tailor–made for our purposes. In particular, if $\chi \in L^2(\mathbb{R})$, χ is associated, for instance, with the analytic function

$$\mu \mapsto \frac{1}{2i\pi} \int_{-\infty}^{\infty} \chi(\rho) \frac{d\rho}{\rho - \mu} , \qquad \mu \in \mathbb{C}\backslash\mathbb{R} ; \qquad (7.24)$$

the Dirac mass at $\lambda = \lambda_0$ is associated with the function $\mu \mapsto (2i\pi (\lambda_0 - \mu))^{-1}$.

In all cases, setting

$$\varepsilon(\mu) = \begin{cases} 1 \text{ if Im } \mu > 0 \\ 0 \text{ if Im } \mu < 0 \end{cases}, \qquad \bar{\varepsilon}(\mu) = \begin{cases} 0 \;\; \text{ if Im } \mu > 0 \\ -1 \text{ if Im } \mu < 0 \end{cases}, \qquad (7.25)$$

one may write $\{c(\mu)\}_{\mu=\lambda} = \{c(\mu).\varepsilon(\mu)\}_{\mu=\lambda} - \{c(\mu).\bar{\varepsilon}(\mu)\}_{\mu=\lambda}$ in the sense of hyperfunctions.

THEOREM 7.3. *Let* $f \in L^2(\Gamma\backslash\Pi)$, *with* $\Delta f \in L^2(\Gamma\backslash\Pi)$ *too, have the Fourier expansion (7.6) and the Roelcke–Selberg expansion (7.2). The function*

$$C_0^+(\mu) = \frac{1}{8\pi} \int_0^1 a_0(y) \, y^{-\frac{3}{2}} \frac{(\pi y)^{-\frac{i\mu}{2}}}{\Gamma(-\frac{i\mu}{2})} \, dy \qquad (7.26)$$

is holomorphic in the half-plane Im $\mu > 1$ *and extends as a meromorphic function in the half-plane* Im $\mu > 0$ *with an only possible (simple) pole at* $\mu = i$. *The residue of* C_0^+ *at this point is* $(-4i\pi)^{-1} \Phi^0$, *so that, in particular,* $f \in L^2_{[\frac{1}{4},\infty[}(\Gamma\backslash\Pi)$ *if and only if* C_0^+ *is holomorphic throughout the half-plane* Im $\mu > 0$. *The function*

$$C_0^-(\mu) = -\frac{1}{8\pi} \int_1^{\infty} a_0(y) \, y^{-\frac{3}{2}} \frac{(\pi y)^{-\frac{i\mu}{2}}}{\Gamma(-\frac{i\mu}{2})} \, dy \qquad (7.27)$$

is holomorphic in the half-plane Im $\mu < 0$. *Finally, the distribution* C_0 *given, in the sense of hyperfunction theory, as*

$$C_0(\lambda) = \{C_0^+(\mu).\varepsilon(\mu)\}_{\mu=\lambda} - \{C_0^-(\mu).\bar{\varepsilon}(\mu)\}_{\mu=\lambda} \qquad (7.28)$$

coincides with the function

$$C_0(\lambda) = \frac{1}{8\pi} \frac{\pi^{-\frac{i\lambda}{2}}}{\Gamma(-\frac{i\lambda}{2})} \Phi(-\lambda) = \frac{1}{8\pi} \frac{\zeta(-i\lambda)}{\zeta^*(1-i\lambda)} \Phi(\lambda). \qquad (7.29)$$

PROOF. As a consequence of proposition 7.1 and (7.19), one has

$$a_0(y) = \Phi^0 + \frac{1}{4\pi} \int_{-\infty}^{\infty} \Phi(\rho) \, y^{\frac{1-i\rho}{2}} \, d\rho , \qquad (7.30)$$

which implies, when Im $\mu > 1$,

$$C_0^+(\mu) = -\frac{1}{4\pi}(1+i\mu)^{-1}\frac{\pi^{-\frac{i\mu}{2}}}{\Gamma(-\frac{i\mu}{2})}\Phi^0 + \frac{i}{16\pi^2}\frac{\pi^{-\frac{i\mu}{2}}}{\Gamma(-\frac{i\mu}{2})}\int_{-\infty}^{\infty}\Phi(\rho)\frac{d\rho}{\rho+\mu}.$$

(7.31)

The same formula gives the value of $C_0^-(\mu)$ in the lower half–plane, which is immediately seen to imply theorem 7.3, using (7.24) then (7.3) again. □

Remark. In the case when f is rapidly decreasing at infinity in the fundamental domain, the function C_0^- can be extended from the lower half–plane across the real line, so that $C_0(\lambda)$ is nothing but the function obtained by complex continuation (starting from Im $\mu > 1$) from the function

$$C_0(\mu) = C_0^+(\mu) - C_0^-(\mu)$$

$$= \frac{1}{8\pi}\frac{\pi^{-\frac{i\mu}{2}}}{\Gamma(-\frac{i\mu}{2})}\int_0^{\infty}a_0(y)\,y^{-\frac{3}{2}-\frac{i\mu}{2}}\,dy$$

(7.32)

and, from (7.29), $\Phi(-\lambda)$ is the value, continued at $\mu = \lambda$ from above, of the integral

$$\int_0^{\infty}a_0(y)\,y^{-\frac{3}{2}-\frac{i\mu}{2}}\,dy.$$

(7.33)

On the other hand, $\Phi(-\lambda) = (E_{\frac{1+i\lambda}{2}}|f)$ is the integral, over the fundamental domain, of $f(z)E_{\frac{1-i\lambda}{2}}(z)\,d\mu(z)$, so that the result of theorem 7.3 coincides in this case with that obtained by the Rankin–Selberg method (cf. (3) in [49], p.268). The hyperfunction concept, on the other hand, cannot be dispensed with in the cases (for instance that discussed in section 9) when there are no a priori reasons why the continuation of C_0^+ or C_0^- should be possible across the real line: this is even more so when theorem 7.4 is considered, since unless there are no discrete summands in the Roelcke-Selberg expansion of f, $\tilde{c}_n^+$ will *always* get poles on the real line.

THEOREM 7.4. *Let $f \in L^2(\Gamma\backslash\Pi)$, with $\Delta^N f \in L^2(\Gamma\backslash\Pi)$ for all N too, have the Fourier expansion (7.6): also assume that its Roelcke–Selberg expansion (7.2) reduces to*

$$f(z) = \sum_{k\geq 1}\Phi^k\,M_k(z).$$

(7.34)

Given $n \neq 0$, define

$$\tilde{c}_n^+(\mu) = \frac{1}{8\pi}\int_0^{\infty}a_n(y)\,y^{-\frac{3}{2}}\frac{(\pi y)^{-\frac{i\mu}{2}}}{\Gamma(-\frac{i\mu}{2})}\,dy$$

(7.35)

when Im $\mu > 0$. Then $\tilde{c}_n^+(\mu)$ extends as a meromorphic function in the entire plane: it has no pole with Im $\mu > -4$ except the real points $\pm\lambda_k$ ($k \geq 1$); all its poles are simple. If $\{k, k+1, \ldots, k+l\}$ $(l \geq 0)$ is a complete set *of indices with λ_{k+j} independent of j (recall that $k \mapsto \lambda_k$ is not decreasing), one has*

$$\sum_{0 \le j \le l} \Phi^{k+j} M_{k+j}(z) = y^{\frac{1}{2}} \sum_{n \ne 0} d_n K_{\frac{i\lambda_k}{2}}(2\pi|n|y)\, e^{-2i\pi nx} \tag{7.36}$$

with

$$d_n = -8i\pi\, |n|^{-\frac{i\lambda_k}{2}} \times \text{residue of } \tilde{c}_n^+(\mu) \text{ at } \mu = \lambda_k: \tag{7.37}$$

here, of course, we mean the complex function–theoretic residue, not that introduced in definition 3.5.

Remark. The left–hand side of (7.36) is just the orthogonal projection of f onto the eigenspace of Δ for the eigenvalue $\frac{1}{4}(1+\lambda_k^2)$.

PROOF. We first show that $\tilde{c}_n^+(\mu)$ makes sense for Im $\mu > 0$, and that it can be obtained by performing a term–by–term integration in (7.35), starting from

$$a_n(y) = y^{\frac{1}{2}} \sum_{k \ge 1} \Phi^k\, b_n^k\, K_{\frac{i\lambda_k}{2}}(2\pi|n|y)\,, \tag{7.38}$$

a consequence of (7.4). This has to be integrated against $y^{-\frac{3}{2}-\frac{i\mu}{2}}\, dy$ from 0 to ∞ (Im $\mu > 0$): estimating the integral from 0 to 1 follows from the argument at the end of the proof of proposition 7.1, so all we have to do is showing that

$$\sum_{k \ge 1} |\Phi^k|\, |b_n^k| \int_1^\infty |K_{\frac{i\lambda_k}{2}}(2\pi|n|y)|\, y^{-1+\frac{\text{Im }\mu}{2}}\, dy < \infty\,. \tag{7.39}$$

From ([**27**], p.101 and 13), one has when $0 < \text{Im }\mu < \mu_0$

$$\int_0^\infty (K_{\frac{i\lambda_k}{2}}(2\pi|n|y))^2\, y^{\text{Im }\mu}\, dy \le C_{n,\mu_0}\, |\Gamma(\frac{1+\text{Im }\mu+i\lambda_k}{2})\,\Gamma(\frac{1+\text{Im }\mu-i\lambda_k}{2})|$$

$$\le C_{n,\mu_0}\, |\lambda_k|^{\text{Im }\mu}\, e^{-\frac{\pi}{2}|\lambda_k|} \tag{7.40}$$

so that, by the Cauchy–Schwarz inequality, the series on the left–hand side of (7.39) is majorized by

$$C_{n,\mu_0} \sum_{k \ge 1} |\Phi^k|\, |b_n^k|\, |\lambda_k|^{\frac{\text{Im }\mu}{2}}\, e^{-\frac{\pi}{4}|\lambda_k|}\,, \tag{7.41}$$

a convergent series thanks to (7.5), since $\sum_k |\Phi^k|^2\, (1+\lambda_k^2)^N < \infty$ for all N.

After a term–by–term integration, we get, using ([**27**], p.91),

$$\tilde{c}_n^+(\mu) = \frac{1}{32\,\pi}\, \frac{|n|^{\frac{i\mu}{2}}}{\Gamma(-\frac{i\mu}{2})} \sum_{k \ge 1} \Phi^k\, b_n^k\, \Gamma(\frac{i(-\mu+\lambda_k)}{4})\,\Gamma(\frac{i(-\mu-\lambda_k)}{4})\,. \tag{7.42}$$

In view of the estimate

$$|\Gamma(\frac{i(-\mu+\lambda_k)}{4})\Gamma(\frac{i(-\mu-\lambda_k)}{4})| \le C\,|\lambda_k|^{\frac{\mathrm{Im}\,\mu}{2}-1}\,e^{-\frac{\pi|\lambda_k|}{4}},$$

$$(7.43)$$

valid, say, when μ stays within any compact subset of the complex plane excluding the poles of these two Gamma factors, it is clear that the only possible poles of $\tilde{c}_n^+(\mu)$ are given by

$$\mu \pm \lambda_k = 0,\ -4i,\ -8i,\ \dots \qquad\qquad (7.44)$$

and that the contribution to the residue at $\mu = \lambda_k$ that comes from the k-th term is

$$-\frac{1}{8i\pi}\,|n|^{\frac{i\lambda_k}{2}}\,\Phi^k\,b_n^k\,.$$

□

Remark. If one only assumes that f and Δf lie in $L^2(\Gamma\backslash\Pi)$, (7.41) still ensures that we can at least get the meromorphic continuation of the function $\tilde{c}_n$ in a strip of comfortable width across the real line, thus sufficient for tracking the real poles.

8. The "product" of two Eisenstein distributions

The first short part of this section is heuristic in character. Its aim is to show that a formal application of the bilinear machine $L_{\nu_1,\nu_2;\nu}$ introduced in (1.4) to a pair of Eisenstein distributions brings into consideration, formally, an interesting modular distribution $\mathfrak{T}$, the existence of which, however, is unclear. Out of the arithmetic context, we then transfer the bilinear operation to Π, i.e., as a bilinear operation on pairs of eigenfunctions of Δ. We shall then use the result as a new definition, thus ensuring by indirect means, in the next section, the significance of $\mathfrak{T}$.

This is the distribution, or function, suggested by an application of (1.4) with $\nu_1 = i\lambda_1$, $\nu_2 = i\lambda_2$, $\nu = i\lambda$:

$$\mathfrak{T}(s) = \int \chi_{i\lambda_1, i\lambda_2; i\lambda}(s_1, s_2; s)\, \mathfrak{E}_{i\lambda_1}(s_1)\, \mathfrak{E}_{i\lambda_2}(s_2)\, ds_1\, ds_2\,, \qquad (8.1)$$

in which the Eisenstein distributions were introduced in (3.4). Making the integral kernel explicit, this could also be written as

$$< \mathfrak{T}, u > = \int u(s)\, ds < \mathfrak{E}_{i\lambda_1} \otimes \mathfrak{E}_{i\lambda_2}, (s_1, s_2) \mapsto$$

$$|s_1 - s_2|^{\frac{-1+i(\lambda+\lambda_1+\lambda_2)}{2}} |s_1 - s|^{\frac{-1+i(-\lambda+\lambda_1-\lambda_2)}{2}} |s_2 - s|^{\frac{-1+i(-\lambda-\lambda_1+\lambda_2)}{2}} >,$$

$$u \in C^\infty_{-i\lambda}. \quad (8.2)$$

We expect $\mathfrak{T}$ not to be a genuine (distribution-valued) function of λ, but to involve Dirac-like discontinuities for the values of λ which correspond to the discrete part of the spectrum of Δ in $L^2(\Gamma\backslash\Pi)$.

Let us set

$$\alpha = \frac{1 + i(-\lambda + \lambda_1 - \lambda_2)}{2}, \quad \beta = \frac{1 + i(-\lambda - \lambda_1 + \lambda_2)}{2}, \quad \gamma = \frac{1 + i(\lambda + \lambda_1 + \lambda_2)}{2}$$

and, for later use,

$$\delta = \frac{1 + i(-\lambda + \lambda_1 + \lambda_2)}{2} \qquad (8.3)$$

so that $\alpha + \beta + \gamma - \delta = 1$, and forget, in the integral (8.2), the meaningless terms that would correspond to $s_1 - s_2 = 0$ (but of course, we do not prevent the singularities $s = s_1$ or $s = s_2$): we formally get, for $\mathfrak{T}$, the function

$$\mathfrak{T}(s) = \frac{1}{4} \sum_{n_1 m_2 - m_1 n_2 \neq 0} |m_1|^{i\lambda_1 - 1} |m_2|^{i\lambda_2 - 1} \left|\frac{n_1}{m_1} - \frac{n_2}{m_2}\right|^{\gamma - 1} \left|\frac{n_1}{m_1} - s\right|^{\alpha - 1} \left|\frac{n_2}{m_2} - s\right|^{\beta - 1}.$$
$$(8.4)$$

This may also be written as

$$\mathfrak{T}(s) = \frac{1}{4} \sum_{n_1 m_2 - m_1 n_2 \neq 0} |n_1 m_2 - m_1 n_2|^{\gamma-1} |n_1 - m_1 s|^{\alpha-1} |n_2 - m_2 s|^{\beta-1}, \tag{8.5}$$

which would be a very interesting new modular distribution if it did converge. However, it does not.

Given $N \geq 1$, define the Hecke operator T_N^λ on $\mathbf{Z}$–periodic distributions in the space $\mathcal{D}'_{i\lambda}$ through

$$< T_N^\lambda \mathfrak{T}, u > = N^{-\frac{1}{2}} \sum_{\substack{ad = N, d > 0 \\ b \bmod d}} (\frac{a}{d})^{\frac{i\lambda-1}{2}} < \mathfrak{T}, s \mapsto u(\frac{ds-b}{a}) >, \qquad u \in C_{-i\lambda}^\infty \tag{8.6}$$

or, if the distribution one is dealing with is a function,

$$(T_N^\lambda \mathfrak{T})(s) = N^{-\frac{1}{2}} \sum_{\substack{ad = N, d > 0 \\ b \bmod d}} (\frac{a}{d})^{\frac{1+i\lambda}{2}} \mathfrak{T}(\frac{as+b}{d}). \tag{8.7}$$

If $f = \Theta_{i\lambda} \mathfrak{T}$ as defined in (2.19), it is immediate that

$$(\Theta_{i\lambda} T_N^\lambda \mathfrak{T})(z) = N^{-\frac{1}{2}} \sum_{\substack{ad = N, d > 0 \\ b \bmod d}} f(\frac{az+b}{d}) = (T_N f)(z), \tag{8.8}$$

where T_N denotes the usual Hecke operator ([36], p.241) acting on (non-holomorphic) modular forms.

There are too many terms in (8.5). Concentrating on those with $n_1 m_2 - m_1 n_2 = \pm 1$, consider instead the (still non–convergent) sum

$$\mathfrak{T}_0(s) = \frac{1}{2} \sum_{n_1 m_2 - m_1 n_2 = 1} |n_1 - m_1 s|^{\alpha-1} |n_2 - m_2 s|^{\beta-1}, \tag{8.9}$$

also *formally* a modular distribution. Applying formally the definition above, we would get (using $\alpha + \beta = 1 - i\lambda$)

$$(T_N^\lambda \mathfrak{T}_0)(s) = \frac{1}{2} N^{\frac{1-\alpha-\beta}{2}} \times$$

$$\sum_{\substack{ad = N, d > 0 \\ b \bmod d}} \sum_{n_1 m_2 - m_1 n_2 = 1} |dn_1 - bm_1 - am_1 s|^{\alpha-1} |dn_2 - bm_2 - am_2 s|^{\beta-1}. \tag{8.10}$$

Since

$$\begin{pmatrix} m_2 & n_2 \\ m_1 & n_1 \end{pmatrix} \begin{pmatrix} a & -b \\ 0 & d \end{pmatrix} = \begin{pmatrix} am_2 & dn_2 - bm_2 \\ am_1 & dn_1 - bm_1 \end{pmatrix}, \tag{8.11}$$

it follows from the decomposition ([36], p.238)

$$M_N(\mathbf{Z}) = \bigcup_{\substack{ad = N, d > 0 \\ b \bmod d}} \Gamma \begin{pmatrix} a & -b \\ 0 & d \end{pmatrix}, \tag{8.12}$$

where $M_N(\mathbf{Z})$ denotes the set of matrices with integer coefficients and determinant N, that

$$(T_N^\lambda \mathfrak{T}_0)(s) = \frac{1}{2} N^{\frac{i\lambda}{2}} \sum_{n_1' m_2' - m_1' n_2' = N} |n_1' - m_1' s|^{\alpha - 1} |n_2' - m_2' s|^{\beta - 1}. \tag{8.13}$$

From (8.5) and (8.13) it follows (not forgetting that in the sum (8.5), the sign of $n_1 m_2 - m_1 n_2$ was not fixed) that, in a still formal sense,

$$\mathfrak{T} = \sum_{N \geq 1} N^{\frac{-1 + i(\lambda_1 + \lambda_2)}{2}} T_N^\lambda \mathfrak{T}_0, \tag{8.14}$$

a formula which, to start with, makes the study of the sum (8.5) a two–step problem. We stop our heuristic considerations here, and turn to the problem of transferring the bilinear operation $\mathbf{L}_{i\lambda_1, i\lambda_2; i\lambda}$ to Π.

PROPOSITION 8.1. *Given λ_1, λ_2 and $\lambda \in \mathbf{R}$, and $u_1 \in C_{i\lambda_1}^\infty$, $u_2 \in C_{i\lambda_2}^\infty$, set, in accordance with (4.35),*

$$u_j^\sharp(\xi) = |\xi_2|^{-1 - i\lambda_j} u_j\left(\frac{\xi_1}{\xi_2}\right). \tag{8.15}$$

Then, with the notation from section 4 (in particular (4.34)), one has for almost all s

$$(TV((V^*T^*u_1^\sharp).(V^*T^*u_2^\sharp)))_\lambda^\flat(s) =$$

$$2^{-\frac{9}{2}} \pi^{-2} \frac{\Gamma(\frac{1 - i(\lambda + \lambda_1 + \lambda_2)}{4}) \Gamma(\frac{1 + i(\lambda - \lambda_1 + \lambda_2)}{4}) \Gamma(\frac{1 + i(\lambda + \lambda_1 - \lambda_2)}{4}) \Gamma(\frac{1 + i(\lambda - \lambda_1 - \lambda_2)}{4})}{\Gamma(\frac{-i\lambda_1}{2}) \Gamma(\frac{-i\lambda_2}{2}) \Gamma(\frac{i\lambda}{2})}$$

$$\int_{-\infty}^\infty \int_{-\infty}^\infty \chi_{i\lambda_1, i\lambda_2; i\lambda}(s_1, s_2; s) \, u_1(s_1) \, u_2(s_2) \, ds_1 \, ds_2, \tag{8.16}$$

where the function $\chi_{i\lambda_1, i\lambda_2; i\lambda}$ was defined in (1.5).

PROOF. Starting from (4.34), (4.36),

$$(TV((V^*T^*u_1^\sharp).(V^*T^*u_2^\sharp)))_\lambda^\flat(s) = \frac{1}{2} (2\pi)^{-\frac{5}{2}} \frac{\Gamma(\frac{1 + i\lambda}{2}) \Gamma(\frac{1 - i\lambda_1}{2}) \Gamma(\frac{1 - i\lambda_2}{2})}{\Gamma(-\frac{i\lambda_1}{2}) \Gamma(-\frac{i\lambda_2}{2}) \Gamma(\frac{i\lambda}{2})}$$

$$\int_{-\infty}^\infty \int_{-\infty}^\infty A_{\lambda_1, \lambda_2; \lambda}(s_1, s_2; s) \, u_1(s_1) \, u_2(s_2) \, ds_1 \, ds_2 \tag{8.17}$$

with

$$A_{\lambda_1,\lambda_2;\lambda}(s_1,s_2;s)$$

$$= \int_\Pi \left(\frac{|z-s|^2}{y}\right)^{-\frac{1}{2}-\frac{i\lambda}{2}} \left(\frac{|z-s_1|^2}{y}\right)^{-\frac{1}{2}+\frac{i\lambda_1}{2}} \left(\frac{|z-s_2|^2}{y}\right)^{-\frac{1}{2}+\frac{i\lambda_2}{2}} d\mu(z). \tag{8.18}$$

Using the easily proven identity

$$A_{\lambda_1,\lambda_2;\lambda}(\frac{as_1+b}{cs_1+d},\frac{as_2+b}{cs_2+d};\frac{as+b}{cs+d})$$

$$= |cs+d|^{1+i\lambda} |cs_1+d|^{1-i\lambda_1} |cs_2+d|^{1-i\lambda_2} A_{\lambda_1,\lambda_2;\lambda}(s_1,s_2;s), \tag{8.19}$$

and noting that, if s_1,s_2,s are the images of $0,1,\infty$ under the fractional-linear transformation associated with $\left(\begin{smallmatrix} a & b \\ c & d \end{smallmatrix}\right)$, then

$$\chi_{i\lambda_1,i\lambda_2;i\lambda}(s_1,s_2,;s) = |c|^{1+i\lambda} |d|^{1-i\lambda_1} |c+d|^{1-i\lambda_2}, \tag{8.20}$$

one gets

$$A_{\lambda_1,\lambda_2;\lambda}(s_1,s_2;s) = I(\lambda_1,\lambda_2;\lambda)\,\chi_{i\lambda_1,i\lambda_2;i\lambda}(s_1,s_2;s) \tag{8.21}$$

with

$$I(\lambda_1,\lambda_2;\lambda) = \int_\Pi \left(\frac{|z|^2}{y}\right)^{-\frac{1}{2}+\frac{i\lambda_1}{2}} \left(\frac{|z-1|^2}{y}\right)^{-\frac{1}{2}+\frac{i\lambda_2}{2}} y^{\frac{1}{2}+\frac{i\lambda}{2}} d\mu(z), \tag{8.22}$$

a convergent integral. The justification of all that precedes is based on the easily proven inequality

$$\int_{\mathbf{R}^3} |(s_1-s_2)(s_2-s)(s-s_1)|^{-\frac{1}{2}} ((1+s_1^2)(1+s_2^2)(1+s^2))^{-\frac{1}{2}} ds_1\,ds_2\,ds < \infty.$$

Using (4.39) and the Plancherel formula for the dx–integration, we get

$$I(\lambda_1,\lambda_2;\lambda) = \frac{8\,\pi^{1-\frac{1}{2}(\lambda_1+\lambda_2)}}{\Gamma(\frac{1-i\lambda_1}{2})\Gamma(\frac{1-i\lambda_2}{2})} \int_0^\infty y^{-\frac{1}{2}+\frac{i\lambda}{2}}\,dy \int_0^\infty \sigma^{-\frac{i}{2}(\lambda_1+\lambda_2)}\,\cos(2\pi\sigma)$$
$$K_{\frac{i\lambda_1}{2}}(2\pi\sigma y)\,K_{\frac{i\lambda_2}{2}}(2\pi\sigma y)\,d\sigma, \tag{8.23}$$

in which the $d\sigma$–integration has to be carried first. Integrating with respect to dy first so as to take advantage of [27], p.101, one would formally obtain

$$I(\lambda_1,\lambda_2;\lambda) = \frac{\pi^{\frac{1}{2}}}{2\,\Gamma(\frac{1-i\lambda_1}{2})\,\Gamma(\frac{1-i\lambda_2}{2})\,\Gamma(\frac{1+i\lambda}{2})} \times \Gamma(\frac{1-i(\lambda+\lambda_1+\lambda_2)}{4})$$
$$\Gamma(\frac{1+i(\lambda-\lambda_1+\lambda_2)}{4})\Gamma(\frac{1+i(\lambda+\lambda_1-\lambda_2)}{4})\Gamma(\frac{1+i(\lambda-\lambda_1-\lambda_2)}{4}), \tag{8.24}$$

and the process can be justified if one first inserts under the right–hand side of (8.23) the factor $h(\varepsilon\sigma)$ for some $h \in \mathcal{S}(\mathbf{R})$ with $h(0) = 1$, letting at the end ε go to zero.   □

PROPOSITION 8.2. *Given f_1 and $f_2 \in C^\infty(\Pi)$, define their Poisson bracket as*

$$\{f_1, f_2\} := y^2 \left(-\frac{\partial f_1}{\partial y}\frac{\partial f_2}{\partial x} + \frac{\partial f_1}{\partial x}\frac{\partial f_2}{\partial y} \right). \tag{8.25}$$

Under the assumptions of proposition 8.1, one has

$$(TV(\{V^*T^*u_1^\sharp, V^*T^*u_2^\sharp\}))_\lambda^\flat(s) =$$

$$2^{-\frac{7}{2}}\pi^{-2}\frac{\Gamma(\frac{3-i(\lambda+\lambda_1+\lambda_2)}{4})\Gamma(\frac{3+i(\lambda-\lambda_1+\lambda_2)}{4})\Gamma(\frac{3+i(\lambda+\lambda_1-\lambda_2)}{4})\Gamma(\frac{3+i(\lambda-\lambda_1-\lambda_2)}{4})}{\Gamma(\frac{-i\lambda_1}{2})\Gamma(\frac{-i\lambda_2}{2})\Gamma(\frac{i\lambda}{2})}$$

$$\int_{-\infty}^{\infty}\int_{-\infty}^{\infty}\chi_{i\lambda_1, i\lambda_2; i\lambda}^{-}(s_1, s_2; s)\, u_1(s_1)\, u_2(s_2)\, ds_1\, ds_2, \tag{8.26}$$

where the integral kernel was defined in (5.47).

PROOF. In the same way as in the proof of proposition 8.1, we are led to a formula fully similar to (8.17), in which the pointwise product on the left-hand side has to be replaced by the "Poisson bracket" just defined, and $A_{\lambda_1, \lambda_2; \lambda}(s_1, s_2; s)$ on the right-hand side has to be replaced by

$$B_{\lambda_1, \lambda_2; \lambda}(s_1, s_2; s) =$$

$$\int_\Pi \left(\frac{|z-s|^2}{y}\right)^{-\frac{1}{2}-\frac{i\lambda}{2}}\left\{ \left(\frac{|z-s_1|^2}{y}\right)^{-\frac{1}{2}+\frac{i\lambda_1}{2}}, \left(\frac{|z-s_2|^2}{y}\right)^{-\frac{1}{2}+\frac{i\lambda_2}{2}} \right\}\, d\mu(z). \tag{8.27}$$

Since, obviously,

$$B_{\lambda_1, \lambda_2; \lambda}(s_1, s_2; s) = -B_{\lambda_2, \lambda_1; \lambda}(s_2, s_1; s) \tag{8.28}$$

and

$$\chi_{i\lambda_1, i\lambda_2; i\lambda}(s_1, s_2; s) = \chi_{i\lambda_2, i\lambda_1; i\lambda}(s_2, s_1; s), \tag{8.29}$$

one may be satisfied with computing the kernel $B_{\lambda_1, \lambda_2; \lambda}(s_1, s_2; s)$ in the case when $\frac{s_2-s_1}{(s_1-s)(s_2-s)} > 0$, which we assume from now on.

If $g \in G$, one has whenever f_1 and $f_2 \in C^\infty(\Pi)$ the identity

$$\{f_1, f_2\} \circ g^{-1} = \{f_1 \circ g^{-1}, f_2 \circ g^{-1}\}, \tag{8.30}$$

which permits to extend the invariance property (8.19) to the case when $A_{\lambda_1, \lambda_2; \lambda}(s_1, s_2; s)$ is replaced by $B_{\lambda_1, \lambda_2; \lambda}(s_1, s_2; s)$. Setting $z = \frac{aw+b}{cw+d}$ and choosing $g = \left(\begin{smallmatrix} a & b \\ c & d \end{smallmatrix}\right)$ so that $(g.s_1, g.s_2; g.s)$ should be near $(0, 1, \infty)$, one easily gets that

$$B_{\lambda_1, \lambda_2; \lambda}(s_1, s_2; s) = J(\lambda_1, \lambda_2; \lambda)\, \chi_{i\lambda_1, i\lambda_2; i\lambda}^{-}(s_1, s_2; s) \tag{8.31}$$

with

$$
J(\lambda_1, \lambda_2; \lambda) = \int_\Pi \left\{ \left(\frac{|z|^2}{y} \right)^{-\frac{1}{2} + \frac{i\lambda_1}{2}}, \left(\frac{|z-1|^2}{y} \right)^{-\frac{1}{2} + \frac{i\lambda_2}{2}} \right\} y^{\frac{1}{2} + \frac{i\lambda}{2}} \, d\mu(z)
$$

$$
= \int_\Pi \left(\frac{|z-1|^2}{y} \right)^{-\frac{1}{2} + \frac{i\lambda_2}{2}} \left\{ y^{\frac{1}{2} + \frac{i\lambda}{2}}, \left(\frac{|z|^2}{y} \right)^{-\frac{1}{2} + \frac{i\lambda_2}{2}} \right\} d\mu(z)
$$

$$
= -\frac{1 + i\lambda}{2} \int_\Pi \left(\frac{|z-1|^2}{y} \right)^{-\frac{1}{2} + \frac{i\lambda_2}{2}} y^{\frac{3}{2} + \frac{i\lambda}{2}} \frac{\partial}{\partial x} (\frac{x^2 + y^2}{y})^{-\frac{1}{2} + \frac{i\lambda_1}{2}} \, d\mu(z).
$$

$$(8.32)$$

Again, (4.39) and the Plancherel formula for the dx–integration yield the formula (to be compared to (8.22))

$$
J(\lambda_1, \lambda_2; \lambda) = \frac{8 \, \pi^{2 - \frac{i}{2}(\lambda_1 + \lambda_2)} \, (1 + i\lambda)}{\Gamma(\frac{1 - i\lambda_1}{2}) \Gamma(\frac{1 - i\lambda_2}{2})} \times
$$

$$
\int_0^\infty y^{\frac{1}{2} + \frac{i\lambda}{2}} \, dy \int_0^\infty \sigma^{1 - \frac{i}{2}(\lambda_1 + \lambda_2)} \, \sin(2\pi\sigma) \, K_{\frac{i\lambda_1}{2}}(2\pi\sigma y) \, K_{\frac{i\lambda_2}{2}}(2\pi\sigma y) \, d\sigma.
$$

$$(8.33)$$

Using ([27], p. 101):

$$
\int_0^\infty y^{\frac{1 + i\lambda}{2}} K_{\frac{i\lambda_1}{2}}(2\pi\sigma y) K_{\frac{i\lambda_2}{2}}(2\pi\sigma y) \, dy = \frac{1}{8} \, \pi^{-\frac{3}{2} - \frac{i\lambda}{2}} \, \sigma^{-\frac{3}{2} - \frac{i\lambda}{2}} \times
$$

$$
\frac{\Gamma(\frac{3 + i(\lambda + \lambda_1 + \lambda_2)}{4}) \, \Gamma(\frac{3 + i(\lambda - \lambda_1 + \lambda_2)}{4}) \, \Gamma(\frac{3 + i(\lambda + \lambda_1 - \lambda_2)}{4}) \, \Gamma(\frac{3 + i(\lambda - \lambda_1 - \lambda_2)}{4})}{\Gamma(\frac{3 + i\lambda}{2})} \quad (8.34)
$$

and

$$
\int_0^\infty \sigma^{\frac{-1 - i(\lambda + \lambda_1 + \lambda_2)}{2}} \, \sin(2\pi\sigma) \, d\sigma = \frac{1}{2} \, \pi^{\frac{i(\lambda + \lambda_1 + \lambda_2)}{2}} \, \frac{\Gamma(\frac{3 - i(\lambda + \lambda_1 + \lambda_2)}{4})}{\Gamma(\frac{3 + i(\lambda + \lambda_1 + \lambda_2)}{4})},
$$

$$(8.35)$$

we get

$$
B_{\lambda_1, \lambda_2; \lambda}(s_1, s_2; s) = \pi^{\frac{1}{2}} \times
$$

$$
\frac{\Gamma(\frac{3 - i(\lambda + \lambda_1 + \lambda_2)}{4}) \Gamma(\frac{3 + i(\lambda - \lambda_1 + \lambda_2)}{4}) \Gamma(\frac{3 + i(\lambda + \lambda_1 - \lambda_2)}{4}) \Gamma(\frac{3 + i(\lambda - \lambda_1 - \lambda_2)}{4})}{\Gamma(\frac{1 - i\lambda_1}{2}) \, \Gamma(\frac{1 - i\lambda_2}{2}) \, \Gamma(\frac{1 + i\lambda}{2})} \times
$$

$$
\chi^-_{i\lambda_1, i\lambda_2; i\lambda}(s_1, s_2; s),
$$

$$(8.36)$$

which we only have to plug in the right–hand side of the (unwritten) formula analogous to (8.17), mentioned in the beginning of the present proof. □

With the definitions (2.19) and (4.45) of $\Theta_{i\lambda}$ and $\Theta_{i\lambda}^{-1}$, the result of proposition 8.1 may also be written as

$$\Theta_{i\lambda}^{-1}\left(\Theta_{i\lambda_1}u_1 \cdot \Theta_{i\lambda_2}u_2\right) = \frac{1}{16\,\pi}\,\pi^{-1+\frac{i(-\lambda+\lambda_1+\lambda_2)}{2}} \times$$

$$\frac{\Gamma(\frac{1-i(\lambda+\lambda_1+\lambda_2)}{4})\Gamma(\frac{1+i(\lambda-\lambda_1+\lambda_2)}{4})\Gamma(\frac{1+i(\lambda+\lambda_1-\lambda_2)}{4})\Gamma(\frac{1+i(\lambda-\lambda_1-\lambda_2)}{4})}{\Gamma(\frac{i\lambda}{2})\Gamma(-\frac{i\lambda}{2})} \times$$

$$\mathbf{L}_{i\lambda_1,i\lambda_2;i\lambda}(u_1, u_2). \quad (8.37)$$

Also, defining $\mathbf{L}_{i\lambda_1,i\lambda_2;i\lambda}^{-}$ just like $\mathbf{L}_{i\lambda_1,i\lambda_2;i\lambda}$ with $\chi_{i\lambda_1,i\lambda_2;i\lambda}^{-}$ substituted for $\chi_{i\lambda_1,i\lambda_2;i\lambda}$, one gets from proposition 8.2 the formula

$$\Theta_{i\lambda}^{-1}\left(\{\Theta_{i\lambda_1}u_1, \Theta_{i\lambda_2}u_2\}\right) = \frac{1}{8\,\pi}\,\pi^{-1+\frac{i(-\lambda+\lambda_1+\lambda_2)}{2}} \times$$

$$\frac{\Gamma(\frac{3-i(\lambda+\lambda_1+\lambda_2)}{4})\Gamma(\frac{3+i(\lambda-\lambda_1+\lambda_2)}{4})\Gamma(\frac{3+i(\lambda+\lambda_1-\lambda_2)}{4})\Gamma(\frac{3+i(\lambda-\lambda_1-\lambda_2)}{4})}{\Gamma(\frac{i\lambda}{2})\Gamma(-\frac{i\lambda}{2})} \times$$

$$\mathbf{L}_{i\lambda_1,i\lambda_2;i\lambda}^{-}(u_1, u_2). \quad (8.38)$$

Our job in the remainder of this paper will include giving the left–hand side of (8.16) (or (8.26)) a meaning in the case when $u_j = \mathfrak{E}_{i\lambda_j}$. On the other hand, in the case when the non–class one principal series is substituted for the class–one series throughout this work, no such trick as that expressed by (8.16),(8.26) is available without heavy modifications (since no dual Radon transform valued into a space of *functions* on G/K exists any more): if conceptually simple, these modifications may entail such computational complications as to make the facing head–on of the analogue of (8.1) preferable.

Finally, let us observe, for the sake of completeness, that linear combinations of the two bilinear operations $\mathbf{L}_{\nu_1,\nu_2;\nu}$ and $\mathbf{L}_{\nu_1,\nu_2;\nu}^{-}$ are the only ones that satisfy the covariance property (1.6). For if $K(s_1, s_2; s)$ is the integral kernel of such an operation, it is easily seen, using nothing more than the definition of the representations π_{ν_1}, π_{ν_2} and π_ν, that K must satisfy the identity

$$K(\frac{as_1+b}{cs_1+d}, \frac{as_2+b}{cs_2+d}; \frac{as+b}{cs+d}) =$$

$$|cs+d|^{1+\nu}\,|cs_1+d|^{1-\nu_1}\,|cs_2+d|^{1-\nu_2}\,K(s_1, s_2; s), \quad (8.39)$$

analogous to (8.19), for every $\begin{pmatrix} a & b \\ c & d \end{pmatrix} \in G$, so that everything boils down (since the kernel $\chi_{i\lambda_1,i\lambda_2;i\lambda}$ satisfies this property), to remarking, as was done in the proof of proposition 8.2, that G acts in a transitive way on the set of triples $(s_1, s_2; s)$ for which the sign of $\frac{s_1-s_2}{(s-s_1)(s_2-s)}$ is fixed.

9. The Roelcke–Selberg expansion of the product of two Eisenstein series: the continuous part

Define

$$\text{height}(z) = \max_{\gamma \in \Gamma} \text{Im} \, (\gamma.z) \tag{9.1}$$

for all $z \in \Pi$: in other words, $\text{height}(z)$ is the Γ–invariant function on Π that coincides with Im z in the standard fundamental domain D of Π defined by the inequalities $|z| > 1$, $|\text{Re } z| < \frac{1}{2}$. It is easily seen that, given z and w in Π, one has the inequality

$$\text{height}(z) \leq 2 \cosh d(z, w) \, \text{height}(w). \tag{9.2}$$

The function $z \mapsto (\text{height}(z))^{\frac{1}{2}}$ is a good order of magnitude of Γ-invariant functions that just fail to lie in $L^2(\Gamma \backslash \Pi)$, which is precisely the case with Eisenstein series $E_{\frac{1-i\lambda}{2}}$ since, as a consequence of (3.29),

$$|E_{\frac{1-i\lambda}{2}}(z)| \leq C \, (\text{height}(z))^{\frac{1}{2}}. \tag{9.3}$$

Indeed, f will lie in $L^2(\Gamma \backslash \Pi)$ if it satisfies the same kind of estimate, with an extra factor of $(\log(\text{height}(w)))^{\alpha}$ $(\alpha < -\frac{1}{2})$ on the right–hand side. Also,

LEMMA 9.1. *If f and Δf lie in $L^2(\Gamma \backslash \Pi)$, one has for some constant $C > 0$ the estimate*

$$|f(z)| \leq C \, (\text{height}(z))^{\frac{1}{2}}. \tag{9.4}$$

PROOF. Standard elliptic estimates yield

$$|f(z)|^2 \leq C \, [\, \|f\|^2_{L^2(\Gamma \backslash \Pi)} + \|\Delta f\|^2_{L^2(\Gamma \backslash \Pi)} \,] \tag{9.5}$$

for $1 \leq \text{Im } z \leq 2$. If $g(z) = f(nz)$ for some $n \geq 1$ one may note that

$$\|g\|^2_{L^2(\Gamma \backslash \Pi)} \leq n \, \|f\|^2_{L^2(\Gamma \backslash \Pi)} \tag{9.6}$$

since, as z describes the standard fundamental domain D, nz lies within the union of n translates of D; also, Δ commutes with the map $z \mapsto nz$. $\quad\square$

From (8.37) our interest lies with the Γ–invariant function

$$(\Theta_{i\lambda_1} \mathfrak{E}_{i\lambda_1} \cdot \Theta_{i\lambda_2} \mathfrak{E}_{i\lambda_2})(z) = \zeta^*(1 - i\lambda_1) \, \zeta^*(1 - i\lambda_2) \, E_{\frac{1-i\lambda_1}{2}}(z) \, E_{\frac{1-i\lambda_2}{2}}(z)$$

$$= E^*_{\frac{1-i\lambda_1}{2}}(z) \, E^*_{\frac{1-i\lambda_2}{2}}(z), \tag{9.7}$$

where, in accordance with ([36], p.208), we have set

$$E^*_s(z) = \zeta^*(2s) \, E_s(z). \tag{9.8}$$

From the Fourier expansion (3.29) of Eisenstein series, it is clear that one only has

$$|(\Theta_{i\lambda_1} \mathcal{E}_{i\lambda_1})(z)\,(\Theta_{i\lambda_2} \mathcal{E}_{i\lambda_2})(z)| \leq C\,\text{height}(z), \qquad (9.9)$$

so that this product of Eisenstein series is far from lying in $L^2(\Gamma\backslash\Pi)$: to be in a position to apply it the Roelcke–Selberg theory, we shall discard some well–chosen Eisenstein series, but first, we shall change $i\lambda_1$ and $i\lambda_2$ to more general complex numbers ν_1 and ν_2.

We shall set, in all that follows, $\text{Re}\,\nu_1 = \ell_1$, $\text{Re}\,\nu_2 = \ell_2$, $\max(\ell_1, \ell_2) = \ell$. We shall assume that $\ell < -1$, that $|\ell_1 - \ell_2| < 1$ and (for convenience only) that $\nu_1 \neq \nu_2$.

As shown by (3.29), the product $E^*_{\frac{1-\nu_1}{2}}(z)\,E^*_{\frac{1-\nu_2}{2}}(z)$ coincides, up to an error term that remains bounded in the standard fundamental domain D, with

$$\sum_{\substack{\epsilon_1 = \pm 1 \\ \epsilon_2 = \pm 1}} \zeta^*(1 - \epsilon_1\nu_1)\,\zeta^*(1 - \epsilon_2\nu_2)\, y^{1 - \frac{\epsilon_1\nu_1 + \epsilon_2\nu_2}{2}},$$

where we can delete the term with $\epsilon_1 = \epsilon_2 = -1$ since $\ell < -1$: in turn, as a consequence of (3.29) again and of the condition $|\ell_1 - \ell_2| < 1$, this is just the same, up to a term that lies in $L^2(D)$, as

$$\sum_{\substack{\epsilon_1^2 = \epsilon_2^2 = 1 \\ (\epsilon_1, \epsilon_2) \neq (-1, -1)}} \zeta^*(1 - \epsilon_1\nu_1)\,\zeta^*(1 - \epsilon_2\nu_2)\, E_{1 - \frac{\epsilon_1\nu_1 + \epsilon_2\nu_2}{2}}(z).$$

Thus the Γ–invariant function

$$f(z) := E^*_{\frac{1-\nu_1}{2}}(z)\,E^*_{\frac{1-\nu_2}{2}}(z) - \zeta^*(1 - \nu_1)\,\zeta^*(1 - \nu_2)\,E_{1 - \frac{\nu_1 + \nu_2}{2}}(z)$$
$$- \zeta^*(1 + \nu_1)\,\zeta^*(1 - \nu_2)\,E_{1 + \frac{\nu_1 - \nu_2}{2}}(z) - \zeta^*(1 - \nu_1)\,\zeta^*(1 + \nu_2)\,E_{1 + \frac{\nu_2 - \nu_1}{2}}(z)$$
$$(9.10)$$

lies in $L^2(\Gamma\backslash\Pi)$.

Remark. As is well–known (it follows from (3.29)), the Γ–invariant function $z \mapsto E_s(z) - \frac{3}{\pi}(s - 1)^{-1}$ is still well–defined when $s = 1$. Since, as ν_1 and ν_2 ($\nu_1 \neq \nu_2$) go to the same ν with $\text{Re}\,\nu < -1$, the singular parts that come from the last two terms on the right–hand side of (9.10) cancel out, the function f remains well–defined and bounded even when $\nu_1 = \nu_2$ if $\text{Re}\,\nu_1 < -1$.

Since (propositions 3.3 and 3.4)

$$E^*_{\frac{1-\nu_1}{2}}(z) = \frac{1}{2} \sum_{|m_1| + |n_1| \neq 0} \pi^{\frac{\nu_1 - 1}{2}}\,\Gamma(\frac{1 - \nu_1}{2})\left(\frac{|m_1 z - n_1|^2}{y}\right)^{\frac{\nu_1 - 1}{2}}, \qquad (9.11)$$

one has

$$E^*_{\frac{1-\nu_1}{2}}(z)\, E^*_{\frac{1-\nu_2}{2}}(z) = \frac{1}{4}\,\pi^{\frac{\nu_1+\nu_2-2}{2}}\,\Gamma(\frac{1-\nu_1}{2})\,\Gamma(\frac{1-\nu_2}{2})$$

$$\sum_{\substack{|m_1|+|n_1|\neq 0 \\ |m_2|+|n_2|\neq 0}} \left(\frac{|m_1 z - n_1|^2}{y}\right)^{\frac{\nu_1-1}{2}} \left(\frac{|m_2 z - n_2|^2}{y}\right)^{\frac{\nu_2-1}{2}}. \quad (9.12)$$

Now the terms with $n_1 m_2 - n_2 m_1 = 0$ in the last sum are associated with indices that can be uniquely written as

$$\binom{n_1}{m_1} = k_1 \binom{n}{m}, \qquad \binom{n_2}{m_2} = k_2 \binom{n}{m}, \qquad (9.13)$$

where $k_1 \geq 1$, $k_2 \in \mathbf{Z}^\times = \mathbf{Z}\backslash\{0\}$ and $(n, m) = 1$. It follows that the sum of all terms on the right–hand side of (9.12) with $n_1 m_2 - n_2 m_1 = 0$ is

$$\frac{1}{4}\,\pi^{\frac{\nu_1+\nu_2-2}{2}}\,\Gamma(\frac{1-\nu_1}{2})\,\Gamma(\frac{1-\nu_2}{2}) \sum_{\substack{k_1 \geq 1 \\ k_2 \in \mathbf{Z}^\times}} \sum_{(m,n)=1} k_1^{\nu_1-1}|k_2|^{\nu_2-1} \left(\frac{|mz-n|^2}{y}\right)^{\frac{\nu_1+\nu_2-2}{2}}$$

$$= \zeta^*(1-\nu_1)\,\zeta^*(1-\nu_2)\,E_{1-\frac{\nu_1+\nu_2}{2}}(z), \quad (9.14)$$

thus

$$f_{\text{main}}(z) := E^*_{\frac{1-\nu_1}{2}}(z)\, E^*_{\frac{1-\nu_2}{2}}(z) - \zeta^*(1-\nu_1)\,\zeta^*(1-\nu_2)\,E_{1-\frac{\nu_1+\nu_2}{2}}(z) \quad (9.15)$$

coincides with the same sum as the right–hand side of (9.12), in which the condition $n_1 m_2 - n_2 m_1 \neq 0$ is substituted for $(|m_1|+|n_1| \neq 0, |m_2|+|n_2| \neq 0)$.

Using when $n_1 m_2 - n_2 m_1 \neq 0$ the formula (cf. (8.18))

$$\int_\Pi \left(\frac{|z-s|^2}{y}\right)^{-\frac{1}{2}-\frac{i\lambda}{2}} \left(\frac{|m_1 z - n_1|^2}{y}\right)^{\frac{\nu_1-1}{2}} \left(\frac{|m_2 z - n_2|^2}{y}\right)^{\frac{\nu_2-1}{2}} d\mu(z)$$

$$= \frac{\pi^{\frac{1}{2}}}{2}\,\frac{\Gamma(\frac{1-\alpha}{2})\Gamma(\frac{1-\beta}{2})\Gamma(\frac{1-\gamma}{2})\Gamma(\frac{1-\delta}{2})}{\Gamma(\frac{1-\nu_1}{2})\Gamma(\frac{1-\nu_2}{2})\Gamma(\frac{1+i\lambda}{2})} \times$$

$$|n_1 m_2 - n_2 m_1|^{\gamma-1}\,|n_1 - m_1 s|^{\alpha-1}\,|n_2 - m_2 s|^{\beta-1} \quad (9.16)$$

with

$$\alpha = \frac{1 - i\lambda + \nu_1 - \nu_2}{2}, \qquad \beta = \frac{1 - i\lambda - \nu_1 + \nu_2}{2},$$

$$\gamma = \frac{1 + i\lambda + \nu_1 + \nu_2}{2}, \qquad \delta = \frac{1 - i\lambda + \nu_1 + \nu_2}{2} \quad (9.17)$$

and the formula $((4.45), (4.34))$

$$(\Theta_{i\lambda}^{-1} f)(s) = \frac{1}{8}\, \pi^{-\frac{3}{2}-\frac{i\lambda}{2}}\, \frac{\Gamma(\frac{1+i\lambda}{2})}{\Gamma(\frac{i\lambda}{2})\Gamma(-\frac{i\lambda}{2})} \int_{\Pi} \left(\frac{|z-s|^2}{y}\right)^{-\frac{1}{2}-\frac{i\lambda}{2}} f(z)\, d\mu(z), \tag{9.18}$$

one would be led into believing that

$$(\Theta_{i\lambda}^{-1} f_{\mathrm{main}})(s) = \frac{1}{16\pi}\, \pi^{\frac{\alpha+\beta+\gamma+\delta}{2}-2}\, \frac{\Gamma(\frac{1-\alpha}{2})\Gamma(\frac{1-\beta}{2})\Gamma(\frac{1-\gamma}{2})\Gamma(\frac{1-\delta}{2})}{\Gamma(\frac{i\lambda}{2})\Gamma(\frac{-i\lambda}{2})}\, \mathfrak{T}(s), \tag{9.19}$$

as a consequence of a comparison with (8.5). However, $\mathfrak{T}(s)$ (actually thought of as a distribution with respect to s and λ) lacked a bona fide definition in section 8, so that the last identity should really be thought of as defining $\mathfrak{T}(s)$. We now introduce an automorphic function $(f_0)_{\mathrm{main}}$ which bears the same relationship to the (tentative) distribution $\mathfrak{T}_0$ (cf. (8.9)) as that between f_{main} and $\mathfrak{T}$. This is

$$(f_0)_{\mathrm{main}}(z) = \frac{1}{2}\pi^{\frac{\nu_1+\nu_2-2}{2}}\, \Gamma(\frac{1-\nu_1}{2})\,\Gamma(\frac{1-\nu_2}{2}) \times$$

$$\sum_{n_1 m_2 - n_2 m_1 = 1} \left(\frac{|m_1 z - n_1|^2}{y}\right)^{\frac{\nu_1 - 1}{2}} \left(\frac{|m_2 z - n_2|^2}{y}\right)^{\frac{\nu_2 - 1}{2}}. \tag{9.20}$$

Using (8.8) and (8.12), it is easily seen that the image of $(f_0)_{\mathrm{main}}$ under the Hecke operator T_N (cf. ([36], p.241) is

$$(T_N(f_0)_{\mathrm{main}})(z) = \frac{1}{2}\pi^{\frac{\nu_1+\nu_2-2}{2}}\, \Gamma(\frac{1-\nu_1}{2})\,\Gamma(\frac{1-\nu_2}{2}) \times$$

$$N^{\frac{1-\nu_1-\nu_2}{2}} \sum_{n_1 m_2 - n_2 m_1 = N} \left(\frac{|m_1 z - n_1|^2}{y}\right)^{\frac{\nu_1 - 1}{2}} \left(\frac{|m_2 z - n_2|^2}{y}\right)^{\frac{\nu_2 - 1}{2}}, \tag{9.21}$$

so that

$$f_{\mathrm{main}} = \sum_{N \geq 1} N^{\frac{\nu_1+\nu_2-1}{2}} T_N(f_0)_{\mathrm{main}}, \tag{9.22}$$

a formula which should be substituted to the tentative (8.14).

Still under the assumption $\ell = \max(\mathrm{Re}\,\nu_1, \mathrm{Re}\,\nu_2) < -1$, we define

$$f_0(z) := (f_0)_{\mathrm{main}}(z) - \pi^{\frac{\nu_1+\nu_2-1}{2}}\, \Gamma(-\frac{\nu_1}{2})\,\Gamma(\frac{1-\nu_2}{2})\, E_{1+\frac{\nu_1-\nu_2}{2}}(z)$$

$$- \pi^{\frac{\nu_1+\nu_2-1}{2}}\, \Gamma(\frac{1-\nu_1}{2})\,\Gamma(-\frac{\nu_2}{2})\, E_{1+\frac{\nu_2-\nu_1}{2}}(z). \tag{9.23}$$

Since ([36], p.246)

$$T_N E_s = N^{\frac{1}{2}-s}\, \sigma_{2s-1}(N)\, E_s \tag{9.24}$$

and ([16], p. 250)

$$\pi^{\frac{\nu_1+\nu_2-1}{2}} \Gamma(-\frac{\nu_1}{2}) \Gamma(\frac{1-\nu_2}{2}) \sum_{N \geq 1} N^{\frac{\nu_1+\nu_2-1}{2}} N^{\frac{-1+\nu_2-\nu_1}{2}} \sigma_{1+\nu_1-\nu_2}(N)$$

$$= \pi^{\frac{\nu_1+\nu_2-1}{2}} \Gamma(-\frac{\nu_1}{2}) \Gamma(\frac{1-\nu_2}{2}) \zeta(1-\nu_2) \zeta(-\nu_1)$$

$$= \zeta^*(1-\nu_2) \zeta^*(-\nu_1)$$

$$= \zeta^*(1-\nu_2) \zeta^*(1+\nu_1), \tag{9.25}$$

it follows from (9.10), (9.15) and (9.23), (9.25) that

$$f = \sum_{N \geq 1} N^{\frac{\nu_1+\nu_2-1}{2}} T_N f_0 \tag{9.26}$$

as well.

In order to be in a position to compute the Roelcke–Selberg expansion of f_0, we prove:

LEMMA 9.2. *Assume that* Re $\nu_1 = \ell_1$, Re $\nu_2 = \ell_2$, $\ell = \max(\ell_1, \ell_2) < -1$, $|\ell_1 - \ell_2| < 1$, $\nu_1 \neq \nu_2$. *Then the* Γ*-invariant function* f_0 *introduced in* (9.23) *lies in* $L^2(\Gamma \backslash \Pi)$.

PROOF. The function

$$\frac{1}{2} \sum_{m \neq 0, n \in \mathbb{Z}} \left(\frac{|mz-n|^2}{y} \right)^{\frac{\ell_1-1}{2}} = \zeta(1-\ell_1) [E_{\frac{1-\ell_1}{2}}(z) - y^{\frac{1-\ell_1}{2}}] \tag{9.27}$$

is bounded in the fundamental domain D if $\ell_1 < -1$, as a consequence of the Fourier expansion (3.29) of $E_{\frac{1-\ell_1}{2}}$, and so is the sum of all terms on the right–hand side of (9.20) with $m_1 m_2 \neq 0$, since it is taken from the product of two absolutely convergent series such as (9.27), with ν_1 and ν_2 substituted for ℓ_1. Thus, up to some error term that remains bounded in D, one has

$$(f_0)_{\text{main}}(z) \sim \pi^{\frac{\nu_1+\nu_2-2}{2}} \Gamma(\frac{1-\nu_1}{2}) \Gamma(\frac{1-\nu_2}{2})$$

$$\left[\sum_{n_2 \in \mathbb{Z}} y^{\frac{1-\nu_1}{2}} \left(\frac{|z-n_2|^2}{y} \right)^{\frac{\nu_2-1}{2}} + \sum_{n_1 \in \mathbb{Z}} y^{\frac{1-\nu_2}{2}} \left(\frac{|z-n_1|^2}{y} \right)^{\frac{\nu_1-1}{2}} \right], \tag{9.28}$$

for the right–hand side of this identity is the sum of all terms with $m_1 m_2 = 0$ taken from the right–hand side of (9.20). Also,

$$\pi^{\frac{\nu_1+\nu_2-2}{2}} \Gamma(\frac{1-\nu_1}{2}) \Gamma(\frac{1-\nu_2}{2}) y^{\frac{1-\nu_1}{2}} \int_{-\infty}^{\infty} \left(\frac{|z-t|^2}{y} \right)^{\frac{\nu_2-1}{2}} dt$$

$$= \pi^{\frac{\nu_1+\nu_2-2}{2}} \Gamma(\frac{1-\nu_1}{2}) \Gamma(\frac{1-\nu_2}{2}) \pi^{\frac{1}{2}} \frac{\Gamma(-\frac{\nu_2}{2})}{\Gamma(\frac{1-\nu_2}{2})} y^{\frac{2-\nu_1+\nu_2}{2}} \tag{9.29}$$

differs by an error term that lies in $L^2(D)$ from

$$\pi^{\frac{\nu_1+\nu_2-1}{2}} \, \Gamma(\frac{1-\nu_1}{2}) \Gamma(-\frac{\nu_2}{2}) \, E_{1+\frac{\nu_2-\nu_1}{2}}(z),$$

the last term on the right–hand side of (9.23). To conclude the proof of the lemma, all that remains to be done is proving that

$$\mathrm{Err}_2(z): \, = \sum_{n\in\mathbf{Z}} |z-n|^{\nu_2-1} - \int_{-\infty}^{\infty} |z-t|^{\nu_2-1} \, dt \qquad (9.30)$$

is a $O(y^{\ell_2-1})$ as $y \to \infty$ while $z \in D$. Now

$$\mathrm{Err}_2(z) = \sum_{n\in\mathbf{Z}} \int_{n}^{n+1} \left[\, |z-n|^{\nu_2-1} - |z-t|^{\nu_2-1} \right] dt$$

$$= (1-\nu_2) \sum_{n\in\mathbf{Z}} \int_{n}^{n+1} dt \int_{n}^{t} (s-x) \left[(s-x)^2 + y^2 \right]^{\frac{\nu_2-3}{2}} ds$$

$$= (1-\nu_2) \sum_{n\in\mathbf{Z}} \int_{n}^{n+1} (s-x)(n+1-s) \left[(s-x)^2 + y^2 \right]^{\frac{\nu_2-3}{2}} ds \, ,$$

$$(9.31)$$

so that

$$|\mathrm{Err}_2(z)| \le C \int_{-\infty}^{\infty} |s-x| \left[(s-x)^2 + y^2 \right]^{\frac{\ell_2-3}{2}} ds$$

$$\le C \, y^{\ell_2-1}. \qquad (9.32)$$

$\square$

LEMMA 9.3. *Under the assumptions of lemma 9.2, the function f_0 is C^∞ on Π; for every N, $\Delta^N f_0$ lies in $L^2(\Gamma\backslash\Pi)$.*

PROOF. Consider first the sum of all terms with $m_1 m_2 \ne 0$ in the definition (9.20) of $(f_0)_{\mathrm{main}}$. To show that one can apply it Δ^N and end up with a function bounded in D, it suffices to prove — using the same argument as in the beginning of the proof of lemma 9.2 — that, for every pair (j,k) of non–negative integers, the series

$$\sum_{m\ne 0, n\in\mathbf{Z}} |(y\frac{\partial}{\partial y})^j (y\frac{\partial}{\partial x})^k \left(\frac{|mz-n|^2}{y} \right)^{\frac{\nu_1-1}{2}} |$$

is bounded in D. One shows by induction on $j+k$ that

$$(y\frac{\partial}{\partial y})^j (y\frac{\partial}{\partial x})^k \left(\frac{|mz-n|^2}{y} \right)^{\frac{\nu_1-1}{2}}$$

is a linear combination, with coefficients depending only on j, k, i, p, q, of expressions

$$[m(mx - n)]^p \, [m^2 y]^q \left(\frac{|mz - n|^2}{y}\right)^{\frac{\nu_1 - 1 - 2i}{2}}$$

with $0 \le i \le j + k$, $p \ge 0$, $q \ge 0$, $p + q \le i$. This is majorized by

$$C \, [m|mx - n| + m^2 y]^i \left(\frac{|mz - n|^2}{y}\right)^{\frac{\ell - 1 - 2i}{2}}$$

$$\le C \, [\tfrac{3}{2} m^2 y + \tfrac{1}{2y}(mx - n)^2]^i \left(\frac{|mz - n|^2}{y}\right)^{\frac{\ell - 1 - 2i}{2}}$$

$$\le C \left(\frac{|mz - n|^2}{y}\right)^{\frac{\ell - 1}{2}}, \tag{9.33}$$

so we are done for this part of f_0.

What remains to be done, using (9.23) and (9.29) again, is getting a uniform bound in D for the $(y\frac{\partial}{\partial y}, y\frac{\partial}{\partial x})$-derivatives of all order of

$$y^{\frac{1 - \nu_1}{2}} \left[\sum_{n \in \mathbf{Z}} \left(\frac{|z - n|^2}{y}\right)^{\frac{\nu_2 - 1}{2}} - \int_{-\infty}^{\infty} \left(\frac{|z - t|^2}{y}\right)^{\frac{\nu_2 - 1}{2}} dt\right]$$

$$= y^{1 - \frac{\nu_1 + \nu_2}{2}} \operatorname{Err}_2(z). \tag{9.34}$$

We thus have to show that

$$|(y\frac{\partial}{\partial y})^j \, (y\frac{\partial}{\partial x})^k \operatorname{Err}_2(z)| \le C \, y^{\ell_2 - 1}. \tag{9.35}$$

Starting from (9.31), we may write

$$(y\frac{\partial}{\partial y})^j \, (y\frac{\partial}{\partial x})^k \operatorname{Err}_2(z)$$

as a linear combination of sums

$$\sum_{n \in \mathbf{Z}} \int_n^{n+1} (n + 1 - s) \, (s - x)^p \, y^q \, [(s - x)^2 + y^2]^{\frac{\nu_2 - 3 - 2i}{2}} \, ds$$

with $0 \le i \le j + k$, $p \ge 0$, $q \ge 0$, $p + q \le 1 + 2i$, so we are done. $\square$

We prepare for the problem of changing some order of summation, which will occur here and in sections 11 and 12, with the following lemma.

LEMMA 9.4. *Assume that* $\max(\operatorname{Re} \nu_1, \operatorname{Re} \nu_2) = \ell < -1$. *For every* $n \in \mathbf{Z}$, *one has for some* $C > 0$ *the estimate*

$$\left| \int_{-\infty}^{\infty} |(n - \sigma)y|^{-\frac{\nu_1}{2}} |\sigma y|^{-\frac{\nu_2}{2}} e^{-2i\pi\sigma\xi} K_{\frac{\nu_1}{2}}(2\pi y|n - \sigma|) K_{\frac{\nu_2}{2}}(2\pi y|\sigma|) \, d\sigma \right|$$

$$\le C \, y^{-\ell} \, (y^2 + \xi^2)^{\frac{\ell - 1}{2}}. \tag{9.36}$$

PROOF. From (3.25),

$$\int_{-\infty}^{\infty} |\sigma y|^{-\frac{\nu_2}{2}} e^{-2i\pi\sigma\xi} K_{\frac{\nu_2}{2}}(2\pi y|\sigma|) \, d\sigma = C(\nu_2) \, y^{-1} \, (1 + \frac{\xi^2}{y^2})^{\frac{\nu_2-1}{2}},$$
(9.37)

where the real part of the last exponent is $\frac{\ell_2-1}{2} \le \frac{\ell-1}{2} < -1$. Now one can easily majorize the convolution integral

$$\int_{-\infty}^{\infty} y^{-2} \, (1 + \frac{(\xi-\eta)^2}{y^2})^{\frac{\ell-1}{2}} (1 + \frac{\eta^2}{y^2})^{\frac{\ell-1}{2}} \, d\eta \le C \, y^{-1} \, (1 + \frac{\xi^2}{y^2})^{\frac{\ell-1}{2}}.$$
(9.38)

□

THEOREM 9.5. *Assume that* Re $\nu_1 = \ell_1$, Re $\nu_2 = \ell_2$, $\ell = \max(\ell_1, \ell_2) <$ -1, $|\ell_1 - \ell_2| < 1$, *and that* $\nu_1 \ne \nu_2$. *The image under the spectral projection corresponding to the continuous part of the spectrum of* Δ *of the function*

$$f_0(z) = \frac{1}{2} \pi^{\frac{\nu_1+\nu_2-2}{2}} \, \Gamma(\frac{1-\nu_1}{2}) \Gamma(\frac{1-\nu_2}{2}) \times$$

$$\sum_{n_1 m_2 - n_2 m_1 = 1} \left(\frac{|m_1 z - n_1|^2}{y} \right)^{\frac{\nu_1-1}{2}} \left(\frac{|m_2 z - n_2|^2}{y} \right)^{\frac{\nu_2-1}{2}}$$

$$- \pi^{\frac{\nu_1+\nu_2-1}{2}} \, \Gamma(-\frac{\nu_1}{2}) \Gamma(\frac{1-\nu_2}{2}) \, E_{1+\frac{\nu_1-\nu_2}{2}}(z)$$

$$- \pi^{\frac{\nu_1+\nu_2-1}{2}} \, \Gamma(\frac{1-\nu_1}{2}) \Gamma(-\frac{\nu_2}{2}) \, E_{1+\frac{\nu_2-\nu_1}{2}}(z)$$
(9.39)

is given by the integral

$$z \mapsto \frac{1}{8\pi} \int_{-\infty}^{\infty} \Phi_0(\lambda) \, E_{\frac{1-i\lambda}{2}}(z) \, d\lambda$$
(9.40)

with

$$\Phi_0(\lambda) = \pi^{\frac{\gamma+\delta}{2}-1} \, \Gamma(\frac{1-\gamma}{2}) \Gamma(\frac{1-\delta}{2}) \frac{\zeta^*(1-\alpha) \, \zeta^*(1-\beta)}{\zeta^*(-i\lambda)},$$
(9.41)

where $\alpha, \beta, \gamma, \delta$ *were defined in* (9.17).
    The function f_0 *is orthogonal to the constants in* $L^2(\Gamma \backslash \Pi)$.

PROOF. Split f_0 as

$$f_0 = (f_0)_{\text{maj}} + (f_0)_{\text{min}},$$
(9.42)

where $(f_0)_{\text{maj}}$ is the sum of all terms with $m_1 m_2 \ne 0$ in the main term on the right–hand side of (9.39). Let

$$a_0 = (a_0)_{\text{maj}} + (a_0)_{\text{min}}$$
(9.43)

be the corresponding decomposition of

$$a_0(y) = \int_{-\frac{1}{2}}^{\frac{1}{2}} f_0(x + iy) \, dx.$$
(9.44)

Using the same trick as that which led to (9.28), and the known constant term in the Fourier expansion (3.29) of Eisenstein series, one easily finds

$(a_0)_{\min}(y)$

$$= \pi^{\frac{\nu_1+\nu_2-2}{2}} \Gamma(\frac{1-\nu_1}{2})\Gamma(\frac{1-\nu_2}{2}) \int_{-\infty}^{\infty} y^{\frac{2-\nu_1-\nu_2}{2}} [(t^2+y^2)^{\frac{\nu_1-1}{2}} + (t^2+y^2)^{\frac{\nu_2-1}{2}}]\, dt$$

$$- \pi^{\frac{\nu_1+\nu_2-1}{2}} \Gamma(-\frac{\nu_1}{2})\Gamma(\frac{1-\nu_2}{2}) [y^{\frac{2+\nu_1-\nu_2}{2}} + y^{\frac{\nu_2-\nu_1}{2}}\frac{\zeta^*(\nu_2-\nu_1)}{\zeta^*(2+\nu_1-\nu_2)}]$$

$$- \pi^{\frac{\nu_1+\nu_2-1}{2}} \Gamma(\frac{1-\nu_1}{2})\Gamma(-\frac{\nu_2}{2}) [y^{\frac{2-\nu_1+\nu_2}{2}} + y^{\frac{\nu_1-\nu_2}{2}}\frac{\zeta^*(\nu_1-\nu_2)}{\zeta^*(2-\nu_1+\nu_2)}] \qquad (9.45)$$

so that, as a consequence of (9.29),

$$(a_0)_{\min}(y) = -\pi^{\frac{\nu_1+\nu_2-1}{2}} [\Gamma(-\frac{\nu_1}{2})\Gamma(\frac{1-\nu_2}{2})\frac{\zeta^*(\nu_2-\nu_1)}{\zeta^*(2+\nu_1-\nu_2)} y^{\frac{\nu_2-\nu_1}{2}}$$

$$+ \Gamma(\frac{1-\nu_1}{2})\Gamma(-\frac{\nu_2}{2})\frac{\zeta^*(\nu_1-\nu_2)}{\zeta^*(2-\nu_1+\nu_2)} y^{\frac{\nu_1-\nu_2}{2}}]. \qquad (9.46)$$

On the other hand, the set of matrices $\begin{pmatrix} n_1 & n_2 \\ m_1 & m_2 \end{pmatrix} \in \Gamma$ with $m_1 m_2 \neq 0$ can be characterized as the set of $\begin{pmatrix} n_1^0+km_1 & . \\ m_1 & m_2 \end{pmatrix}$ with $m_1 m_2 \neq 0$, $k \in \mathbf{Z}$, where n_1^0 is *one* arbitrarily chosen integer with $n_1^0 m_2 \equiv 1 \bmod m_1$: then n_2 is uniquely determined and $\frac{n_2}{m_2} = \frac{n_1}{m_1} - \frac{1}{m_1 m_2}$. One has

$$\sum_{k \in \mathbf{Z}} \int_{-\frac{1}{2}}^{\frac{1}{2}} [(x-\frac{n_1^0+km_1}{m_1})^2 + y^2]^{\frac{\nu_1-1}{2}} [(x-\frac{n_2}{m_2})^2 + y^2]^{\frac{\nu_2-1}{2}}\, dx$$

$$= \int_{-\infty}^{\infty} (x^2+y^2)^{\frac{\nu_1-1}{2}} [(x+\frac{1}{m_1 m_2})^2 + y^2]^{\frac{\nu_2-1}{2}}\, dx : \qquad (9.47)$$

using

$$\int_{-\infty}^{\infty} (x^2+y^2)^{\frac{\nu_1-1}{2}} e^{-2i\pi x\sigma}\, dx = \frac{2\pi^{\frac{1-\nu_1}{2}}}{\Gamma(\frac{1-\nu_1}{2})} y^{\frac{\nu_1}{2}} |\sigma|^{-\frac{\nu_1}{2}} K_{\frac{\nu_1}{2}}(2\pi y|\sigma|) \qquad (9.48)$$

and Plancherel's formula, one then gets

$$(a_0)_{\mathrm{maj}}(y) = 2y \sum_{\substack{m_1 m_2 \neq 0 \\ (m_1, m_2)=1}} |m_1|^{\nu_1-1} |m_2|^{\nu_2-1}$$

$$\int_{-\infty}^{\infty} |\sigma|^{-\frac{\nu_1+\nu_2}{2}} e^{-\frac{2i\pi\sigma}{m_1 m_2}} K_{\frac{\nu_1}{2}}(2\pi y|\sigma|) K_{\frac{\nu_2}{2}}(2\pi y|\sigma|)\, d\sigma. \qquad (9.49)$$

We now turn to the estimation of

$$C_0^{\pm}(\mu) = (C_0)_{\mathrm{maj}}^{\pm}(\mu) + (C_0)_{\min}^{\pm}(\mu) \qquad (9.50)$$

with $C_0^{\pm}(\mu)$ as defined in theorem 7.3, and where the two terms in the decomposition of the right–hand side have an obvious meaning (but a smaller domain of holomorphy than their sum: they do not correspond to automorphic functions). It is immediate that $(C_0)^+_{\min}(\mu)$, defined when $\operatorname{Im} \mu > 1$ by

$$(C_0)^+_{\min}(\mu) = \frac{1}{8\pi} \frac{\pi^{-\frac{i\mu}{2}}}{\Gamma(-\frac{i\mu}{2})} \int_0^1 (a_0)_{\min}(y)\, y^{-\frac{3}{2}-\frac{i\mu}{2}}\, dy\,, \qquad (9.51)$$

extends as a meromorphic function on the entire plane with only simple poles at $\mu = i\,[1 \pm (\nu_1 - \nu_2)]$: the residue at $\mu = i\,(1 + \nu_1 - \nu_2)$ is

$$\frac{1}{4i\pi}\, \pi^{\nu_1} \frac{\Gamma(-\frac{\nu_1}{2})\,\Gamma(\frac{1-\nu_2}{2})}{\Gamma(\frac{1+\nu_1-\nu_2}{2})} \frac{\zeta^*(1+\nu_1-\nu_2)}{\zeta^*(2+\nu_1-\nu_2)}\,, \qquad (9.52)$$

and of course ν_1 and ν_2 have to be traded for each other at the other pole.

Next, performing the change of variables $\sigma \mapsto \frac{\sigma}{y}$ for $y \geq 1$, and remembering that $K_{\frac{\nu_1}{2}}(2\pi|\sigma|)$ is rapidly decreasing at infinity and is $O(|\sigma|^{\frac{\ell_1}{2}})$ near $\sigma = 0$, one sees that

$$(a_0)_{\mathrm{maj}}(y) = O(y^{\frac{\ell_1+\ell_2}{2}}), \qquad y \to \infty \qquad (9.53)$$

so that

$$(C_0)^-_{\mathrm{maj}}(\mu) = -\frac{1}{8\pi} \frac{\pi^{-\frac{i\mu}{2}}}{\Gamma(-\frac{i\mu}{2})} \int_1^\infty (a_0)_{\mathrm{maj}}(y)\, y^{-\frac{3}{2}-\frac{i\mu}{2}}\, dy \qquad (9.54)$$

extends as a holomorphic function in the half–plane $\operatorname{Im} \mu < 1-\ell_1-\ell_2$: in the strip $0 < \operatorname{Im} \mu < 1-\ell_1-\ell_2$, a fortiori in the strip $0 < \operatorname{Im} \mu < 1 - 2\ell$, $C_0^+(\mu)$ thus differs from

$$\tilde{C}_0^+(\mu) = (C_0)^+_{\min}(\mu) + \frac{1}{8\pi} \frac{\pi^{-\frac{i\mu}{2}}}{\Gamma(-\frac{i\mu}{2})} \int_0^\infty (a_0)_{\mathrm{maj}}(y)\, y^{-\frac{3}{2}-\frac{i\mu}{2}}\, dy \qquad (9.55)$$

by a holomorphic term.

The second term on the right–hand side of this equation is

$$\frac{1}{4\pi} \frac{\pi^{-\frac{i\mu}{2}}}{\Gamma(-\frac{i\mu}{2})} \int_0^\infty y^{\frac{-1-i\mu+\nu_1+\nu_2}{2}}\, dy \sum_{\substack{m_1 m_2 \neq 0 \\ (m_1, m_2)=1}} |m_1|^{\nu_1 - 1}\, |m_2|^{\nu_2 - 1}$$

$$\int_{-\infty}^\infty |\sigma y|^{-\frac{\nu_1+\nu_2}{2}}\, e^{-2i\pi \frac{\sigma}{m_1 m_2}}\, K_{\frac{\nu_1}{2}}(2\pi y|\sigma|)\, K_{\frac{\nu_2}{2}}(2\pi y|\sigma|)\, d\sigma\,, \quad (9.56)$$

where the integrations or summations have to be performed from the right to the left. Now, the case $n = 0$ of lemma 9.4 shows that the last integral is less than

$$C\, y^{-1}\, (1 + \frac{1}{(m_1 m_2 y)^2})^{\frac{\ell-1}{2}} \leq C\, y^{2\epsilon-1}\, |m_1 m_2|^{2\epsilon} \qquad (9.57)$$

where we choose when $y < 1$, under the assumption Im $\mu > 1$,

$$\max(0, \frac{1}{4} - \frac{\text{Im }\mu}{4} - \frac{\ell}{2}) < \varepsilon < -\frac{\ell}{2}. \qquad (9.58)$$

When $y > 1$, we majorize the left–hand side of (9.57) by $C\,y^{-1}$. Since

$$\left(\int_0^1 y^{\frac{-1+\text{Im }\mu}{2}+\ell+2\varepsilon-1}\,dy \right) \sum_{m_1 m_2 \neq 0} |m_1|^{\ell_1-1+2\varepsilon}\,|m_2|^{\ell_2-1+2\varepsilon}$$

$$+\left(\int_1^\infty y^{\frac{-1+\text{Im }\mu}{2}+\ell-1}\,dy \right) \sum_{m_1 m_2 \neq 0} |m_1|^{\ell_1-1}\,|m_2|^{\ell_2-1} < \infty, \qquad (9.59)$$

we may, if $1 < \text{Im }\mu < 1 - 2\ell$, perform in (9.56) the dy–integration before the summation with respect to m_1, m_2 without changing the result.

Next it is clear, as one can see by integrating both sides against some function of ξ in the Schwartz space $\mathcal{S}(\mathbb{R})$, that in the case when, moreover, Im $\mu > -1-\ell_1-\ell_2$,

$$\int_0^\infty y^{-\frac{1}{2}-\frac{i\mu}{2}}\,dy \int_{-\infty}^\infty |\sigma|^{-\frac{\nu_1+\nu_2}{2}} e^{-2i\pi\sigma\xi}\,K_{\frac{\nu_1}{2}}(2\pi y|\sigma|)\,K_{\frac{\nu_2}{2}}(2\pi y|\sigma|)\,d\sigma$$

$$(9.60)$$

agrees, as a tempered distribution, with

$$\mathcal{F}\left\{ \sigma \mapsto |\sigma|^{-\frac{\nu_1+\nu_2}{2}} \int_0^\infty y^{-\frac{1}{2}-\frac{i\mu}{2}} K_{\frac{\nu_1}{2}}(2\pi y|\sigma|)\,K_{\frac{\nu_2}{2}}(2\pi y|\sigma|)\,dy \right\}(\xi);$$

$$(9.61)$$

also, since

$$\int_0^\infty y^{-\frac{1}{2}+\frac{\text{Im }\mu}{2}+\frac{\ell_1+\ell_2}{2}-\ell} \times (y^2 + \xi^2)^{\frac{\ell-1}{2}}\,dy < \infty \qquad (9.62)$$

if $-1-\ell_1-\ell_2 < \text{Im }\mu < 1 - 2\ell$ and $\xi \neq 0$, it follows from lemma 9.4 that the integral (9.61) is a continuous function of ξ for $\xi \neq 0$.

These lengthy, but necessary, justifications make it possible, under the condition

$$-1 - \ell_1 - \ell_2 < \text{Im }\mu < 1 - 2\ell, \qquad (9.63)$$

(yes, this defines a non–void strip since $|\ell_1 - \ell_2| < 1$), to write

$$\tilde{C}_0^+(\mu) - (C_0)_{\min}^+(\mu) = \frac{1}{4\pi}\,\frac{\pi^{-\frac{i\mu}{2}}}{\Gamma(-\frac{i\mu}{2})} \sum_{\substack{m_1 \neq 0, m_2 \neq 0 \\ (m_1, m_2) = 1}} |m_1|^{\nu_1-1}\,|m_2|^{\nu_2-1}$$

$$\mathcal{F}\left\{ \sigma \mapsto |\sigma|^{-\frac{\nu_1+\nu_2}{2}} \int_0^\infty y^{-\frac{1}{2}-\frac{i\mu}{2}} K_{\frac{\nu_1}{2}}(2\pi y|\sigma|)\,K_{\frac{\nu_2}{2}}(2\pi y|\sigma|)\,dy \right\}(\frac{1}{m_1 m_2}).$$

$$(9.64)$$

Using the Weber–Schafheitlin integral ([27], p.101 or (6.10)), and the fact that the Fourier transform, evaluated at $\frac{1}{m_1 m_2}$, of $|\sigma|^{\frac{-1+i\mu-\nu_1-\nu_2}{2}}$, is

$$\pi^{\frac{\nu_1+\nu_2-i\mu}{2}} \frac{\Gamma(\frac{1+i\mu-\nu_1-\nu_2}{4})}{\Gamma(\frac{1-i\mu+\nu_1+\nu_2}{4})} |m_1 m_2|^{\frac{1+i\mu-\nu_1-\nu_2}{2}}, \tag{9.65}$$

we get

$$\tilde{C}_0^+(\mu) - (C_0)_{\min}^+(\mu) = 2^{-5}\,\pi^{\frac{-3+\nu_1+\nu_2-i\mu}{2}} \times$$

$$\frac{\Gamma(\frac{1+i\mu-\nu_1-\nu_2}{4})\,\Gamma(\frac{1-i\mu+\nu_1-\nu_2}{4})\,\Gamma(\frac{1-i\mu-\nu_1+\nu_2}{4})\,\Gamma(\frac{1-i\mu-\nu_1-\nu_2}{4})}{\Gamma(-\frac{i\mu}{2})\,\Gamma(\frac{1-i\mu}{2})} \times$$

$$\sum_{\substack{m_1 m_2 \neq 0 \\ (m_1,m_2)=1}} |m_1|^{\frac{-1+i\mu+\nu_1-\nu_2}{2}} |m_2|^{\frac{-1+i\mu-\nu_1+\nu_2}{2}}, \tag{9.66}$$

where the sum on the right–hand side is just

$$4\,\frac{\zeta(\frac{1-i\mu-\nu_1+\nu_2}{2})\,\zeta(\frac{1-i\mu+\nu_1-\nu_2}{2})}{\zeta(1-i\mu)}.$$

Observe that the expression (9.66), or, rather, its analytic continuation to the domain $0 < \operatorname{Im}\mu < 1-2\ell$, has a simple pole at $\mu = i(1+\nu_1-\nu_2)$, arising from the first zeta factor: one can check that the corresponding residue agrees with that of $-(C_0)_{\min}^+(\mu)$ given in (9.52), and the same thing goes for the other pole $\mu = i(1 - \nu_1 + \nu_2)$. This is of course only a verification since, as a consequence of theorem 7.3, C_0^+ must be holomorphic in the entire upper half–plane with the sole possible exception of $\mu = i$, and $\tilde{C}_0^+$ in the strip $0 < \operatorname{Im}\mu < 1 - 2\ell$.

From what precedes, $C_0^+(\mu) = [C_0^+(\mu)-\tilde{C}_0^+(\mu)]+[\tilde{C}_0^+(\mu)-(C_0)_{\min}^+(\mu)]+ (C_0)_{\min}^+(\mu)$ has actually no pole at $\mu = i$, which proves that f_0 is orthogonal to the constants in $L^2(\Gamma\backslash\Pi)$ according to theorem 7.3.

Since the two functions $(C_0)_{\min}^\pm$, holomorphic near the real line, agree there as a consequence of (9.46), and since the functions $(C_0)_{\mathrm{maj}}^\pm$ extend as holomorphic functions near $\mathbb{R}$, one has

$$C_0(\lambda) = (C_0)_{\mathrm{maj}}^+(\lambda) - (C_0)_{\mathrm{maj}}^-(\lambda), \tag{9.67}$$

which is just the second term on the right–hand side of (9.55). Thus, from (9.66),

$$C_0(\lambda) = \frac{1}{8\,\pi}\,\pi^{\frac{\alpha+\beta+\gamma+\delta-3}{2}} \frac{\Gamma(\frac{1-\delta}{2})\,\Gamma(\frac{\alpha}{2})\,\Gamma(\frac{\beta}{2})\,\Gamma(\frac{1-\gamma}{2})}{\Gamma(\frac{\alpha+\beta-1}{2})\,\Gamma(\frac{\alpha+\beta}{2})} \frac{\zeta(\alpha)\,\zeta(\beta)}{\zeta(\alpha+\beta)}, \tag{9.68}$$

with $\alpha,\beta,\gamma,\delta$ as in (9.17).

As a consequence of theorem 7.3 again,

$$\Phi_0(\lambda) = 8\pi \frac{\zeta^*(1-i\lambda)}{\zeta(-i\lambda)} C_0(\lambda)$$

$$= 8\pi^{\frac{3-\alpha-\beta}{2}} \Gamma(\frac{\alpha+\beta-1}{2}) \frac{\zeta^*(\alpha+\beta)}{\zeta^*(\alpha+\beta-1)} C_0(\lambda)$$

$$= \pi^{-1+\frac{\gamma+\delta}{2}} \Gamma(\frac{1-\gamma}{2}) \Gamma(\frac{1-\delta}{2}) \frac{\zeta^*(1-\alpha)\zeta^*(1-\beta)}{\zeta^*(\alpha+\beta-1)}, \quad (9.69)$$

which is just the same as (9.41). $\qquad\square$

THEOREM 9.6. *Let* $\mathrm{Re}\,\nu_1 < -1$, $\mathrm{Re}\,\nu_2 < -1$, $|\mathrm{Re}\,(\nu_1 - \nu_2)| < 1$, *and* $\nu_1 \neq \nu_2$. *The image under the spectral projection corresponding to the continuous part of the spectrum of* Δ *of the function* f *defined in* (9.10) *is given by the integral*

$$z \mapsto \frac{1}{8\pi} \int_{-\infty}^{\infty} \Phi(\lambda)\, E_{\frac{1-i\lambda}{2}}(z)\, d\lambda \qquad (9.70)$$

with

$$\Phi(\lambda) = \frac{\zeta^*(\frac{1+i\lambda-\nu_1+\nu_2}{2})\zeta^*(\frac{1+i\lambda+\nu_1-\nu_2}{2})\zeta^*(\frac{1-i\lambda-\nu_1-\nu_2}{2})\zeta^*(\frac{1+i\lambda-\nu_1-\nu_2}{2})}{\zeta^*(-i\lambda)}. \qquad (9.71)$$

PROOF. From ([**36**], p.241), the coefficients $C_0(\lambda)$ and $(C_0(\lambda))_N$ relative to f_0 and $T_N f_0$ respectively are related by

$$(C_0(\lambda))_N = N^{\frac{i\lambda}{2}} \sigma_{-i\lambda}(N)\, C_0(\lambda). \qquad (9.72)$$

Since ([**16**], p.250)

$$\sum_{N\geq 1} N^{\frac{-1+i\lambda+\nu_1+\nu_2}{2}} \sigma_{-i\lambda}(N)$$

$$= \sum_{N\geq 1} N^{\gamma-1} \sigma_{\alpha+\beta-1}(N)$$

$$= \zeta(1-\gamma)\zeta(2-\alpha-\beta-\gamma)$$

$$= \zeta(1-\gamma)\zeta(1-\delta), \qquad (9.73)$$

it follows from (9.26) that

$$\Phi(\lambda) = \zeta(1-\gamma)\zeta(1-\delta)\, \Phi_0(\lambda), \qquad (9.74)$$

so that theorem 9.6 follows from theorem 9.5 and (9.17). $\qquad\square$

Remark. The condition $\nu_1 \neq \nu_2$ can be dispensed with as a consequence of the remark which follows (9.10). More precisely, when $\nu_1 = \nu_2 = \nu$, $\mathrm{Re}\,\nu < -1$, the theorem will hold if f is defined as

$$f(z) = (E_{\frac{1-\nu}{2}}(z))^2 - (\zeta^*(1-\nu))^2 E_{1-\nu}(z) - 2\zeta^*(1+\nu)\zeta^*(1-\nu) E_1^0(z)$$

$$+ \frac{6}{\pi} [\zeta^*(1+\nu)\frac{d}{d\nu}\zeta^*(1-\nu) - \zeta^*(1-\nu)\frac{d}{d\nu}\zeta^*(1+\nu)], \qquad (9.75)$$

where

$$\frac{\pi}{6} E_1^0(z) = \lim_{s \to 1} [E_s^*(z) - \frac{1}{2(s-1)}]$$

$$= \frac{\pi y}{6} + \gamma + \sum_{n \neq 0} \frac{\sigma_1(|n|)}{|n|} e^{-2\pi|n|y} e^{2i\pi nx} \qquad (9.76)$$

and γ is Euler's constant.

10. A digression on Kloosterman sums

Recall from (2.16) that introducing

$$\omega_\nu = \pi^{\nu - \frac{1}{2}} \frac{\Gamma(\frac{1-\nu}{2})}{\Gamma(\frac{\nu}{2})} \tag{10.1}$$

permits to write

$$(\mathcal{F}(|x|^{\nu-1}))(\sigma) = \omega_\nu^{-1} |\sigma|^{-\nu}; \tag{10.2}$$

also,

$$\omega_\nu \, \omega_{1-\nu} = 1 \quad \text{and} \quad \zeta(s) = \omega_s \, \zeta(1-s). \tag{10.3}$$

Recall ([**33, 7, 12, 20**]) that one defines the Kloosterman sum

$$S(j, n; m) = \sum_{\substack{\xi, \eta \in \mathbf{Z}/m\mathbf{Z} \\ \xi\eta = 1}} e^{\frac{2i\pi}{m}(j\xi + n\eta)} \tag{10.4}$$

for every triple of integers with $m \geq 1$; in particular, $S(j, n; 1) = 1$.

DEFINITION 10.1. For every pair (s, t) of complex numbers with Re $s > 1$, Re $t > 1$, we define

$$\zeta_n(s, t) = \frac{1}{4} \sum_{\substack{m_1 m_2 \neq 0 \\ (m_1, m_2) = 1}} |m_1|^{-s} |m_2|^{-t} e^{2i\pi n \frac{m_2}{m_1}},$$

$$(\overline{m}_2 m_2 \equiv 1 \bmod m_1). \tag{10.5}$$

Note that $\zeta_n(s, t) = \zeta_{-n}(s, t)$ and that the sum for $m_1 \neq 0$ is also twice the sum for $m_1 \geq 1$ (but one cannot, then, fix the sign of m_2).

PROPOSITION 10.2.

$$\zeta_0(s, t) = \frac{\zeta(s)\,\zeta(t)}{\zeta(s + t)}. \tag{10.6}$$

PROOF. We have already used this right after (9.66): the proof is obvious. □

DEFINITION 10.3. For $0 < \xi < 1$, recall that the Hurwitz zeta function is defined when Re $s > 1$ by

$$\zeta(s, \xi) = \sum_{n \geq 0} (\xi + n)^{-s}: \tag{10.7}$$

it extends as a meromorphic function in the whole plane, with only a simple pole at $s = 1$. We also set

$$\tilde{\zeta}(s,\xi) = \zeta(s,\xi) + \zeta(s, 1 - \xi), \tag{10.8}$$

or

$$\tilde{\zeta}(s,\xi) = \sum_{n \in \mathbf{Z}} |\xi + n|^{-s} \tag{10.9}$$

in the case when Re $s > 1$: it is natural to extend $\tilde{\zeta}(s,\xi)$, for $\xi \notin \mathbf{Z}$, so that it should depend only on ξ mod 1.

LEMMA 10.4. *Given* $\xi \in \mathbf{R} \backslash \mathbf{Z}$, *one has*

$$\tilde{\zeta}(s,\xi) = \omega_s \sum_{j \neq 0} e^{2i\pi j\xi} |j|^{s-1} \tag{10.10}$$

whenever Re $s < 0$.

PROOF. If Re $s > 1$,

$$\tilde{\zeta}(s,\xi) = |\xi|^{-s} + 2\zeta(s) + \frac{1}{2} \sum_{n \neq 0} [\,|n + \xi|^{-s} + |n - \xi|^{-s} - 2\,|n|^{-s}\,]. \tag{10.11}$$

Since the bracket is $O(|n|^{-\operatorname{Re} s - 2})$ as $|n| \to \infty$, this formula makes sense whenever Re $s > -1$.

If $a > 0$ and Re $s > 0$,

$$a^{-\frac{s}{2}} = \frac{2\pi^{\frac{s}{2}}}{\Gamma(\frac{s}{2})} \int_0^\infty e^{-\pi a x^2} x^{s-1} \, dx, \tag{10.12}$$

hence, for $n \neq 0$,

$$\frac{1}{2} [\,|n + \xi|^{-s} + |n - \xi|^{-s} - 2\,|n|^{-s}\,] =$$

$$\frac{\pi^{\frac{s}{2}}}{\Gamma(\frac{s}{2})} \int_0^\infty x^{s-1} [\, e^{-\pi(n+\xi)^2 x^2} + e^{-\pi(n-\xi)^2 x^2} - 2\, e^{-\pi n^2 x^2} \,] \, dx, \tag{10.13}$$

a formula that remains true for Re $s > -2$ since the bracket inside the integral is $O(x^2)$ as $x \to 0$. If $-2 < $ Re $s < 0$, the last formula is also true for $n = 0$ (substitute ε for n, then let ε go to zero).

Hence, in the case when $-1 < $ Re $s < 0$,

$$\tilde{\zeta}(s,\xi) = 2\zeta(s) + \frac{1}{2} \sum_{n \in \mathbf{Z}} [\,|n + \xi|^{-s} + |n - \xi|^{-s} - 2\,|n|^{-s}\,]$$

$$= 2\zeta(s) + \frac{\pi^{\frac{s}{2}}}{\Gamma(\frac{s}{2})} \sum_{n \in \mathbf{Z}} \int_0^\infty x^{s-1} [\, e^{-\pi(n+\xi)^2 x^2} + e^{-\pi(n-\xi)^2 x^2} - 2\, e^{-\pi n^2 x^2} \,] \, dx.$$

$$\tag{10.14}$$

Using Poisson's formula,

$$\tilde{\zeta}(s,\xi) = 2\zeta(s) + \frac{\pi^{\frac{s}{2}}}{\Gamma(\frac{s}{2})} \int_0^\infty x^{s-2}\, dx \sum_j e^{-\frac{\pi j^2}{x^2}} (e^{2i\pi j\xi} + e^{-2i\pi j\xi} - 2), \tag{10.15}$$

where we can drop the term with $j = 0$ since it is zero. Thus

$$\tilde{\zeta}(s,\xi) = 2\zeta(s) + \frac{\pi^{\frac{s}{2}}}{\Gamma(\frac{s}{2})} \int_0^\infty x^{-s}\, dx \sum_{j\neq 0} e^{-\pi j^2 x^2} (e^{2i\pi j\xi} + e^{-2i\pi j\xi} - 2)$$

$$= 2\zeta(s) + \frac{\pi^{\frac{s}{2}}}{\Gamma(\frac{s}{2})} \sum_{j\neq 0} \frac{1}{2} \frac{\Gamma(\frac{1-s}{2})}{\pi^{\frac{1-s}{2}}} |j|^{s-1} (e^{2i\pi j\xi} + e^{-2i\pi j\xi} - 2)$$

$$= 2\zeta(s) + \frac{1}{2}\omega_s \sum_{j\neq 0} (e^{2i\pi j\xi} + e^{-2i\pi j\xi} - 2) |j|^{s-1}, \tag{10.16}$$

so we are done thanks to (10.3). $\qquad\square$

THEOREM 10.5. $\zeta_n(s,t)$ extends as a holomorphic function, first to the domain $(\mathrm{Re}\,(s+t) > 2, \mathrm{Re}\,t > -1)$, next to the domain $(\mathrm{Re}\,(s+t) > \frac{3}{2}, \mathrm{Re}\,t < 0)$. If (s,t) lies in the latter one, one has

$$\zeta_n(s,t) = \frac{1}{2}\pi^{t-\frac{1}{2}} \frac{\Gamma(\frac{1-t}{2})}{\Gamma(\frac{t}{2})} \sum_{m\geq 1} m^{-s-t} \sum_{j\neq 0} S(j,n;m) |j|^{t-1}. \tag{10.17}$$

PROOF. In the case when $n = 0$, starting from the identification

$$S(j,0;m) = c_m(j) \tag{10.18}$$

([16], p.237-238) of this special Kloosterman sum with a Ramanujan sum $c_m(j)$, we get (loc.cit., p.250)

$$\sum_{m\geq 1} m^{-s-t} S(j,0;m) = (\zeta(s+t))^{-1} \sum_{d|j} d^{1-s-t}, \tag{10.19}$$

thus, if $\mathrm{Re}\,t < 0, \mathrm{Re}\,s > 1$,

$$\frac{1}{2}\omega_t \sum_{m\geq 1} m^{-s-t} \sum_{j\neq 0} S(j,0;m) |j|^{t-1} = \omega_t (\zeta(s+t))^{-1} \sum_{j\geq 1} j^{t-1} \sum_{d|j} d^{1-s-t}$$

$$= \omega_t (\zeta(s+t))^{-1} \sum_{j\geq 1} j^{t-1} \sigma_{1-s-t}(j)$$

$$= \omega_t \frac{\zeta(s)\zeta(1-t)}{\zeta(s+t)}$$

$$= \frac{\zeta(s)\zeta(t)}{\zeta(s+t)}. \tag{10.20}$$

Assume now $n \neq 0$, and $\mathrm{Re}\ s > 1$, $\mathrm{Re}\ t > 1$ again. Starting from (10.5), one gets

$$\zeta_n(s,t) = \frac{1}{2} \sum_{m_1 > 0} m_1^{-s} \sum_{\substack{m_2 \neq 0 \\ (m_1, m_2) = 1}} |m_2|^{-t} e^{2i\pi n \frac{\overline{m_2}}{m_1}}$$

$$= \zeta(t) + \frac{1}{2} \sum_{m_1 \geq 2} m_1^{-s-t} \sum_{\substack{1 \leq k < m_1 \\ (k, m_1) = 1}} \sum_{q \in \mathbf{Z}} |\frac{k}{m_1} + q|^{-t} e^{2i\pi n \frac{\overline{k}}{m_1}}$$

$$= \zeta(t) + \frac{1}{2} \sum_{m_1 \geq 2} m_1^{-s-t} \sum_{\substack{\xi, \eta \bmod m_1 \\ \xi\eta \equiv 1}} \tilde{\zeta}(t, \frac{\xi}{m_1}) e^{2i\pi n \frac{\eta}{m_1}}. \tag{10.21}$$

When $\mathrm{Re}\ t > -1$ and $0 < \xi < 1$, the integral expression ([27], p.23)

$$\zeta(t, \xi) = \frac{1}{2} \xi^{-t} - \frac{\xi^{1-t}}{1-t} + \frac{1}{\Gamma(t)} \int_0^\infty \left[\frac{1}{e^x - 1} - \frac{1}{x} + \frac{1}{2} \right] e^{-\xi x} x^{t-1}\, dx \tag{10.22}$$

shows that

$$|\zeta(t, \xi)| \leq C\, \xi^{-\mathrm{Re}\ t - 1} \tag{10.23}$$

and

$$|\tilde{\zeta}(t, \xi)| \leq C\, [\xi^{-\mathrm{Re}\ t - 1} + (1 - \xi)^{-\mathrm{Re}\ t - 1}], \tag{10.24}$$

so that

$$\sum_{\substack{\xi \in \mathbf{Z}/m_1\mathbf{Z} \\ \xi \neq 0}} |\tilde{\zeta}(t, \frac{\xi}{m_1})| \leq C\, m_1 \log m_1. \tag{10.25}$$

Thus (10.21) permits to analytically continue $\zeta_n(s,t)$ to the range ($\mathrm{Re}\ t > -1$, $\mathrm{Re}\ (s+t) > 2$). Then, using lemma 10.4, we get when ($\mathrm{Re}\ (s+t) > 2$, $-1 < \mathrm{Re}\ t < 0$) the identity

$$\zeta_n(s,t) = \zeta(t) + \frac{1}{2} \omega_t \sum_{m_1 \geq 2} m_1^{-s-t} \sum_{\substack{\xi, \eta \bmod m_1 \\ \xi\eta \equiv 1}} e^{2i\pi j \frac{\xi}{m_1}} e^{2i\pi n \frac{\eta}{m_1}} |j|^{t-1}$$

$$= \zeta(t) + \frac{1}{2} \omega_t \sum_{m_1 \geq 2} m_1^{-s-t} \sum_{j \neq 0} S(j, n; m_1) |j|^{t-1}$$

$$= \frac{1}{2} \omega_t \sum_{m_1 \geq 1} m_1^{-s-t} \sum_{j \neq 0} S(j, n; m_1) |j|^{t-1}. \tag{10.26}$$

The Weil estimate ([**48, 12**])

$$|S(j,n;m)| \leq \sigma_0(m)\, m^{\frac{1}{2}} \left(|j|,|n|,m\right)^{\frac{1}{2}},\qquad\qquad (10.27)$$

in which $\sigma_0(m)$ is the number of divisors of m, permits to extend the meaning of the right-hand side of (10.17) to the domain indicated. □

PROPOSITION 10.6. *With the shorthand*

$$\langle m \rangle^{-s}: = |m|^{-s}\operatorname{sign}(m),\qquad\qquad (10.28)$$

set, for Re $s > 1$, Re $t > 1$,

$$\zeta_n^-(s,t) = \frac{1}{4}\sum_{\substack{m_1 m_2 \neq 0 \\ (m_1,m_2)=1}} \langle m_1 \rangle^{-s}\,\langle m_2 \rangle^{-t}\, e^{2i\pi n \frac{\overline{m}_2}{m_1}},$$

$$(\overline{m}_2 m_2 \equiv 1 \bmod m_1).\quad (10.29)$$

One has $\zeta_{-n}^-(s,t) = -\zeta_n^-(s,t)$. *The conditions that make theorem* 10.5 *valid extend, yielding the formula*

$$\zeta_n^-(s,t) = \frac{1}{2i}\,\pi^{t-\frac{1}{2}}\,\frac{\Gamma(\frac{2-t}{2})}{\Gamma(\frac{1+t}{2})} \sum_{m \geq 1} m^{-s-t} \sum_{j \neq 0} S(j,n;m)\,\langle j \rangle^{t-1}.$$

$$(10.30)$$

PROOF. One defines, for Re $s > 1$ and $\xi \in \,]0,1[$,

$$\tilde{\zeta}^-(s,\xi) = \zeta(s,\xi) - \zeta(s,1-\xi)$$

$$= \sum_{n \in \mathbf{Z}} \langle n + \xi \rangle^{-s}.\qquad\qquad (10.31)$$

Lemma 10.4 extends (taking the $\frac{\partial}{\partial \xi}$-derivative), yielding for Re $s < 0$ the formula

$$\tilde{\zeta}^-(s,\xi) = \frac{1}{i}\,\pi^{s-\frac{1}{2}}\,\frac{\Gamma(\frac{2-s}{2})}{\Gamma(\frac{1+s}{2})} \sum_{j \neq 0} e^{2i\pi j\xi}\,\langle j \rangle^{s-1}:\qquad (10.32)$$

the proof of theorem 10.5 then extends *mutatis mutandis*. □

One should remember that the so-called Kloosterman-Selberg series analogous to (10.17), though involving no summation with respect to j, have been considered by Selberg [**33**], Deshouillers-Iwaniec [**7**], Goldfeld-Sarnak [**12**], Iwaniec [**20**]: see remark following the proof of theorem 13.6.

Remark. Instead of $\zeta_n(s,t)$ and $\zeta_n^-(s,t)$, one might just as well have considered

$$(\zeta_n)_{\text{sym}}(s,t) = \frac{1}{4}\sum_{\substack{m_1 m_2 \neq 0 \\ (m_1,m_2)=1}} |m_1|^{-s}|m_2|^{-t}\, e^{2i\pi n(\frac{\overline{m}_2}{m_1} - \frac{1}{2\,m_1 m_2})},$$

$$(\overline{m}_2 m_2 \equiv 1 \bmod m_1).\quad (10.33)$$

and

$$(\zeta_n^-)_{\text{sym}}(s,t) \;=\; \frac{1}{4} \sum_{\substack{m_1 m_2 \neq 0 \\ (m_1,m_2)=1}} \langle m_1 \rangle^{-s} \langle m_2 \rangle^{-t} \, e^{2i\pi n(\frac{\overline{m_2}}{m_1} - \frac{1}{2 m_1 m_2})} \quad (10.34)$$

instead.

Indeed, expanding $\exp(-\frac{i\pi n}{m_1 m_2})$, one gets

$$(\zeta_n)_{\text{sym}}(s,t) = \sum_{k \text{ even} \geq 0} \frac{(i\pi n)^k}{k!} \, \zeta_n(s+k, t+k) - \sum_{k \text{ odd} \geq 1} \frac{(i\pi n)^k}{k!} \, \zeta_n^-(s+k, t+k) \quad (10.35)$$

and a similar identity holds with $(\zeta_n^-)_{\text{sym}}(s,t)$. Thus, $(\zeta_n)_{\text{sym}}(s,t) - \zeta_n(s,t)$ and $(\zeta_n^-)_{\text{sym}}(s,t) - \zeta_n^-(s,t)$ extend as holomorphic functions in the domain $\operatorname{Re} s > 0$, $\operatorname{Re} t > 0$: this is the domain in which we shall consider the functions $\zeta_n(s,t)$ and $\zeta_n^-(s,t)$ in what follows.

The advantage of the new functions is that the two equations

$$(\zeta_n)_{\text{sym}}(s,t) = (\zeta_n)_{\text{sym}}(t,s) \quad \text{and} \quad (\zeta_n^-)_{\text{sym}}(s,t) = -(\zeta_n^-)_{\text{sym}}(t,s) \quad (10.36)$$

hold, since

$$\left(\begin{smallmatrix} n_1 & n_2 \\ m_1 & m_2 \end{smallmatrix} \right) \in \Gamma \Rightarrow \frac{n_1}{m_1} - \frac{1}{2\, m_1 m_2} = -\left(\frac{-n_2}{m_2} - \frac{1}{2\, m_1 m_2} \right). \quad (10.37)$$

A problem. From proposition 10.2, it follows that, for $\operatorname{Re} s > 0$, $\operatorname{Re} t > 0$ and $\operatorname{Re}(s+t) < 1$, whether (s,t) lies in the singular set of $\zeta_0(s,t)$ depends only on $s+t$. Is an analogous statement valid in the case of $\zeta_n(s,t)$ or $\zeta_n^-(s,t)$, $n \neq 0$? This will turn out to be true in the open set $\operatorname{Re} s > 0$, $\operatorname{Re} t > 0$, $|\operatorname{Re}(s-t)| \neq 1$, $s \neq 1$, $t \neq 1$. The proof will rely on all results contained in sections 11 to 13. The interesting values of $s+t$, in this case, are not only the non-trivial zeros of the Riemann zeta-function, but also the numbers $1 + i\lambda_k$ for which $\frac{1}{4}(1+\lambda_k^2)$ lies in the discrete part of the spectrum of Δ.

11. The Roelcke–Selberg expansion of the product of two Eisenstein series: the discrete part

In this section, and the following one, we discuss the discrete part, under the Roelcke–Selberg decomposition, of the product, or Poisson bracket, of two Eisenstein series. This will bring the functions $\zeta_n(s,t)$ and $\zeta_n^-(s,t)$, introduced in the preceding section, into the picture: it will be shown that the poles on a certain line of one–variable versions of these two Dirichlet series correspond to the eigenvalues of the modular Laplacian which contribute to the spectral decomposition of the product or Poisson bracket under consideration. The non–trivial zeroes of the zeta function will also contribute to the singularities of the first of these two functions.

On the technical level, the present section will be devoted, for its most part, to analyzing the singularities on the real line of the function

$$\tilde{c}_n^+(\mu) = \frac{1}{8\pi} \int_0^\infty a_n(y)\, y^{-\frac{3}{2}} \frac{(\pi y)^{-\frac{i\mu}{2}}}{\Gamma(-\frac{i\mu}{2})}\, dy\,, \tag{11.1}$$

with $n \neq 0$ and

$$a_n(y) = \int_{-\frac{1}{2}}^{\frac{1}{2}} f_0(x + iy)\, e^{2i\pi nx}\, dx\,. \tag{11.2}$$

As it will turn out, the function $\tilde{c}_n^+$ is holomorphic for $\operatorname{Im} \mu > 0$. However, theorem 7.4 is not directly applicable since f_0 has a continuous part (under the Roelcke–Selberg decomposition), made explicit in theorem 9.5. With obvious notations, we split (11.1) and (11.2) as

$$a_n = (a_n)_{\text{cont}} + (a_n)_{\text{disc}}, \qquad \tilde{c}_n^+ = (\tilde{c}_n)_{\text{cont}}^+ + (\tilde{c}_n)_{\text{disc}}^+\,. \tag{11.3}$$

THEOREM 11.1. *Under the assumptions of theorem 9.5, the function* $(\tilde{c}_n)_{\text{cont}}^+(\mu)$, *initially defined for* $\operatorname{Im} \mu > 0$, *extends as a holomorphic function to the half-plane* $\operatorname{Im} \mu > -1 + |\ell_1 - \ell_2|$, *except for the poles* $-i\omega$, ω *any non-trivial zero of the zeta function.*

PROOF. Using (3.29) and the decomposition (9.40) of the continuous part of f_0, one finds

$$(a_n)_{\text{cont}}(y) = \frac{1}{4\pi} \int_{-\infty}^\infty \Phi_0(\rho)\, (\zeta^*(1 - i\rho))^{-1}\, |n|^{-\frac{i\rho}{2}} \sigma_{i\rho}(|n|)$$

$$y^{\frac{1}{2}}\, K_{\frac{i\rho}{2}}(2\pi|n|y)\, d\rho \tag{11.4}$$

with

$$\Phi_0(\rho) = \pi^{\frac{\gamma+\delta}{2}-1} \Gamma(\frac{1-\gamma}{2}) \Gamma(\frac{1-\delta}{2}) \frac{\zeta^*(1-\alpha)\,\zeta^*(1-\beta)}{\zeta^*(-i\rho)} , \tag{11.5}$$

where

$$\alpha = \frac{1 - i\rho + \nu_1 - \nu_2}{2}, \qquad \beta = \frac{1 - i\rho - \nu_1 + \nu_2}{2},$$

$$\gamma = \frac{1 + i\rho + \nu_1 + \nu_2}{2}, \qquad \delta = \frac{1 - i\rho + \nu_1 + \nu_2}{2} . \tag{11.6}$$

Observe that, now, we know Φ_0 explicitly, thus we are in a much better position —so far as estimates are concerned— than in section 7. Indeed, combining asymptotics for the Gamma function on vertical lines ([27], p.13) with the fact ([37], p.161 or [29], p.100) that $(\zeta(-i\rho))^{-1}$ is bounded by $|\rho|^{-\frac{1}{2}}$ times some power of $\log|\rho|$ as $|\rho| \to \infty$ and with the estimate ([37], p. 147)

$$\zeta(\sigma + it) = O(|t|^{\frac{1-\sigma}{2}+\epsilon}), \qquad |t| \to \infty, \ 0 < \sigma < 1, \tag{11.7}$$

we get the estimates

$$|\Phi_0(\rho)| \le C\, e^{-\frac{\pi}{4}|\rho|} |\rho|^b, \qquad |\rho| \to \infty \tag{11.8}$$

and

$$\left| \frac{\Phi_0(\rho)}{\zeta^*(1-i\rho)} \right| \le C\,|\rho|^b, \tag{11.9}$$

in which the exponent b does not really matter.

We now compute (using (7.30) and (11.4))

$$(\tilde{c}_n)^+_{\mathrm{cont}}(\mu) = \frac{1}{32\,\pi^2} \int_0^\infty y^{-\frac{3}{2}} \frac{(\pi y)^{-\frac{i\mu}{2}}}{\Gamma(-\frac{i\mu}{2})} \, dy$$

$$\int_{-\infty}^\infty \frac{\Phi_0(\rho)}{\zeta^*(1-i\rho)} |n|^{-\frac{i\rho}{2}} \sigma_{i\rho}(|n|) \, y^{\frac{1}{2}} \, K_{\frac{i\rho}{2}}(2\pi|n|y) \, d\rho. \tag{11.10}$$

In order to show that the two integrations may be permuted, it suffices to show that, for $\mathrm{Im}\,\mu > 0$,

$$\int_0^\infty y^{-1+\frac{\mathrm{Im}\,\mu}{2}} \, dy \int_0^\infty (1+\rho)^b \, |K_{\frac{i\rho}{2}}(2\pi|n|y)| \, d\rho < \infty. \tag{11.11}$$

That the contribution to this integral of the domain $(0 < y < 1)$ is finite is a consequence of lemma 7.2. On the other hand, the contribution of the domain $(y > 1)$ is less than

$$\int_0^\infty (1+\rho)^b \, d\rho \left(\int_0^\infty y^{\mathrm{Im}\,\mu} \, (K_{\frac{i\rho}{2}}(2\pi|n|y))^2 \, dy \right)^{\frac{1}{2}}$$

$$= C \int_0^\infty (1+\rho)^b \, [\Gamma(\frac{1+\mathrm{Im}\,\mu+i\rho}{2}) \Gamma(\frac{1+\mathrm{Im}\,\mu-i\rho}{2})]^{\frac{1}{2}} \, d\rho ,$$

where the constant C depends only on n, μ, according to ([27], p.101): this is a convergent integral.

Since ([27], p.91)

$$\frac{1}{8\pi} \int_0^\infty K_{\frac{i\rho}{2}}(2\pi|n|y) \frac{(\pi y)^{-\frac{i\mu}{2}}}{\Gamma(-\frac{i\mu}{2})} \frac{dy}{y}$$

$$= \frac{2^{-5}\pi^{-1}}{\Gamma(-\frac{i\mu}{2})} |n|^{\frac{i\mu}{2}} \Gamma(\frac{i(-\mu+\rho)}{4}) \Gamma(\frac{i(-\mu-\rho)}{4}), \quad (11.13)$$

we thus get when $\operatorname{Im} \mu > 0$

$$(\tilde{c}_n)^+_{\text{cont}}(\mu) = \frac{2^{-7}\pi^{-2}}{\Gamma(-\frac{i\mu}{2})} \int_{-\infty}^\infty |n|^{\frac{i(\mu-\rho)}{2}} \sigma_{i\rho}(|n|) (\zeta^*(1-i\rho))^{-1}$$

$$\Phi_0(\rho) \Gamma(\frac{i(-\mu+\rho)}{4}) \Gamma(\frac{i(-\mu-\rho)}{4}) d\rho, \quad (11.14)$$

a holomorphic function.

Recall from (11.5) that

$$\frac{\Phi_0(\rho)}{\zeta^*(1-i\rho)} = \pi^{\frac{-1-i\rho+\nu_1+\nu_2}{2}} \frac{\Gamma(\frac{1-i\rho-\nu_1-\nu_2}{4}) \Gamma(\frac{1+i\rho-\nu_1-\nu_2}{4}) \Gamma(\frac{1-i\rho-\nu_1+\nu_2}{4}) \Gamma(\frac{1-i\rho+\nu_1-\nu_2}{4})}{\Gamma(-\frac{i\rho}{2}) \Gamma(\frac{1-i\rho}{2})}$$

$$\times \frac{\zeta(\frac{1-i\rho-\nu_1+\nu_2}{2}) \zeta(\frac{1-i\rho+\nu_1-\nu_2}{2})}{\zeta(1-i\rho) \zeta(-i\rho)}. \quad (11.15)$$

This function, meromorphic in the complex plane, has poles only at $\rho = \pm i(1 - \nu_1 - \nu_2 + 4k)$ for some $k = 0, 1, \ldots$ or $\rho = -i[1 \pm (\nu_1 - \nu_2) + 4k]$ or $\rho = i[1 \pm (\nu_1 - \nu_2)]$ or, finally, $\rho = \pm i\omega$, ω a non–trivial zero of the zeta–function: none of these poles, of course, is real, and the only ones with $-2 < \operatorname{Im} \rho < 0$ are $\rho = -i(1 \pm (\nu_1 - \nu_2))$ and the points $-i\omega$, $\zeta^*(\omega) = 0$; the only ones with $0 < \operatorname{Im} \rho < 2$ are $\rho = i(1 \pm (\nu_1 - \nu_2))$ and the points $i\omega$.

Let h satisfy $0 < h < 1 - |\ell_1 - \ell_2| = 1 - |\operatorname{Re}(\nu_1 - \nu_2)|$, and consider the closed solid rectangle $\mathcal{R}$ with vertices $\pm A$, $\pm A + ih$ for some $A > 0$: we assume that h and A are chosen so that the (finite) set S of poles $i\omega$ of the function $\frac{\Phi_0(\rho)}{\zeta^*(1-i\rho)}$ within $\mathcal{R}$ is disjoint from the boundary $\partial\mathcal{R}$ of $\mathcal{R}$. Let C be the contour from $-\infty$ to $-A$ to $-A + ih$ to $A + ih$ to A to ∞, partly along the real line and partly along $\partial\mathcal{R}$. If μ is any number in the interior of $\mathcal{R}$, $\mu \notin S$, one has

$$2^7\,\pi^2\,\Gamma(-\frac{i\mu}{2})\,(\tilde{c}_n)^+_{\text{cont}}(\mu) = \int_C |n|^{\frac{i(\mu-\rho)}{2}}\,\sigma_{i\rho}(|n|)\,\frac{\Phi_0(\rho)}{\zeta^*(1-i\rho)}\,\Gamma(\tfrac{i(-\mu+\rho)}{4})\,\Gamma(\tfrac{i(-\mu-\rho)}{4})\,d\rho$$

$$+\,2i\pi \sum_{i\omega \in S} \text{Res}_{\rho=i\omega}\left\{|n|^{\frac{i(\mu-\rho)}{2}}\,\sigma_{i\rho}(|n|)\,\frac{\Phi_0(\rho)}{\zeta^*(1-i\rho)}\,\Gamma(\tfrac{i(-\mu+\rho)}{4})\,\Gamma(\tfrac{i(-\mu-\rho)}{4})\right.$$

$$+\,8\pi\,\Gamma(-\frac{i\mu}{2})\,\sigma_{i\mu}(|n|)\,\frac{\Phi_0(\mu)}{\zeta^*(1-i\mu)}, \tag{11.1}$$

where the last term arises from the residue of the integrand at $\rho = \mu$.

This expression yields the analytic continuation of the function $(\tilde{c}_n)^+_{\text{cont}}$ to the union Ω of the upper half–plane with the interior of the rectangle $-\mathcal{R} = \{\mu: -\mu \in \mathcal{R}\}$ together with the real interval $\,]-A, A[$. Indeed, the first (integral) term on the right–hand side of (11.16) yields a holomorphic function of μ for $\mu \in \Omega$, and the third term is fully explicit: we now analyze the second one.

LEMMA 11.2. *Let $u = u(z)$ be a meromorphic function in a neighborhood of $z^0 \neq 0$, with a pole at z^0. The function*

$$\mu \mapsto \text{Res}_{z=z^0}\left\{u(z)\,\Gamma(\frac{i(z-\mu)}{4})\,\Gamma(\frac{i(-z-\mu)}{4})\right\} - 4i\,\Gamma(-\frac{i\mu}{2})\,u(\mu)$$

is regular in a neighborhood of z^0.

PROOF. Apply the residue theorem, for $|\mu - z^0| < \varepsilon$, to the integral

$$\frac{1}{2i\pi}\int_{|z-z^0|<\varepsilon} u(z)\,\Gamma(\frac{i(z-\mu)}{4})\,\Gamma(\frac{i(-z-\mu)}{4})\,dz. \tag{11.17}$$

$\square$

End of proof of theorem 11.1.
We use this with the function

$$u(z) = |n|^{-\frac{iz}{2}}\,\sigma_{iz}(|n|)\,\frac{\Phi_0(z)}{\zeta^*(1-iz)}$$

in the case when $z^0 = i\omega$, then in the case when $z^0 = -i\omega$. In the first case we get

$$\text{Res}_{\rho=i\omega}\left\{u(\rho)\,\Gamma(\frac{i(-\mu+\rho)}{4})\,\Gamma(\frac{i(-\mu-\rho)}{4})\right\} \sim 4i\,\Gamma(-\frac{i\mu}{2})\,u(\mu),$$

$$\mu \text{ near } i\omega, \tag{11.18}$$

where the sign $\sim$ means that the difference of the two functions is regular near $i\omega$ as a function of μ.

Next, we apply this with u, and $z^0 = -i\omega$: setting $z = -\rho$, we get

$$- \operatorname{Res}_{\rho=i\omega} \left\{ u(-\rho)\,\Gamma(\frac{i(-\mu+\rho)}{4})\,\Gamma(\frac{i(-\mu-\rho)}{4}) \right\} \sim 4i\,\Gamma(-\frac{i\mu}{2})\,u(\mu),$$

$$\mu \text{ near } -i\omega. \quad (11.19)$$

Now the function u is even, a consequence of (6.69) for its main factor. It follows that, for ν near $-i\omega$, the meromorphic extension to Ω of the function on the left–hand side of (11.16) agrees with

$$2^4\,\pi\,\Gamma(-\frac{i\mu}{2})\,\sigma_{i\mu}(|n|)\,\frac{\Phi_0(\mu)}{\zeta^*(1-i\mu)}$$

up to some error term which is holomorphic in a neighborhood of $-i\omega$.

Thus $(\tilde{c}_n)^+_{\text{cont}}$ admits a meromorphic extension to the half–plane $\operatorname{Im} \mu > -1 + |\ell_1 - \ell_2|$ and

$$(\tilde{c}_n)^+_{\text{cont}}(\mu) \sim \frac{1}{2^4\,\pi}\,\sigma_{i\mu}(|n|)\,\frac{\Phi_0(\mu)}{\zeta^*(1-i\mu)} \quad (11.20)$$

for μ close to $-i\omega$, any solution of $\zeta^*(\omega) = 0$ with $\operatorname{Re} \omega < 1 - |\ell_1 - \ell_2|$. $\qquad\square$

THEOREM 11.3. *Assume that* $-2 < \operatorname{Re} \nu_1 = \ell_1 < -1$, $-2 < \operatorname{Re} \nu_2 = \ell_2 < -1$, *and that* $\nu_1 \neq \nu_2$; *set* $\ell = \max(\ell_1, \ell_2)$. *Recall the definition (9.39) of the function* f_0, *and consider, for every* $n \in \mathbf{Z}^\times$, *the (meromorphic) function defined when* $\operatorname{Im} \mu > 1 + |\ell_1 - \ell_2|$ *as*

$$b_n(\mu) := \frac{1}{8\pi}\,\pi^{\frac{-1-i\mu+\nu_1+\nu_2}{2}} \times$$

$$\frac{\Gamma(\frac{1+i\mu-\nu_1-\nu_2}{4})\,\Gamma(\frac{1-i\mu+\nu_1-\nu_2}{4})\,\Gamma(\frac{1-i\mu-\nu_1+\nu_2}{4})\,\Gamma(\frac{1-i\mu-\nu_1-\nu_2}{4})}{\Gamma(-\frac{i\mu}{2})\,\Gamma(\frac{1-i\mu}{2})} \times$$

$$\zeta_n(\frac{1-i\mu-\nu_1+\nu_2}{2}, \frac{1-i\mu+\nu_1-\nu_2}{2}), \quad (11.21)$$

where the function $\zeta_n(s,t)$ *was introduced in (10.5) and put into some other form in theorem 10.5. It extends as a meromorphic function in the half–plane* $\operatorname{Im} \mu > -1 + |\ell_1 - \ell_2|$. *Besides the actual poles* $\mu = i(1 + \nu_1 - \nu_2)$ *and* $\mu = i(1 - \nu_1 + \nu_2)$ *(which correspond to* $s = 1$ *and* $t = 1$ *respectively), and the poles* $-i\omega$ *where* ω *is any non–trivial zero of the zeta function with* $\operatorname{Re} \omega < 1 - |\ell_1 - \ell_2|$, *its only possible poles in the strip* $-1 + |\ell_1 - \ell_2| < \operatorname{Im} \mu < 1 - 2\ell$ *(note that this strip contains the closed strip* $0 \leq \operatorname{Im} \mu \leq 3$) *are points* λ_k *with* $\frac{1}{4}(1 + \lambda_k^2)$ *in the discrete part of the spectrum of* Δ: *they are simple. Moreover, the Fourier coefficients* d_n *of the orthogonal projection* $(f_0)_{\lambda_k}$ *of* f_0 *onto the eigenspace of* Δ *for some eigenvalue* $\frac{1}{4}(1 + \lambda_k^2)$, *characterized by the identity*

$$(f_0)_{\lambda_k}(z) = y^{\frac{1}{2}} \sum_{n \neq 0} d_n\,K_{\frac{i\lambda_k}{2}}(2\pi|n|y)\,e^{-2i\pi nx}, \quad (11.22)$$

are given by the formula

$$d_n = -8i\pi\,|n|^{-\frac{i\lambda_k}{2}} \times \text{ residue of } b_n(\mu) \text{ at } \mu = \lambda_k.$$

$$(11.23)$$

Near a pole $-i\omega$, one has

$$b_n(\mu) \sim \frac{1}{8\pi}\,\sigma_{i\mu}(|n|)\frac{\Phi_0(\mu)}{\zeta^*(1-i\mu)},$$

$$(11.24)$$

i.e., the difference of the two sides is regular as a function of μ near $-i\omega$: the function $\frac{\Phi_0(\mu)}{\zeta^*(1-i\mu)}$ was made explicit in (11.15).

PROOF. First observe that the Gamma factors can only contribute poles with $\operatorname{Im}\mu \geq 1 - 2\ell$ or $\operatorname{Im}\mu \leq -1 + |\ell_1 - \ell_2|$, so that all the poles in the given strip come from the factor $\zeta_n(\frac{1-i\mu-\nu_1+\nu_2}{2}, \frac{1-i\mu+\nu_1-\nu_2}{2})$.

The estimate (11.8) first shows that the continuous part (under the Roelcke–Selberg decomposition) of f_0, as well as f_0 itself (lemma 9.3), remains in $L^2(\Gamma\backslash\Pi)$ after having been applied any power of Δ: thus, the same holds with the discrete part of f_0, for which theorem 7.4 can consequently be used. The search for the discrete part of the Roelcke–Selberg decomposition of f_0 involves the consideration of the function

$$(\tilde{c}_n)^+_{\text{disc}}(\mu) = \tilde{c}_n^+(\mu) - (\tilde{c}_n)^+_{\text{cont}}(\mu),$$

$$(11.25)$$

where the analytic continuation of the second term on the right-hand side has already been analyzed in theorem 11.1; according to theorem 7.4 the left-hand side is a meromorphic function in the plane, with no pole with $\operatorname{Im}\mu > -4$ except, possibly, the real points λ_k. What we shall do is computing $\tilde{c}_n^+(\mu)$ when $\operatorname{Im}\mu > 1$ and showing that its analytic extension agrees with $b_n(\mu)$ up to some function holomorphic in the strip $-1 + |\ell_1 - \ell_2| < \operatorname{Im}\mu < 1 - 2\ell$. This will prove that the function b_n is meromorphic in this strip. At the same time, we shall get (11.22)–(11.23) as a consequence of the comparison between $b_n(\mu)$ and $(\tilde{c}_n)^+_{\text{disc}}(\mu)$ near a point λ_k; near points $-i\omega$, on the other hand, the singularities of $\tilde{c}_n^+(\mu)$ only come from those of $(\tilde{c}_n)^+_{\text{cont}}(\mu)$, which proves (11.24), starting from (11.20).

We use the same splitting of f_0 as in (9.42), ending up with

$$a_n(y) = (a_n)_{\text{maj}}(y) + (a_n)_{\text{min}}(y),$$

$$(11.26)$$

where

$$(a_n)_{\text{maj}}(y) = \frac{1}{2}\pi^{\frac{\nu_1+\nu_2-2}{2}}\Gamma(\frac{1-\nu_1}{2})\Gamma(\frac{1-\nu_2}{2})\int_{-\frac{1}{2}}^{\frac{1}{2}} e^{2i\pi nx}\,dx$$

$$\sum_{\substack{m_1 m_2 \neq 0 \\ n_1 m_2 - n_2 m_1 = 1}} \left(\frac{|m_1 z - n_1|^2}{y}\right)^{\frac{\nu_1-1}{2}}\left(\frac{|m_2 z - n_2|^2}{y}\right)^{\frac{\nu_2-1}{2}}$$

$$(11.27)$$

and

$$(a_n)_{\min}(y) = \int_{-\frac{1}{2}}^{\frac{1}{2}} e^{2i\pi n x} \, dx$$

$$\{ \pi^{\frac{\nu_1+\nu_2-2}{2}} \Gamma(\frac{1-\nu_1}{2}) \Gamma(\frac{1-\nu_2}{2}) y^{\frac{2-\nu_1-\nu_2}{2}} [\sum_{n_1 \in \mathbf{Z}} |z-n_1|^{\nu_1-1} + \sum_{n_2 \in \mathbf{Z}} |z-n_2|^{\nu_2-1}]$$

$$-\pi^{\frac{\nu_1+\nu_2-1}{2}} \Gamma(-\frac{\nu_1}{2}) \Gamma(\frac{1-\nu_2}{2}) E_{1+\frac{\nu_1-\nu_2}{2}}(z) - \pi^{\frac{\nu_1+\nu_2-1}{2}} \Gamma(\frac{1-\nu_1}{2}) \Gamma(-\frac{\nu_2}{2}) E_{1+\frac{\nu_2-\nu_1}{2}}(z)$$

$$\tag{11.2}$$

Let us first discuss

$$(\tilde{c}_n)^+_{\min}(\mu) = \frac{1}{8\pi} \frac{\pi^{-\frac{i\mu}{2}}}{\Gamma(-\frac{i\mu}{2})} \int_0^{\infty} (a_n)_{\min}(y) \, y^{-\frac{3}{2}-\frac{i\mu}{2}} \, dy \,. \tag{11.29}$$

By (9.48) and Poisson's formula,

$$\sum_{n_1 \in \mathbf{Z}} |z-n_1|^{\nu_1-1} = \frac{2 \pi^{\frac{1-\nu_1}{2}}}{\Gamma(\frac{1-\nu_1}{2})} \sum_{n_1 \in \mathbf{Z}} y^{\frac{\nu_1}{2}} |n_1|^{-\frac{\nu_1}{2}} K_{\frac{\nu_1}{2}}(2\pi|n_1|y) \, e^{-2i\pi n_1 x} \,. \tag{11.30}$$

Thus the contribution of a term such as

$$\int_{-\frac{1}{2}}^{\frac{1}{2}} e^{2i\pi n x} y^{\frac{2-\nu_1-\nu_2}{2}} \sum_{n_1 \in \mathbf{Z}} |z-n_1|^{\nu_1-1} \, dx$$

taken from $(a_n)_{\min}(y)$ (almost: we have not considered here its coefficient, irrelevant to our purpose), to $(\tilde{c}_n)^+_{\min}(\mu)$ is (using (7.28) and [27], p.91 again)

$$\frac{1}{4\pi} \frac{\pi^{\frac{1-i\mu-\nu_1}{2}}}{\Gamma(-\frac{i\mu}{2})\Gamma(\frac{1-\nu_1}{2})} |n|^{-\frac{\nu_1}{2}} \int_0^{\infty} y^{\frac{-1-i\mu-\nu_2}{2}} K_{\frac{\nu_1}{2}}(2\pi|n|y) \, dy$$

$$= 2^{-4} \pi^{\frac{-2-\nu_1+\nu_2}{2}} |n|^{\frac{-1+i\mu+\nu_2-\nu_1}{2}} \frac{\Gamma(\frac{1-i\mu+\nu_1-\nu_2}{4})\Gamma(\frac{1-i\mu-\nu_1-\nu_2}{4})}{\Gamma(-\frac{i\mu}{2})\Gamma(\frac{1-\nu_1}{2})} \,, \tag{11.31}$$

a function whose only singularities $\mu = -i + (\nu_2 \pm \nu_1)i - 4ki$ $(k = 0, 1, \ldots)$ lie below or on the line $\mathrm{Im}\, \mu = -1 + |\ell_1 - \ell_2|$.

By the same argument, starting from the expression (3.29) of the Fourier coefficients of Eisenstein series, a term such as $E_{1+\frac{\nu_2-\nu_1}{2}}(z)$ within the bracket on the right–hand side of (11.28) would contribute to $(\tilde{c}_n)^+_{\min}(\mu)$ the function

$$\frac{1}{16\pi} |n|^{\frac{1+i\mu-\nu_1+\nu_2}{2}} \sigma_{-1+\nu_1-\nu_2}(|n|) \frac{\Gamma(\frac{1-i\mu-\nu_1+\nu_2}{4})\Gamma(\frac{-1-i\mu+\nu_1-\nu_2}{4})}{\Gamma(-\frac{i\mu}{2})\zeta^*(2-\nu_1+\nu_2)} \,,$$

an expression still without poles μ with $\mathrm{Im}\, \mu > -1 + |\ell_1 - \ell_2|$, except for the pole

$$\mu = i(1 - \nu_1 + \nu_2). \tag{11.32}$$

Our last problem, towards proving theorem 11.3, is analyzing the singularities in the strip $-1 + |\ell_1 - \ell_2| < \mathrm{Im}\,\mu < 1 - 2\ell$ of

$$(\tilde{c}_n)^+_{\mathrm{maj}}(\mu) = \frac{1}{16\,\pi}\,\pi^{\frac{-2-i\mu+\nu_1+\nu_2}{2}}\,\frac{\Gamma(\frac{1-\nu_1}{2})\,\Gamma(\frac{1-\nu_2}{2})}{\Gamma(-\frac{i\mu}{2})}\,\times$$

$$\int_0^\infty y^{-\frac{3}{2}-\frac{i\mu}{2}}\,dy \sum_{\substack{m_1 m_2 \neq 0 \\ n_1 m_2 - n_2 m_1 = 1}} |m_1|^{\nu_1 - 1}\,|m_2|^{\nu_2 - 1}$$

$$y^{\frac{2-\nu_1-\nu_2}{2}} \int_{-\frac{1}{2}}^{\frac{1}{2}} e^{2i\pi nx}\,[(x - \frac{n_1}{m_1})^2 + y^2]^{\frac{\nu_1 - 1}{2}}\,[(x - \frac{n_2}{m_2})^2 + y^2]^{\frac{\nu_2 - 1}{2}}\,dx\,.$$

$$(11.33)$$

The end of the proof starts in a way quite similar to that of theorem 9.5. First, using the same matrix argument as before (9.47), together with (9.48), we write the last series of integrals as

$$\sum_{\substack{m_1 m_2 \neq 0 \\ (m_1, m_2) = 1}} |m_1|^{\nu_1 - 1}\,|m_2|^{\nu_2 - 1}\,e^{2i\pi n \frac{n_1^0}{m_1}}\,I_{m_1 m_2}(y)\,, \qquad (11.34)$$

where n_1^0 is chosen so that $n_1^0 m_2 \equiv 1 \bmod m_1$, thus can be replaced by $\overline{m}_2$ (cf. (10.5)) in the exponent, and

$$I_{m_1 m_2}(y) = y^{\frac{2-\nu_1-\nu_2}{2}} \int_{-\infty}^\infty e^{2i\pi nx}\,(x^2 + y^2)^{\frac{\nu_1 - 1}{2}}\,[(x + \frac{1}{m_1 m_2})^2 + y^2]^{\frac{\nu_2 - 1}{2}}\,dx$$

$$= \frac{4\,\pi^{\frac{2-\nu_1-\nu_2}{2}}}{\Gamma(\frac{1-\nu_1}{2})\,\Gamma(\frac{1-\nu_2}{2})}\,y\,\times$$

$$\int_{-\infty}^\infty |n - \sigma|^{-\frac{\nu_1}{2}}\,|\sigma|^{-\frac{\nu_2}{2}}\,e^{-2i\pi \frac{\sigma}{m_1 m_2}}\,K_{\frac{\nu_1}{2}}(2\pi|n - \sigma|y)\,K_{\frac{\nu_2}{2}}(2\pi|\sigma|y)\,d\sigma\,.$$

$$(11.35)$$

We have been careful to state lemma 9.4 in a way that would apply as well to our present situation. Thus, under the assumption $-1 - \ell_1 - \ell_2 < \mathrm{Im}\,\mu < 1 - 2\ell$, one can exchange the order of summations and integrations from $(dy; m_1, m_2; d\sigma)$ to $(m_1, m_2; d\sigma; dy)$ (from the right to the left) in a way fully similar to that expounded between formulas (9.56) and (9.63), ending up with

$$(\tilde{c}_n)^+_{\text{maj}}(\mu) = \frac{1}{4\pi} \frac{\pi^{-\frac{i\mu}{2}}}{\Gamma(-\frac{i\mu}{2})} \sum_{\substack{m_1 m_2 \neq 0 \\ (m_1,m_2)=1}} |m_1|^{\nu_1-1} |m_2|^{\nu_2-1} e^{2i\pi n \frac{m_2}{m_1}} \times$$

$$\mathcal{F}\{\sigma \mapsto |n-\sigma|^{-\frac{\nu_1}{2}} |\sigma|^{-\frac{\nu_2}{2}} \times$$

$$\int_0^\infty y^{-\frac{1}{2}-\frac{i\mu}{2}} K_{\frac{\nu_1}{2}}(2\pi|n-\sigma|y) K_{\frac{\nu_2}{2}}(2\pi|\sigma|y)\, dy\}(\frac{1}{m_1 m_2}), \quad (11.36)$$

where the Fourier transform that occurs on the right–hand side is indeed a continuous function of ξ for $\xi \neq 0$, thus has a well–defined meaning at $\xi = \frac{1}{m_1 m_2}$.

Using the Weber–Schafheitlin integral ([27], p.101) again, we find, if $-1-\ell_1-\ell_2 < \text{Im}\,\mu < 1 - 2\ell$,

$$(\tilde{c}_n)^+_{\text{maj}}(\mu) = 2^{-5}\, \pi^{-\frac{3}{2}} \times$$

$$\frac{\Gamma(\frac{1-i\mu+\nu_1+\nu_2}{4})\,\Gamma(\frac{1-i\mu+\nu_1-\nu_2}{4})\,\Gamma(\frac{1-i\mu-\nu_1+\nu_2}{4})\,\Gamma(\frac{1-i\mu-\nu_1-\nu_2}{4})}{\Gamma(-\frac{i\mu}{2})\,\Gamma(\frac{1-i\mu}{2})} \times$$

$$\sum_{\substack{m_1 m_2 \neq 0 \\ (m_1,m_2)=1}} |m_1|^{\nu_1-1} |m_2|^{\nu_2-1} e^{2i\pi n \frac{m_2}{m_1}} \times \mathcal{F}\{\sigma \mapsto |\sigma|^{\frac{-1+i\mu-\nu_1-\nu_2}{2}}$$

$${}_2F_1(\frac{1-i\mu+\nu_1+\nu_2}{4}, \frac{1-i\mu+\nu_1-\nu_2}{4}; \frac{1-i\mu}{2}; \frac{n(2\sigma-n)}{\sigma^2})\}(\frac{1}{m_1 m_2}).$$

$$(11.37)$$

We know from theorem 11.1 and theorem 7.4 that $(\tilde{c}_n)^+(\mu)$ extends as a holomorphic function for $\text{Im}\,\mu > 0$: applying (11.31) and (11.32) (relative to the behaviour of $(\tilde{c}_n)^+_{\text{min}}(\mu)$), we know that $(\tilde{c}_n)^+_{\text{maj}}(\mu)$ extends as a holomorphic function for $\text{Im}\,\mu > 0$, except for the actual poles $\mu = i(1 \pm (\nu_1 - \nu_2))$. On the other hand, from (11.21) and from the convergence of the series defining $\zeta_n(s,t)$ for $\text{Re}\,s > 1$, $\text{Re}\,t > 1$, it follows that $b_n(\mu)$ is holomorphic in the strip $1 + |\ell_1 - \ell_2| < \text{Im}\,\mu < 1 - 2\ell$, a fortiori in the strip $-1-\ell_1-\ell_2 < \text{Im}\,\mu < 1 - 2\ell$.

Thus, what remains to be done is showing that $(\tilde{c}_n)^+_{\text{maj}}(\mu)$, as defined by (11.37) for $-1-\ell_1-\ell_2 < \text{Im}\,\mu < 1 - 2\ell$, differs from $b_n(\mu)$, holomorphic in the same strip, by a function that extends holomorphically to the strip $-1 + |\ell_1 - \ell_2| < \text{Im}\,\mu < 1 - 2\ell$.

The product of Gamma coefficients in front of the right–hand side of (11.37) presents, when analytically continued to the strip $-1 + |\ell_1 - \ell_2| < \text{Im}\,\mu < 1 - 2\ell$, only one singularity, namely $\mu_0 = -i(1 + \nu_1 + \nu_2)$, with imaginary part $-1-\ell_1-\ell_2$. However, we need not worry about μ_0, which is neither a genuine singularity of the continuation of $(\tilde{c}_n)^+_{\text{maj}}(\mu)$ (thanks to theorem 7.4 and to our previous results concerning $(\tilde{c}_n)^+_{\min}(\mu))$ nor one of $b_n(\mu)$ since $1 + |\ell_1 - \ell_2| < -1 - \ell_1 - \ell_2 < 1 - 2\ell$. Even though μ_0 will appear as a singularity in several terms of the decomposition of (the continuation of) $(\tilde{c}_n)^+_{\text{maj}}(\mu)$ that follows, it disappears from their sum.

Let

$$1 = \chi_0(\sigma) + \chi_1(\sigma) + \chi_\infty(\sigma) \tag{11.38}$$

be a smooth partition of unity on $\mathbf{R}$ such that, with $\tau = \frac{n(2\sigma - n)}{\sigma^2}$ (observe that $\tau < 1$ for $\sigma \neq n$), one should have

$$\begin{cases} -3 \leq \tau \leq \frac{3}{4} & \text{on support } (\chi_0) \\ \tau \geq \frac{1}{2} & \text{on support } (\chi_1) \\ \tau \leq -2 & \text{on support } (\chi_\infty). \end{cases}$$

Let $C \in \mathbf{R}$ be such that

$$_2F_1(\dots) = 1 + C\sigma^{-1} + \psi(\sigma), \tag{11.39}$$

with $\psi(\sigma) = O(\sigma^{-2})$, $|\sigma| \to \infty$. Let us split $_2F_1(\dots)$ as

$$\begin{aligned} _2F_1(\dots) = &\ 1 \\ &+ C\sigma^{-1} \\ &+ \psi(\sigma)\chi_0(\sigma) \\ &- C\sigma^{-1}(\chi_1(\sigma) + \chi_\infty(\sigma)) \\ &+ (_2F_1(\dots) - 1)\chi_\infty(\sigma) \\ &+ (_2F_1(\dots) - 1)\chi_1(\sigma) \end{aligned} \tag{11.40}$$

and examine the contribution to $(\tilde{c}_n)^+_{\text{maj}}(\mu)$ obtained by substituting, in (11.37), each of the preceding six terms for $_2F_1(\dots)$ itself. Recalling (9.65) and definition 10.1, one sees that substituting 1 for $_2F_1(\dots)$ in (11.37) would yield, instead of $(\tilde{c}_n)^+_{\text{maj}}(\mu)$, precisely the function $b_n(\mu)$ which appears in the statement of theorem 11.3. Also,

$$\mathcal{F}\{\sigma \mapsto |\sigma|^{\frac{-1+i\mu-\nu_1-\nu_2}{2}}\sigma^{-1}\}\left(\frac{1}{m_1 m_2}\right)$$

$$= -i\pi^{\frac{2-i\mu+\nu_1+\nu_2}{2}}\frac{\Gamma(\frac{1+i\mu-\nu_1-\nu_2}{4})}{\Gamma(\frac{3-i\mu+\nu_1+\nu_2}{4} + \frac{1}{2})}|m_1 m_2|^{\frac{-1+i\mu-\nu_1-\nu_2}{2}}\text{sign}\left(\frac{1}{m_1 m_2}\right),$$

$$\tag{11.41}$$

where the coefficient is holomorphic for $\text{Im}\,\mu < 1 - 2\ell$; since

$$\nu_1 - 1 + \frac{-1 + i\mu - \nu_1 - \nu_2}{2}$$

(the new exponent of $|m_1|$ in the term under consideration) has a real part < -1 if $\text{Im}\,\mu > -1 + |\ell_1 - \ell_2|$, the term $C\sigma^{-1}$ in (11.39) contributes to

$(\tilde{c}_n)^+_{\mathrm{maj}}(\mu)$ a function of μ holomorphic for $-1 + |\ell_1 - \ell_2| < \operatorname{Im}\mu < 1 - 2\ell$ (except for the above–mentioned pole μ_0). Since

$$\operatorname{Re}\left(\frac{-1 + i\mu - \nu_1 - \nu_2}{2} - 2\right) < 0 \qquad (11.42)$$

under this condition, so does the third term $\psi(\sigma)\,\chi_0(\sigma)$ on the right–hand side of (11.40): indeed, the function $|\sigma|^{\frac{-1+i\mu-\nu_1-\nu_2}{2}}\,\psi(\sigma)\,\chi_0(\sigma)$ tends to 0 as $|\sigma| \to \infty$ and its second derivative is summable.

The remaining three terms all have compact support in σ since $\tau \to 0$ as $|\sigma| \to \infty$. Remembering that $|\sigma|^\rho \operatorname{sign}(\rho)$, as a distribution in σ, is a holomorphic function of ρ for $\rho \neq -2, -4, \ldots$, one can see that

$$\mathcal{F}\{\sigma \mapsto |\sigma|^{\frac{-1+i\mu-\nu_1-\nu_2}{2}}\sigma^{-1}\,(\chi_1(\sigma) + \chi_\infty(\sigma))\}\left(\frac{1}{m_1 m_2}\right)$$

is holomorphic for $\operatorname{Im}\mu < 1 - 2\ell$.

To study the fifth term, in which $-\tau$ is large, we write ([**27**], p.48))

$$_2F_1\left(\tfrac{1-i\mu+\nu_1+\nu_2}{4}, \tfrac{1-i\mu+\nu_1-\nu_2}{4}; \tfrac{1-i\mu}{2}; \tfrac{n(2\sigma-n)}{\sigma^2}\right)$$

$$= \frac{\Gamma(\frac{1-i\mu}{2})\Gamma(-\frac{\nu_2}{2})}{\Gamma(\frac{1-i\mu+\nu_1-\nu_2}{4})\Gamma(\frac{1-i\mu-\nu_1-\nu_2}{4})}\left(\frac{\sigma^2}{n(n-2\sigma)}\right)^{\frac{1-i\mu+\nu_1+\nu_2}{4}}{}_2F_1\left(\tfrac{1-i\mu+\nu_1+\nu_2}{4}, \tfrac{3+i\mu+\nu_1+\nu_2}{4}; 1+\tfrac{\nu_2}{2}; \tfrac{\sigma^2}{n(2\sigma-n)}\right)$$

$$+ \frac{\Gamma(\frac{1-i\mu}{2})\Gamma(\frac{\nu_2}{2})}{\Gamma(\frac{1-i\mu+\nu_1+\nu_2}{4})\Gamma(\frac{1-i\mu-\nu_1+\nu_2}{4})}\left(\frac{\sigma^2}{n(n-2\sigma)}\right)^{\frac{1-i\mu+\nu_1-\nu_2}{4}}{}_2F_1\left(\tfrac{1-i\mu+\nu_1-\nu_2}{4}, \tfrac{3+i\mu+\nu_1-\nu_2}{4}; 1-\tfrac{\nu_2}{2}; \tfrac{\sigma^2}{n(2\sigma-n)}\right),$$

$$(11.43)$$

an expression of which we have to evaluate the Fourier transform at $\frac{1}{m_1 m_2}$, after having multiplied it by $|\sigma|^{\frac{-1+i\mu-\nu_1-\nu_2}{2}}\chi_\infty(\sigma)$, where the last factor is supported near $\sigma = 0$. We only have to observe that the two terms we are interested in start with powers of $|\sigma|$ the exponents of which are 0 or $-\nu_2$, and that the coefficients do not create singularities when $\operatorname{Im}\mu > -1$ since $\frac{\nu_2}{2} \notin \mathbf{Z}$.

Finally, the very last error term is studied in the same way, by means of the decomposition ([**27**], p.47)

$$_2F_1\left(\tfrac{1-i\mu+\nu_1+\nu_2}{4}, \tfrac{1-i\mu+\nu_1-\nu_2}{4}; \tfrac{1-i\mu}{2}; \tfrac{n(2\sigma-n)}{\sigma^2}\right)$$

$$= \frac{\Gamma(\frac{1-i\mu}{2})\Gamma(-\frac{\nu_1}{2})}{\Gamma(\frac{1-i\mu-\nu_1+\nu_2}{4})\Gamma(\frac{1-i\mu-\nu_1-\nu_2}{4})}{}_2F_1\left(\tfrac{1-i\mu+\nu_1+\nu_2}{4}, \tfrac{1-i\mu+\nu_1-\nu_2}{4}; 1+\tfrac{\nu_1}{2}; \tfrac{(\sigma-n)^2}{\sigma^2}\right)$$

$$+ \frac{\Gamma(\frac{1-i\mu}{2})\Gamma(\frac{\nu_1}{2})}{\Gamma(\frac{1-i\mu+\nu_1+\nu_2}{4})\Gamma(\frac{1-i\mu+\nu_1-\nu_2}{4})}|\tfrac{\sigma-n}{\sigma}|^{-\nu_1}{}_2F_1\left(\tfrac{1-i\mu-\nu_1-\nu_2}{4}, \tfrac{1+i\mu-\nu_1+\nu_2}{4}; 1-\tfrac{\nu_1}{2}; \tfrac{(\sigma-n)^2}{\sigma^2}\right).$$

$$(11.44)$$

This time our interest lies in the behaviour of this function near $\sigma = n$, so we are done since $\frac{\nu_1}{2} \notin \mathbf{Z}$.

Before we leave this proof, a remark is in order concerning the singularities of $b_n(\mu)$ associated with poles of its first Gamma factor $\Gamma(\frac{1+i\mu-\nu_1-\nu_2}{4})$, and for which Im μ can be as large as one pleases. These singularities are not present in $\tilde{c}_n^+(\mu)$, as theorem 7.4 indicates: in our present calculations they arise from our having isolated (in (11.39)) the term 1 in $_2F_1(\dots)$, creating a singularity at $\sigma = 0$ after multiplication by $|\sigma|^{\frac{-1+i\mu-\nu_1-\nu_2}{2}}$. The important fact is that $b_n(\mu)$, contrary to $\tilde{c}_n^+(\mu)$, is fully explicit, and that the singularities of these two functions on the real line agree.   □

Remark. One may actually dispense with the condition $\nu_1 \neq \nu_2$, as has been observed at the end of section 9.

COROLLARY 11.4. *For $n \neq 0$, the function $\zeta_n(s,t)$ extends as a meromorphic function for Re $s > 0$, Re $t > 0$, $|\text{Re }(s-t)| < 1$, $s \neq 1$, $t \neq 1$, $s+t \neq 1$, holomorphic outside the set of all points (s,t) with $s+t = 1-i\lambda_k$, $\frac{1+\lambda_k^2}{4}$ in the discrete part of the spectrum of Δ, or $s + t = \omega$, a non-trivial zero of the zeta function.*

PROOF. In the proof of theorem 11.3 that precedes, all claims relative to the holomorphy of some function with respect to μ in some domain can be replaced by the same claim relative to the dependence on μ, ν_1, ν_2: the remaining question that has to be tackled with to prove the corollary is comparing the domains that occur in the statements of theorem 11.3 and of the present corollary. Setting $\sigma = \text{Re } s$, $\tau = \text{Re } t$, one has under our assumptions $-1 < \tau - \sigma < 1$ and $1 - \frac{3\tau}{2} + \frac{\sigma}{2} = 1 - \tau + \frac{\sigma-\tau}{2} > \frac{1}{2} - \tau$. One may assume $\tau < \frac{5}{2}$ and $\sigma + \tau < 6$ since otherwise one has $\sigma > 1$ and $\tau > 1$ and there is nothing to prove (looking back at the original domain of convergence of the series defining $\zeta_n(s,t)$). Thus one may choose ℓ_2 with

$$\max(-2, \sigma - \tau - 2) < \ell_2 < \min(-1, \sigma - \tau - 1, 1 - \frac{3\tau}{2} + \frac{\sigma}{2}) \tag{11.45}$$

and set $\ell_1 = \ell_2 + \tau - \sigma$, so that

$$-2 < \ell_2 < -1, \qquad -2 < \ell_1 < -1. \tag{11.46}$$

Also observe that

$$-1 + \ell_1 - \ell_2 < -1 + \sigma + \tau < 1 - 2\ell_1 \tag{11.47}$$

(indeed, $\ell_1 - \ell_2 = \tau - \sigma < \tau + \sigma$ since $\sigma > 0$; then,

$$
\begin{aligned}
1 - 2\ell_1 &= 1 - 2\ell_2 - 2(\tau - \sigma) \\
&> 1 - (2 - 3\tau + \sigma) - 2(\tau - \sigma) \\
&= -1 + \sigma + \tau \ ).
\end{aligned}
$$

All that remains to be done, to be in a position to apply theorem 11.3, is finding ν_1, ν_2, μ with

$$\frac{1 - i\mu - \nu_1 + \nu_2}{2} = s, \qquad \frac{1 - i\mu + \nu_1 - \nu_2}{2} = t,$$

$$\operatorname{Re} \nu_1 = \ell_1, \qquad \operatorname{Re} \nu_2 = \ell_2,$$

$$\mu \neq i(1 \pm (\nu_1 - \nu_2)),$$

$$-\frac{i\mu}{2} \neq 0, -1, -2, \ldots .$$

$$(11.48)$$

Since $1 - i\mu = s + t$, $\nu_2 - \nu_1 = s - t$, choosing ν_1 with $\operatorname{Re} \nu_1 = \ell_1$, otherwise arbitrary, will do the trick: the last two conditions just mean that $s \neq 1$, $t \neq 1$, $s + t \neq 1$. $\square$

Remark. The singularities at $s = 1$ or $t = 1$ are genuine, but the condition $s + t \neq 1$ will be removed in proposition 15.8.

We do not know yet which Maass cusp–forms do enter the Roelcke–Selberg decomposition of the function f_0 defined in (9.39) or of the function f, essentially the product of two Eisenstein series, defined in (9.10). From the very definition of f or, so far as f_0 is concerned, from theorem 11.3 together with the fact that $\zeta_n(s, t) = \zeta_{-n}(s, t)$, it follows that one can only get cusp–forms which are *even*, i.e., invariant under the symmetry $z \mapsto -\bar{z}$. It will follow later, from (14.39), that for generic values of (ν_1, ν_2), *all* even Maass cusp–forms do enter the decomposition of f_0: more precisely, a given Hecke form $\mathcal{M}$ (*cf.* beginning of section 14) will enter the decomposition (if a basis of Hecke forms be chosen) unless $\frac{1 + \nu_1 - \nu_2}{2}$ is a zero of the associated L–function $L(., \mathcal{M})$.

12. The expansion of the Poisson bracket of two Eisenstein series

Odd cusp-forms, i.e., forms which change to their negatives under this symmetry, are the subject of the present section. With a view towards applying (8.38) rather than (8.37), one may note that, given any two complex numbers ν_1 and ν_2, the C^∞ function

$$\begin{aligned} h(z) : &= \{\Theta_{\nu_1} \mathfrak{E}_{\nu_1}, \Theta_{\nu_2} \mathfrak{E}_{\nu_2}\}(z) \\ &= \{E^*_{\frac{1-\nu_1}{2}}, E^*_{\frac{1-\nu_2}{2}}\}(z), \end{aligned} \tag{12.1}$$

where the Poisson bracket $\{\ \}$ was defined in (8.25), is Γ-invariant and rapidly decreasing as $y \to \infty$ in the fundamental domain. Also, it changes to its negative as $z \mapsto -\bar{z}$, thus has no continuous part in its Roelcke-Selberg decomposition; finally, it is not zero if $\nu_1 \neq \pm\nu_2$.

If $\mathrm{Re}\ \nu_1 < -1$, $\mathrm{Re}\ \nu_2 < -1$, one has

$$h(z) = \frac{1}{4} \pi^{\frac{\nu_1 + \nu_2 - 2}{2}} \Gamma(\frac{1-\nu_1}{2}) \Gamma(\frac{1-\nu_2}{2})$$
$$\sum_{(m_1,n_1)=1} \sum_{(m_2,n_2)=1} \left\{ \left(\frac{|m_1 z - n_1|^2}{y}\right)^{\frac{\nu_1 - 1}{2}}, \left(\frac{|m_2 z - n_2|^2}{y}\right)^{\frac{\nu_2 - 1}{2}} \right\}. \tag{12.2}$$

PROPOSITION 12.1. *The series in* (12.2) *converges absolutely if* $\mathrm{Re}\ \nu_1 < 0$, $\mathrm{Re}\ \nu_2 < 0$. *Defining*

$$h_0(z) = \frac{1}{2} \pi^{\frac{\nu_1 + \nu_2 - 2}{2}} \Gamma(\frac{1-\nu_1}{2}) \Gamma(\frac{1-\nu_2}{2})$$
$$\sum_{n_1 m_2 - n_2 m_1 = 1} \left\{ \left(\frac{|m_1 z - n_1|^2}{y}\right)^{\frac{\nu_1 - 1}{2}}, \left(\frac{|m_2 z - n_2|^2}{y}\right)^{\frac{\nu_2 - 1}{2}} \right\},$$
$$\tag{12.3}$$

one has

$$h = \sum_{N \geq 1} N^{\frac{\nu_1 + \nu_2 - 1}{2}} T_N h_0. \tag{12.4}$$

The Γ-invariant function h_0 changes into its negative under the symmetry $z \mapsto -\bar{z}$.

PROOF. One has

$$\left\{ \left(\frac{|m_1 z - n_1|^2}{y} \right)^{\frac{\nu_1 - 1}{2}}, \left(\frac{|m_2 z - n_2|^2}{y} \right)^{\frac{\nu_2 - 1}{2}} \right\}$$

$$= -(n_1 m_2 - n_2 m_1) \frac{(\nu_1 - 1)(\nu_2 - 1)}{2} \left(\frac{(m_1 x - n_1)(m_2 x - n_2) + m_1 m_2 y^2}{y} \right) \times$$

$$\left(\frac{|m_1 z - n_1|^2}{y} \right)^{\frac{\nu_1 - 3}{2}} \left(\frac{|m_2 z - n_2|^2}{y} \right)^{\frac{\nu_2 - 3}{2}} \quad (12.5)$$

as a shameless computation shows. Since

$$\left| \frac{(m_1 x - n_1)(m_2 x - n_2) + m_1 m_2 y^2}{y} \right| \le \left(\frac{|m_1 z - n_1|^2}{y} \right)^{\frac{1}{2}} \left(\frac{|m_2 z - n_2|^2}{y} \right)^{\frac{1}{2}},$$

$$(12.6)$$

we are done for the first part. We observe next that if we change (m_1, m_2, n_1, n_2) to $(-m_1, m_2, -n_1, n_2)$, the right-hand side of (12.5) is left unchanged though $n_1 m_2 - n_2 m_1$ is changed to its negative. One can then prove (12.4) in just the same way as (9.22), with the simplification that, on the right-hand side of (12.2), only the terms with $n_1 m_2 - n_2 m_1 \ne 0$ are non-zero. For the last part, change (m_1, m_2, n_1, n_2) to $(-m_1, m_2, n_1, -n_2)$ at the same time z is changed to $-\bar{z}$. $\qquad\square$

We combine in our present case the analogues of theorem 11.3 and corollary 11.4. Note that the zeroes of the zeta function will not occur here as poles of $\zeta_n^-(s, t)$ as a function of $s+t$ (for fixed $s-t$): in the even case, these poles arose from the continuous part of the Roelcke-Selberg decomposition of f_0.

THEOREM 12.2. For $n \ne 0$, the function $\zeta_n^-(s, t)$, defined in (10.29) extends as a meromorphic function for Re $s > -1$, Re $t > -1$, $|\text{Re}\,(s-t)| < 2$, $s+t \ne 1$, holomorphic outside the set of points (s, t) with $s+t = 1 - i\lambda_k$, $\frac{1+\lambda_k^2}{4}$ in the odd part of the (discrete) spectrum of Δ. Assume that $-2 < \text{Re}\,\nu_1 < 0$, $-2 < \text{Re}\,\nu_2 < 0$, and consider the function h_0 defined in (12.3). Given any λ_k in the set of points just considered, the Fourier coefficients d_n^- of the projection $(h_0)_{\lambda_k}$ of h_0 onto the eigenspace of Δ for the eigenvalue $\frac{1+\lambda_k^2}{4}$, characterized by the identity

$$(h_0)_{\lambda_k}(z) = y^{\frac{1}{2}} \sum_{n \ne 0} d_n^- K_{\frac{i\lambda_k}{2}}(2\pi|n|y) e^{-2i\pi n x}, \quad (12.7)$$

are given by the formula

$$d_n^- = -8i\pi |n|^{-\frac{i\lambda_k}{2}} \times \text{residue of } b_n^-(\mu) \text{ at } \mu = \lambda_k, \quad (12.8)$$

where b_n^- is the function

$$b_n^-(\mu): = -\frac{1}{4\pi} \pi^{\frac{-1-i\mu+\nu_1+\nu_2}{2}} \times$$

$$\frac{\Gamma(\frac{3+i\mu-\nu_1-\nu_2}{4}) \Gamma(\frac{3-i\mu+\nu_1-\nu_2}{4}) \Gamma(\frac{3-i\mu-\nu_1+\nu_2}{4}) \Gamma(\frac{3-i\mu-\nu_1-\nu_2}{4})}{\Gamma(-\frac{i\mu}{2}) \Gamma(\frac{1-i\mu}{2})} \times$$

$$\zeta_n^-(\frac{1-i\mu-\nu_1+\nu_2}{2}, \frac{1-i\mu+\nu_1-\nu_2}{2}). \quad (12.9)$$

Remark. This means in particular that $(h_0)_{\lambda_k} = 0$ unless λ_k is actually a pole of at least one of the functions b_n^-. On the other hand, one may remark (for the purpose of comparison with theorem 11.3) that the function $b_n^-(\mu)$, initially defined and holomorphic for $1 + |\text{Re } (\nu_1 - \nu_2)| < \text{Im } \mu < 3 - 2 \max (\text{Re } \nu_1, \text{Re } \nu_2)$, extends as a function in the strip $-3 + |\text{Re } (\nu_1 - \nu_2)| < \text{Im } \mu < 3 - 2 \max (\text{Re } \nu_1, \text{Re } \nu_2)$ holomorphic outside the set of points λ_k: this is the way the first part of theorem 12.2 results from the proof which follows.

PROOF. Set $\text{Re } \nu_1 = \ell_1$, $\text{Re } \nu_2 = \ell_2$, $\max(\ell_1, \ell_2) = \ell$. Just like in (9.42), we set aside the terms on the right-hand side of (12.3) with $m_1 m_2 = 0$, decomposing the a_n's that make the formula

$$h_0(x + iy) = \sum_{n \in \mathbf{Z}} a_n(y) e^{-2i\pi n x} \quad (12.10)$$

valid as

$$a_n = (a_n)_{\text{maj}} + (a_n)_{\text{min}} \quad (12.11)$$

with

$$(a_n)_{\text{maj}}(y) = \frac{1}{2} \pi^{\frac{\nu_1+\nu_2-2}{2}} \Gamma(\frac{1-\nu_1}{2}) \Gamma(\frac{1-\nu_2}{2}) \int_{-\frac{1}{2}}^{\frac{1}{2}} e^{2i\pi n x} dx$$

$$\sum_{\substack{m_1 m_2 \neq 0 \\ n_1 m_2 - n_2 m_1 = 1}} \left\{ \left(\frac{|m_1 z - n_1|^2}{y}\right)^{\frac{\nu_1-1}{2}}, \left(\frac{|m_2 z - n_2|^2}{y}\right)^{\frac{\nu_2-1}{2}} \right\} \quad (12.12)$$

and, after some easy trick,

$$(a_n)_{\text{min}}(y) = \pi^{\frac{\nu_1+\nu_2-2}{2}} \Gamma(\frac{1-\nu_1}{2}) \Gamma(\frac{1-\nu_2}{2}) \int_{-\infty}^{\infty} e^{2i\pi n x} dx$$

$$[y^{\frac{1-\nu_2}{2}} \{y^{\frac{1-\nu_1}{2}}, (x^2 + y^2)^{\frac{\nu_2-1}{2}}\} + y^{\frac{1-\nu_1}{2}} \{(x^2 + y^2)^{\frac{\nu_2-1}{2}}, y^{\frac{1-\nu_1}{2}}\}]. \quad (12.13)$$

Since

$$\{y^{\frac{1-\nu_1}{2}}, (x^2 + y^2)^{\frac{\nu_2-1}{2}}\} = \frac{\nu_1 - 1}{2} y^{\frac{3-\nu_1}{2}} \frac{\partial}{\partial x} (x^2 + y^2)^{\frac{\nu_2-1}{2}},$$

$$(12.14)$$

one finds, with the help of (9.48) again,

$$(a_n)_{\min}(y) = 8i\pi\, n\left[\pi^{\frac{\nu_1-1}{2}}\,\Gamma(\frac{3-\nu_1}{2})\,|n|^{-\frac{\nu_2}{2}}\,y^{\frac{4-\nu_1}{2}}\,K_{\frac{\nu_2}{2}}(2\pi|n|y)\right.$$
$$\left. -\,\pi^{\frac{\nu_2-1}{2}}\,\Gamma(\frac{3-\nu_2}{2})\,|n|^{-\frac{\nu_1}{2}}\,y^{\frac{4-\nu_2}{2}}\,K_{\frac{\nu_1}{2}}(2\pi|n|y)\right].$$

$$(12.15)$$

Our present function $(\tilde{c}_n)^+_{\min}(\mu)$, linked to $(a_n)_{\min}(y)$ by (11.29), can be computed with the help of ([27], p.91), resulting in

$$(\tilde{c}_n)^+_{\min}(\mu) = \frac{i\,n}{4\,\pi^2}\left[\pi^{\nu_1}\,\frac{\Gamma(\frac{3-\nu_1}{2})\,\Gamma(\frac{3-i\mu-\nu_1+\nu_2}{4})\,\Gamma(\frac{3-i\mu-\nu_1-\nu_2}{4})}{\Gamma(-\frac{i\mu}{2})}\,|n|^{\frac{-3+i\mu+\nu_1-\nu_2}{2}}\right.$$
$$\left. -\,\pi^{\nu_2}\,\frac{\Gamma(\frac{3-\nu_2}{2})\,\Gamma(\frac{3-i\mu+\nu_1-\nu_2}{4})\,\Gamma(\frac{3-i\mu-\nu_1-\nu_2}{4})}{\Gamma(-\frac{i\mu}{2})}\,|n|^{\frac{-3+i\mu-\nu_1+\nu_2}{2}}\right],$$

$$(12.16)$$

a function of μ holomorphic for $\operatorname{Im}\mu > -3 + |\ell_1 - \ell_2|$.

We start our computation of $(\tilde{c}_n)^+_{\mathrm{maj}}(\mu)$ just like in (11.33), (11.34), getting this time instead of the function $I_{m_1 m_2}(y)$ as computed in (11.35) the function

$$J_{m_1 m_2}(y) = \int_{-\infty}^{\infty} e^{2i\pi nx}\left\{\left(\frac{x^2+y^2}{y}\right)^{\frac{\nu_1-1}{2}}, \left(\frac{(x+\frac{1}{m_1 m_2})^2+y^2}{y}\right)^{\frac{\nu_2-1}{2}}\right\}\,dx.$$

$$(12.17)$$

Applying (12.5) and noting that

$$\frac{x(x+\frac{1}{m_1 m_2})+y^2}{y} = \frac{(x+\frac{1}{m_1 m_2})^2+y^2}{2y} + \frac{x^2+y^2}{2y} - \frac{1}{2\,m_1^2 m_2^2\,y},$$

$$(12.18)$$

one may express the Poisson bracket on the right-hand side of (12.17) as

$$\{\ \} = \frac{(\nu_1-1)(\nu_2-1)}{4\,m_1 m_2}\times\left[(m_1 m_2)^{-2}\,y^{-1}\left(\frac{x^2+y^2}{y}\right)^{\frac{\nu_1-3}{2}}\left(\frac{(x+\frac{1}{m_1 m_2})^2+y^2}{y}\right)^{\nu_2}\right.$$
$$-\left(\frac{x^2+y^2}{y}\right)^{\frac{\nu_1-1}{2}}\left(\frac{(x+\frac{1}{m_1 m_2})^2+y^2}{y}\right)^{\frac{\nu_2-3}{2}}$$
$$\left. -\left(\frac{x^2+y^2}{y}\right)^{\frac{\nu_1-3}{2}}\left(\frac{(x+\frac{1}{m_1 m_2})^2+y^2}{y}\right)^{\frac{\nu_2-1}{2}}\right].$$

$$(12$$

Denoting as $I_{m_1 m_2}(\nu_1, \nu_2; y)$ the function denoted as $I_{m_1 m_2}(y)$ in (11.35), we thus get

$$J_{m_1 m_2}(y) = \frac{(\nu_1 - 1)(\nu_2 - 1)}{4\, m_1 m_2} \left[(m_1 m_2)^{-2}\, y^{-1}\, I_{m_1 m_2}(\nu_1 - 2, \nu_2 - 2; y) \right.$$

$$\left. - I_{m_1 m_2}(\nu_1, \nu_2 - 2; y) - I_{m_1 m_2}(\nu_1 - 2, \nu_2; y) \right]. \quad (12.20)$$

Using definition (7.30), we get from (12.12) and (12.20) (remembering our shorthand (10.28)) the expression

$$(\tilde{c}_n)^+_{\text{maj}}(\mu) = \frac{1}{16\,\pi}\, \pi^{\frac{-2-i\mu+\nu_1+\nu_2}{2}}\, \frac{\Gamma(\frac{3-\nu_1}{2})\,\Gamma(\frac{3-\nu_2}{2})}{\Gamma(-\frac{i\mu}{2})}$$

$$\sum_{\substack{m_1 m_2 \neq 0 \\ (m_1, m_2) = 1}} \langle m_1 \rangle^{\nu_1 - 2}\, \langle m_2 \rangle^{\nu_2 - 2}\, e^{2i\pi n\, \frac{m_2}{m_1}} \int_0^\infty y^{\frac{-3-i\mu}{2}}$$

$$\left[(m_1 m_2)^{-2}\, y^{-1}\, I_{m_1 m_2}(\nu_1 - 2, \nu_2 - 2; y) - I_{m_1 m_2}(\nu_1, \nu_2 - 2; y) - I_{m_1 m_2}(\nu_1 - 2, \nu_2; y) \right]\, dy\,.$$

$$(12.21)$$

The same argument as that developed between (9.56) and (9.59), starting again from lemma 9.4, permits to change the order of the Σ_{m_1, m_2}-summation and the dy-integration, provided that $3 < \text{Im } \mu < 3 - 2\ell$. Things become slighly more involved so far as exchanging the order of the dy- and $d\sigma$-integrations is concerned. However, we know from lemma 9.4 and (11.35) that the integral

$$\int_0^\infty y^{\frac{-5-i\mu}{2}}\, I_{m_1 m_2}(\nu_1 - 2, \nu_2 - 2; y)\, dy \quad (12.22)$$

is a holomorphic function of μ for $1 + |\ell_1 - \ell_2| < \text{Im } \mu < 7 - 2\ell$, and the two other analogous integrals which come from (12.18) are holomorphic for $1 + |\ell_1 - \ell_2| < \text{Im } \mu < 3 - 2\ell$. This permits to compute each of these integrals by analytic continuation, starting from a range of values of μ ($5 - \ell_1 - \ell_2 < \text{Im } \mu < 7 - 2\ell$ in the first case, $1 - \ell_1 - \ell_2 < \text{Im } \mu < 3 - 2\ell$ in the other ones) for which we can exchange the order of the $dy\, d\sigma$-integrations just like in (11.36). We then get, from (12.21) and (11.31), with the understanding that $3 < \text{Im } \mu < 3 - 2\ell$ but that the integral defining ϕ in what follows is meant in the sense of analytic continuation theory,

$$(\tilde{c}_n)^+_{\text{maj}}(\mu) = \frac{1}{4\,\pi}\, \frac{\pi^{\frac{2-i\mu}{2}}}{\Gamma(-\frac{i\mu}{2})} \sum_{\substack{m_1 m_2 \neq 0 \\ (m_1, m_2) = 1}} \langle m_1 \rangle^{\nu_1 - 2}\, \langle m_2 \rangle^{\nu_2 - 2}\, e^{2i\pi n\, \frac{m_2}{m_1}}\, \hat{\phi}\left(\frac{1}{m_1 m_2}\right),$$

$$(12.23)$$

where $\hat{\phi}$ is the Fourier transform of

$$
\phi(\sigma) = \int_0^\infty y^{\frac{-3-i\mu}{2}} \left[\frac{\pi}{(m_1 m_2)^2} |n-\sigma|^{\frac{2-\nu_1}{2}} |\sigma|^{\frac{2-\nu_2}{2}} K_{\frac{\nu_1-2}{2}}(2\pi|n-\sigma|y) K_{\frac{\nu_2-2}{2}}(2\pi|\sigma| \right.
$$
$$
+ \frac{\nu_1-1}{2} |n-\sigma|^{-\frac{\nu_1}{2}} |\sigma|^{\frac{2-\nu_2}{2}} y K_{\frac{\nu_1}{2}}(2\pi|n-\sigma|y) K_{\frac{\nu_2-2}{2}}(2\pi|\sigma|y)
$$
$$
\left. + \frac{\nu_2-1}{2} |n-\sigma|^{\frac{2-\nu_1}{2}} |\sigma|^{-\frac{\nu_2}{2}} y K_{\frac{\nu_1-2}{2}}(2\pi|n-\sigma|y) K_{\frac{\nu_2}{2}}(2\pi|\sigma|y) \right] dy. \qquad (12.2
$$

Using one last time the Weber-Schafheitlin integral ([27], p.101), and taking as many common factors as is possible, we get

$$
\phi(\sigma) = \frac{2^{-3}\pi^{\frac{-1+i\mu}{2}}}{\Gamma(\frac{1-i\mu}{2})} |\sigma|^{\frac{1+i\mu-\nu_1-\nu_2}{2}} \times
$$
$$
\Gamma(\tfrac{-1-i\mu+\nu_1+\nu_2}{4}) \Gamma(\tfrac{-1-i\mu+\nu_1-\nu_2}{4}) \Gamma(\tfrac{-1-i\mu-\nu_1+\nu_2}{4}) \Gamma(\tfrac{3-i\mu-\nu_1-\nu_2}{4})
$$
$$
\left[\left(\tfrac{\pi\sigma}{m_1 m_2}\right)^2 \left(\tfrac{-1-i\mu}{2}\right)\left(\tfrac{-5-i\mu+\nu_1+\nu_2}{4}\right)^{-1} {}_2F_1\left(\tfrac{-5-i\mu+\nu_1+\nu_2}{4}, \tfrac{-1-i\mu+\nu_1-\nu_2}{4}; \tfrac{-1-i\mu}{2}; \tfrac{n(2\sigma-n)}{\sigma^2}\right) \right.
$$
$$
+ \left(\tfrac{\nu_1-1}{2}\right)\left(\tfrac{-1-i\mu+\nu_1-\nu_2}{4}\right) {}_2F_1\left(\tfrac{-1-i\mu+\nu_1+\nu_2}{4}, \tfrac{3-i\mu+\nu_1-\nu_2}{4}; \tfrac{1-i\mu}{2}; \tfrac{n(2\sigma-n)}{\sigma^2}\right)
$$
$$
\left. + \left(\tfrac{\nu_2-1}{2}\right)\left(\tfrac{-1-i\mu-\nu_1+\nu_2}{4}\right) {}_2F_1\left(\tfrac{-1-i\mu+\nu_1+\nu_2}{4}, \tfrac{-1-i\mu+\nu_1-\nu_2}{4}; \tfrac{1-i\mu}{2}; \tfrac{n(2\sigma-n)}{\sigma^2}\right) \right]. \qquad (12.25)
$$

The end of the proof is fully analogous to that of theorem 11.3, and the main contribution of $\hat{\phi}(\frac{1}{m_1 m_2})$ to $(\tilde{c}_n)^+_{\mathrm{maj}}(\mu)$ is provided by substituting 1 for each of the ${}_2F_1$ functions. With the help of (9.65), we get as a main term in $\hat{\phi}(\frac{1}{m_1 m_2})$ the sum

$$
\frac{2^{-3}\pi^{\frac{-1+i\mu}{2}}}{\Gamma(\frac{1-i\mu}{2})} \Gamma(\tfrac{-1-i\mu+\nu_1+\nu_2}{4}) \Gamma(\tfrac{-1-i\mu+\nu_1-\nu_2}{4}) \Gamma(\tfrac{-1-i\mu-\nu_1+\nu_2}{4}) \Gamma(\tfrac{3-i\mu-\nu_1-\nu_2}{4})
$$
$$
\left[\frac{\pi^2}{(m_1 m_2)^2} \left(\tfrac{-1-i\mu}{2}\right)\left(\tfrac{-5-i\mu+\nu_1+\nu_2}{4}\right)^{-1} \pi^{\frac{-6-i\mu+\nu_1+\nu_2}{2}} \frac{\Gamma(\frac{7+i\mu-\nu_1-\nu_2}{4})}{\Gamma(\frac{-5-i\mu+\nu_1+\nu_2}{4})} |m_1 m_2|^{\frac{7+i\mu-\nu_1-\nu_2}{4}} \right.
$$
$$
+ \frac{\nu_1-1}{8}(-1-i\mu+\nu_1-\nu_2) \pi^{\frac{-2-i\mu+\nu_1+\nu_2}{2}} \frac{\Gamma(\frac{3+i\mu-\nu_1-\nu_2}{4})}{\Gamma(\frac{-1-i\mu+\nu_1+\nu_2}{4})} |m_1 m_2|^{\frac{3+i\mu-\nu_1-\nu_2}{4}}
$$
$$
\left. + \frac{\nu_2-1}{8}(-1-i\mu-\nu_1+\nu_2) \pi^{\frac{-2-i\mu+\nu_1+\nu_2}{2}} \frac{\Gamma(\frac{3+i\mu-\nu_1-\nu_2}{4})}{\Gamma(\frac{-1-i\mu+\nu_1+\nu_2}{4})} |m_1 m_2|^{\frac{3+i\mu-\nu_1-\nu_2}{4}} \right], \qquad (12.26)
$$

which reduces to

$$
\frac{2^{-3}\pi^{\frac{-3+\nu_1+\nu_2}{2}}}{\Gamma(\frac{1-i\mu}{2})} \Gamma(\tfrac{3+i\mu-\nu_1-\nu_2}{4}) \Gamma(\tfrac{-1-i\mu+\nu_1-\nu_2}{4}) \Gamma(\tfrac{-1-i\mu-\nu_1+\nu_2}{4}) \Gamma(\tfrac{3-i\mu-\nu_1-\nu_2}{4})
$$
$$
\left[\left(\tfrac{-1-i\mu}{2}\right)\left(\tfrac{3+i\mu-\nu_1-\nu_2}{4}\right) + \left(\tfrac{\nu_1-1}{8}\right)(-1-i\mu+\nu_1-\nu_2) + \left(\tfrac{\nu_2-1}{8}\right)(-1-i\mu-\nu_1+\nu_2) \right] |m_1 m_2|^{\frac{3+i\mu-\nu_1-\nu_2}{2}}
$$

$$
\qquad (12.27)
$$

or to

$$-\frac{\pi^{\frac{-3+\nu_1+\nu_2}{2}}}{4\,\Gamma(\frac{1-i\mu}{2})}\,\Gamma(\tfrac{3+i\mu-\nu_1-\nu_2}{4})\,\Gamma(\tfrac{3-i\mu+\nu_1-\nu_2}{4})\,\Gamma(\tfrac{3-i\mu-\nu_1+\nu_2}{4})\,\Gamma(\tfrac{3-i\mu-\nu_1-\nu_2}{4})\ \times$$

$$|m_1 m_2|^{\frac{3+i\mu-\nu_1-\nu_2}{2}}. \tag{12.28}$$

This leads to the function $b_n(\mu)$ which occurs in the statement of theorem 12.2, and the other terms can be analyzed just like the corresponding ones in the proof of theorem 11.3.   □

Remark. At least in the case when $|\mathrm{Re}\,(s-t)| < 1$, the condition $s+t \neq 1$ in the statement of theorem 12.2 can be removed, as will be shown in proposition 15.8.

13. Automorphic distributions on $\mathbf{R}^2$

This section plays an essential role in the part of section 18 that deals with the Lax–Phillips scattering theory [25] for the automorphic wave equation. It also plays a mild one in section 16. On the other hand, it is certainly useful to realize that the problem of defining a Hilbert space of automorphic distributions on $\mathbf{R}^2$ is not one that could easily be tackled with in any direct way; also, it is to our belief interesting to see how the Weyl calculus plays a (possibly not anticipated) role in this context.

The definition of the Weyl calculus of operators on $L^2(\mathbf{R})$ is the rule $h \mapsto \mathrm{Op}(h)$ that has been recalled in (5.1). In this section, we show (though the discussion is only tentative at some point since the expansion (13.23) that follows is only given a rather weak meaning) why the problem of decomposing an automorphic distribution $\mathfrak{S}$ on $\mathbf{R}^2$ into its homogeneous terms is best handled through the consideration of the operator with Weyl symbol $\mathfrak{S}$. So is the Hilbert space structure on what constitutes the "right" space of automorphic distributions.

The Weyl calculus is covariant under the Heisenberg representation, but we shall have no use of this in what follows: our interest lies with the metaplectic representation instead. Recall that there exists a certain twofold covering $\tilde{G} = \widetilde{SL}(2, \mathbf{R})$ of G and a unitary representation Met of $\tilde{G}$ in $L^2(\mathbf{R})$ such that, for every $\tilde{g} \in \tilde{G}$ lying above some point $g \in G$, and every tempered distribution f on $\mathbf{R}^2$ (then, $\mathrm{Op}(f)$ acts from $\mathcal{S}(\mathbf{R})$ to $\mathcal{S}'(\mathbf{R})$), the covariance rule

$$\mathrm{Met}(\tilde{g})\,\mathrm{Op}(h)\,\mathrm{Met}(\tilde{g})^{-1} = \mathrm{Op}(h \circ g^{-1}) \tag{13.1}$$

should hold: this is of course very classical.

The covariance formula (13.1) gives an operator–theoretic meaning to the set of symbols h which are Γ-invariant: as a consequence, if the composition of any two such operators can be given a meaning, the resulting symbol is also Γ-invariant.

But first, let us note that the group Γ does not have any fundamental domain in $\mathbf{R}^2$ (in which G acts in a linear way): indeed, it does not act in a discontinuous way for, if $\frac{\xi_1}{\xi_2}$ is irrational, the set of transforms under Γ of the point $\xi = (\xi_1, \xi_2)$ is dense in $\mathbf{R}^2$. It is therefore not at all obvious, at first sight, how one should proceed in order to get an analogue of the Petersson scalar product of two (suitable) Γ-invariant distributions on $\mathbf{R}^2$. To that effect, we shall use *sets of coherent states.*

Recall that the set of all unitary operators $\mathrm{Met}(\tilde{g})$, $\tilde{g} \in \tilde{G}$, is generated as a group by the operators of the following three species: (i) transformations $u \mapsto v$, $v(x) = a^{-\frac{1}{2}} u(a^{-1} x)$, $a > 0$; (ii) multiplications by exponentials $\exp i \pi c x^2$, c real; (iii) $e^{-\frac{i\pi}{4}}$ times the Fourier transformation. These three transformations are associated with points $\tilde{g}$ that lie above the points

$$g = \begin{pmatrix} a & 0 \\ 0 & a^{-1} \end{pmatrix}, \qquad \begin{pmatrix} 1 & 0 \\ c & 1 \end{pmatrix} \qquad \text{and} \qquad \begin{pmatrix} 0 & 1 \\ -1 & 0 \end{pmatrix} \qquad (13.2)$$

of G. As a consequence, Met is not an irreducible transformation, but acts within $L^2_{\mathrm{even}}(\mathbb{R})$ and $L^2_{\mathrm{odd}}(\mathbb{R})$ separately: the two terms can then be shown to be acted upon in an irreducible way. Remembering that Op is an isometry from $L^2(\mathbb{R}^2)$ onto the space of Hilbert–Schmidt operators on $L^2(\mathbb{R})$, one sees from (13.1) that the quasi–regular representation of G in $L^2(\mathbb{R}^2)$ can be splitted into four parts (the *even–even, even–odd, odd–even* and *odd–odd* parts respectively), where, say, the even–odd part of $L^2(\mathbb{R}^2)$ is the set of Weyl symbols of all operators that send the whole of $L^2(\mathbb{R})$ into $L^2_{\mathrm{odd}}(\mathbb{R})$, and vanish on $L^2_{\mathrm{odd}}(\mathbb{R})$. Since, with $\check{u}(x) = u(-x)$, the symbol of the operator $u \mapsto \mathrm{Op}(h)\check{u}$ is (*cf.* [41], p.262) the function $\mathcal{G}h$, where the transformation $\mathcal{G}$ was introduced in (4.12), and thanks to the remark that just follows (5.10), it is clear that the four species of symbols defined above can be characterized, in the same order, as follows: (i) the even ones invariant under $\mathcal{G}$; (ii) the odd ones invariant under $\mathcal{G}$; (iii) the odd ones changing to their negatives under $\mathcal{G}$; (iv) the even ones changing to their negatives under $\mathcal{G}$. Note that only even functions h on $\mathbb{R}^2$ have been considered in the present work, with the exception of section 5. For simplicity (as the Radon transform, viewed as valued into a space of functions on G/K, does not exist in the odd case), we shall limit ourselves, in what follows, to the consideration of even (*i.e.,* even–even or odd–odd) symbols.

For any point $z \in \Pi$, consider the two functions (normalized in $L^2(\mathbb{R})$)

$$u_z(x) = 2^{\frac{1}{4}} \left(\mathrm{Im}\, \frac{1}{z} \right)^{\frac{1}{4}} \exp \frac{i\pi x^2}{\bar{z}},$$

$$u_z^1(x) = 2^{\frac{5}{4}} \pi^{\frac{1}{2}} \left(\mathrm{Im}\, \frac{1}{z} \right)^{\frac{3}{4}} x \exp \frac{i\pi x^2}{\bar{z}} \qquad (13.3)$$

(here, of course, x has nothing to do with the real part of z!). Using when $b \neq 0$ the decomposition

$$\begin{pmatrix} a & b \\ c & d \end{pmatrix} = \begin{pmatrix} 1 & 0 \\ \frac{a}{b} & 1 \end{pmatrix} \begin{pmatrix} b & 0 \\ 0 & \frac{1}{b} \end{pmatrix} \begin{pmatrix} 0 & 1 \\ -1 & 0 \end{pmatrix} \begin{pmatrix} 1 & 0 \\ \frac{d}{b} & 1 \end{pmatrix}, \qquad (13.4)$$

we find after some computation that, if $\tilde{g} \in \tilde{G}$ lies above $\left(\begin{smallmatrix} a & b \\ c & d \end{smallmatrix}\right) \in G$ and $b > 0$, one has the formulas

$$\text{Met}(\tilde{g})\, u_z = \pm \left(\frac{a + \frac{b}{\bar{z}}}{|a + \frac{b}{\bar{z}}|}\right)^{-\frac{1}{2}} u_{\frac{az+b}{cz+d}},$$

$$\text{Met}(\tilde{g})\, u_z^1 = \pm \left(\frac{a + \frac{b}{\bar{z}}}{|a + \frac{b}{\bar{z}}|}\right)^{-\frac{3}{2}} u_{\frac{az+b}{cz+d}}^1, \tag{13.5}$$

in which the important fact is that the coefficients in front of the right–hand sides are constants (*i.e.*, depend only on z) of modulus one. The formulas (13.3) appear as slightly more complicated than the corresponding ones introduced in [41], p.260 (in view of the change $z \mapsto -\frac{1}{\bar{z}}$). This is due to our using the Poincaré upper half–plane in the present work, whereas life is somewhat simpler in the right half–plane $-i\Pi$ (compare the formula $|z|^{2\lambda} = z^\lambda \bar{z}^\lambda$ to its upper half–plane version $|z|^{2\lambda} = e^{-i\pi\lambda} z^\lambda (-\bar{z})^\lambda$): however, in view of the arithmetic environment, there was simply no choice here but to comply with the tradition. It is easily noted that the functions u_z and u_z^1 are normalized in $L^2(\mathbb{R})$, and one can prove (with the help of the discrete series of representations of $\tilde{G}$) that the set of all functions u_z (resp. u_z^1), as z describes Π, is total in $L^2_{\text{even}}(\mathbb{R})$ (resp. $L^2_{\text{odd}}(\mathbb{R})$). In view of the transformation formulas (13.5) and of this last–reported property, these sets of functions constitute, with a standard terminology borrowed from physicists, families of coherent states; in the odd case (only), one can also write ([41], p.269)

$$\|v\|^2_{L^2(\mathbb{R})} = (8\pi)^{-1} \int_\Pi |(u_z^1|v)|^2 \, d\mu(z), \tag{13.6}$$

from which one gets a *resolution of the identity* in the space $L^2_{\text{odd}}(\mathbb{R})$.

On $L^2(\mathbb{R})$, we define the inner product $(v|u)$ to be linear with respect to the *second* variable. The Wigner function $W(v, u)$ of any two functions u and $v \in L^2(\mathbb{R})$ is the function on $\mathbb{R}^2$ that makes the formula

$$(v|\text{Op}(h)u) = \int_{\mathbb{R}^2} h(\xi_1, \xi_2)\, W(v, u)(\xi_1, \xi_2)\, d\xi_1 \, d\xi_2 \tag{13.7}$$

valid for every $h \in L^2(\mathbb{R}^2)$ (we switched to the coordinates (ξ_1, ξ_2) on $\mathbb{R}^2$ in conformity with section 4 rather than section 5). The Wigner function just defined is also the Weyl symbol of the rank–one operator $\phi \mapsto (v, \phi)u$, and can be computed as

$$W(v, u)(\xi_1, \xi_2) = 2 \int_{-\infty}^{\infty} \bar{v}(\xi_1 + t)\, u(\xi_1 - t)\, e^{4i\pi t \xi_2}\, dt. \tag{13.8}$$

It is then easy to find

$$W(u_z, u_z)(\xi) = 2 \exp\left(-\frac{2\pi}{\text{Im } z}|\xi_1 - z\,\xi_2|^2\right), \tag{13.9}$$

since the covariance formula (13.1) and the second interpretation of the Wigner function make it possible to reduce the computation to the case when $z = i$. In [41], p.261, we also computed

$$W(u_z^1, u_z^1)(\xi) = 2\,[\frac{4\pi}{\text{Im } z}|\xi_1 - z\xi_2|^2 - 1]\exp\left(-\frac{2\pi}{\text{Im } z}|\xi_1 - z\xi_2|^2\right). \tag{13.10}$$

Using the fact that $u_z(x)$ is $(\text{Im }\frac{1}{z})^{\frac{1}{4}}$ times an antiholomorphic function of z, one easily finds (*cf. loc.cit.*) the formulas

$$(u_w|u_z) = 2^{\frac{1}{2}}\,(\text{Im }\frac{1}{z}\,\text{Im }\frac{1}{w})^{\frac{1}{4}}\,[i(-\frac{1}{z} + \frac{1}{w})]^{-\frac{1}{2}},$$

$$(u_w^1|u_z^1) = (u_w|u_z)^3. \tag{13.11}$$

We shall also set

$$p_{z,z}(\xi) := W(u_z, u_z)(\xi),$$

$$p_{z,z}^1(\xi) := W(u_z^1, u_z^1)(\xi). \tag{13.12}$$

Extending the formula (4.20) (which provides the decomposition of an even function on $\mathbf{R}^2$ as a continuous sum of homogeneous ones) as

$$h_{i\nu}^{\flat}(s) = \frac{1}{2\pi}\int_0^\infty t^{-\nu}h(ts, t)\,dt, \tag{13.13}$$

where ν does not have to be pure imaginary any longer, we get from (13.9)

$$(p_{z,z})_{i\nu}^{\flat}(s) = (2\pi)^{\frac{\nu-3}{2}}\,\Gamma(\frac{1-\nu}{2})\left(\frac{|s - z|^2}{\text{Im } z}\right)^{\frac{\nu-1}{2}}, \tag{13.14}$$

a formula that may be compared to the definition (2.6). Doing the same with $p_{z,z}^1$ substituted for $p_{z,z}$, we end up with the pair of formulas

$$(p_{z,z})_{i\nu}^{\flat} = \frac{2^{\frac{\nu-3}{2}}}{\pi}\,\phi_z^{-\nu},$$

$$(p_{z,z}^1)_{i\nu}^{\flat} = -\frac{2^{\frac{\nu-3}{2}}}{\pi}\,\nu\,\phi_z^{-\nu}. \tag{13.15}$$

One should not mistake the set of functions (u_z) with the set of functions $(\phi_z^{-\nu})$: both the metaplectic representation and the full non–unitary principal series of representations of G do occur in the present context, the first one since one considers operators which act on $L^2(\mathbf{R})$, on which it operates, the latter since it permits to decompose the symbols of such.

One may write

$$(1 + s^2)^{\frac{1-\nu}{2}}\,h_{i\nu}^{\flat}(s) = \frac{1}{2\pi}\int_0^\infty t^{-\nu}h(t\xi)\,dt \tag{13.16}$$

with $\xi = \begin{pmatrix} \frac{s}{\sqrt{1+s^2}} \\ \frac{1}{\sqrt{1+s^2}} \end{pmatrix} \in \mathbf{R}^2$, a unit vector. This shows that if h lies in Schwartz's space $\mathcal{S}_{\text{even}}(\mathbf{R}^2)$ and Re $\nu < 1$, then $h^\flat_{i\nu}$ is C^∞ on $\mathbf{R}$; using also

$$(\pi_{-\nu}(\left(\begin{smallmatrix} 0 & -1 \\ 1 & 0 \end{smallmatrix}\right))h^\flat_{i\nu})(s) = |s|^{\nu-1} h^\flat_{i\nu}(-s^{-1})$$

$$= \frac{1}{2\pi} \int_0^\infty t^{-\nu} h(-t, st)\, dt, \qquad (13.17)$$

one sees that $h^\flat_{i\nu} \in C^\infty_{-\nu}$. This makes it possible to define

$$< \mathfrak{T}^\sharp_\nu, h >: \ = 4\pi < \mathfrak{T}_\nu, h^\flat_{i\nu} > \qquad (13.18)$$

if $\mathfrak{T}_\nu \in \mathcal{D}'_\nu$, Re $\nu < 1$ and $h \in \mathcal{S}_{\text{even}}(\mathbf{R}^2)$, thus defining a linear map $\mathfrak{T}_\nu \mapsto \mathfrak{T}^\sharp_\nu$ from $\mathcal{D}'_\nu$ to $\mathcal{S}'_{\text{even}}(\mathbf{R}^2)$ (by convention, we set $< \mathfrak{T}^\sharp_\nu, h >= 0$ if h is odd). It is readily seen that, in the case when $\nu = i\lambda$, this map extends the map $u_\lambda \mapsto u^\sharp_\lambda$ introduced in (4.35).

Since a Wigner function $W(v, u)$ lies in $\mathcal{S}(\mathbf{R}^2)$ if u and v lie in $\mathcal{S}(\mathbf{R})$, this permits to associate with every $\mathfrak{T}_\nu \in \mathcal{D}'_\nu$ (Re $\nu < 1$) the operator $\text{Op}_\nu(\mathfrak{T}_\nu) = \text{Op}(\mathfrak{T}^\sharp_\nu) \colon \mathcal{S}(\mathbf{R}) \to \mathcal{S}'(\mathbf{R})$: it commutes with the symmetry $u \mapsto \check{u}$. One has in particular (using in succession (13.7), (13.12), (13.18) and (13.15))

$$(u_z | \text{Op}_\nu(\mathfrak{T}_\nu) u_z) = < \mathfrak{T}^\sharp_\nu, W(u_z, u_z) >$$

$$= < \mathfrak{T}^\sharp_\nu, p_{z,z} >$$

$$= 4\pi < \mathfrak{T}_\nu, (p_{z,z})^\flat_{i\nu} >$$

$$= 2^{\frac{\nu+1}{2}} < \mathfrak{T}_\nu, \phi^{-\nu}_z >$$

$$= 2^{\frac{\nu+1}{2}} (\Theta_\nu \mathfrak{T}_\nu)(z) \qquad (13.19)$$

according to definition 2.4. Also,

$$(u^1_z | \text{Op}_\nu(\mathfrak{T}_\nu) u^1_z) = 4\pi < \mathfrak{T}_\nu, (p^1_{z,z})^\flat_{i\nu} >$$

$$= -2^{\frac{\nu+1}{2}} \nu (\Theta_\nu \mathfrak{T}_\nu)(z). \qquad (13.20)$$

As the set of functions $(u_z)_{z\in\Pi}$ $(resp.(u^1_z)_{z\in\Pi})$ generates a dense subspace of $\mathcal{S}_{\text{even}}(\mathbf{R})$ $(resp.\ \mathcal{S}_{\text{odd}}(\mathbf{R}))$, it is clear that an operator $Op(\mathfrak{S})$ with an even symbol in $\mathcal{S}'(\mathbf{R}^2)$ (it is then a weakly continuous operator from $\mathcal{S}(\mathbf{R})$ to $\mathcal{S}'(\mathbf{R})$) is characterized by the values of the functions $(w, z) \mapsto (u_w | Op(\mathfrak{S}) u_z)$ and $(w, z) \mapsto (u^1_w | Op(\mathfrak{S}) u^1_z)$ as z and w vary in Π. These may also be written as $< \mathfrak{S}, W(u_w, u_z) >$ and $< \mathfrak{S}, W(u^1_w, u^1_z) >$.

Let $(\mathcal{M}_k)_{k\geq 1}$ be a complete orthonormal set of Maass cusp forms, $\mathcal{M}_k$ corresponding to the eigenvalue $\frac{1}{4}(1 + \lambda^2_k)$ of Δ. We assume $0 < \lambda_1 \leq \lambda_2 \leq$

... If, for some k,

$$\mathcal{M}_k(z) = \sum_{n \neq 0} b_n \, y^{\frac{1}{2}} \, K_{\frac{i\lambda_k}{2}}(2\pi|n|y) \, e^{-2i\pi nx}, \qquad (13.21)$$

we define $\mathfrak{M}_k \in \mathcal{D}'_{i\lambda_k}$ by

$$< \mathfrak{M}_k, v > = \frac{1}{2} \sum_{n \neq 0} |n|^{\frac{i\lambda_k}{2}} \, b_n \, \hat{v}(n), \qquad v \in C^\infty_{-i\lambda_k} \qquad (13.22)$$

as in theorem 6.4, so that

$$< \mathfrak{M}_k, \phi_z^{-i\lambda_k} > = \mathcal{M}_k(z), \qquad (13.23)$$

i.e., $\mathcal{M}_k = \Theta_{i\lambda_k} \mathfrak{M}_k$. We also set

$$\mathfrak{M}_{-k} := \theta_{i\lambda_k} \mathfrak{M}_k \in \mathcal{D}'_{-i\lambda_k}, \qquad (13.24)$$

so that $\mathcal{M}_k = \Theta_{-i\lambda_k} \mathfrak{M}_{-k}$ as well. We set $\lambda_{-k} = -\lambda_k$.

We make the (even, tempered) distributions $\mathfrak{E}_\nu^\natural$ and $\mathfrak{M}_k^\natural$, as associated by (13.18) to $\mathfrak{E}_\nu$ and $\mathfrak{M}_k$, explicit, at the same time analyzing the complex continuation of the distribution–valued function $\nu \mapsto \mathfrak{E}_\nu^\natural$.

PROPOSITION 13.1. *For* Re $\nu < -1$, $h \in \mathcal{S}(\mathbb{R})$,

$$< \mathfrak{E}_\nu^\natural, h > = \frac{1}{2} \sum_{|m|+|n| \neq 0} \int_{-\infty}^\infty |t|^{-\nu} h(tn, tm) \, dt. \qquad (13.25)$$

The function $\nu \mapsto \mathfrak{E}_\nu^\natural$ *extends as a holomorphic function of* ν *for* $\nu \neq \pm 1$, *with simple poles at* $\nu = \pm 1$; *the residues there are given as* Res$_{\nu=-1} \, \mathfrak{E}_\nu^\natural = -1$ *and* Res$_{\nu=1} \, \mathfrak{E}_\nu^\natural = \delta$. *One has the identity* $\mathcal{F}\mathfrak{E}_\nu^\natural = \mathfrak{E}_{-\nu}^\natural$. *One may also write, for* Re $\nu < 0$, $\nu \neq -1$,

$$< \mathfrak{E}_\nu^\natural, h > = \zeta(-\nu) \int_{-\infty}^\infty |t|^{-\nu-1}(\mathcal{F}_1 h)(0,t) \, dt + \zeta(1-\nu) \int_{-\infty}^\infty |t|^{-\nu} h(t,0) \, dt$$

$$+ \sum_{n \neq 0} \sigma_\nu(|n|) \int_{-\infty}^\infty |t|^{-\nu-1} (\mathcal{F}_1 h)(\frac{n}{t}, t) \, dt, \qquad (13.26)$$

where $\mathcal{F}_1 h$ *denotes the Fourier transform of* h *with respect to the first variable.*

If $\mathcal{M}_k$ *is the Maass cusp–form characterized by the Fourier expansion* (13.21) *and* $\mathfrak{M}_k$ *is the associated modular distribution in* (13.22), *one has*

$$< \mathfrak{M}_k^\natural, h > = \frac{1}{2} \sum_{n \neq 0} |n|^{\frac{i\lambda_k}{2}} b_n \int_{-\infty}^\infty |t|^{-i\lambda_k-1} (\mathcal{F}_1 h)(\frac{n}{t}, t) \, dt, \qquad (13.27)$$

and the same goes for $\mathfrak{M}_{-k}^\natural$, *substituting* $-\lambda_k$ *for* λ_k.

PROOF. Let $h \in \mathcal{S}_{\text{even}}(\mathbb{R}^2)$, and assume $\text{Re } \nu < -1$. From (13.13), one gets

$$h_{i\nu}^{\flat}(s) = \frac{1}{2\pi} |s|^{\nu-1} \int_0^\infty t^{-\nu} h(t, \frac{t}{s}) \, dt \tag{13.28}$$

so that (2.4) yields

$$(h_{i\nu}^{\flat})_\infty = \frac{1}{2\pi} \int_0^\infty t^{-\nu} h(t, 0) \, dt. \tag{13.29}$$

From (13.18) together with (3.4), one thus gets

$$< \mathfrak{E}_\nu^\natural, h > = 4\pi < \mathfrak{E}_\nu, h_{i\nu}^{\flat} >$$

$$= \sum_{\substack{m \neq 0 \\ n \in \mathbb{Z}}} |m|^{\nu-1} \int_0^\infty t^{-\nu} h(\frac{tn}{m}, t) \, dt + \sum_{n \neq 0} (h_{i\nu}^{\flat})_\infty |n|^{\nu-1}, \tag{13.30}$$

which immediately leads to (13.25).

Decompose then $\mathfrak{E}_\nu^\natural$ as

$$\mathfrak{E}_\nu^\natural = (\mathfrak{E}_\nu^\natural)_{\text{princ}} + (\mathfrak{E}_\nu^\natural)_{\text{res}}, \tag{13.31}$$

where

$$< (\mathfrak{E}_\nu^\natural)_{\text{princ}}, h > := \sum_{|n|+|m| \neq 0} \int_0^1 t^{-\nu} h(tn, tm) \, dt \tag{13.32}$$

and

$$< (\mathfrak{E}_\nu^\natural)_{\text{res}}, h > := \sum_{|n|+|m| \neq 0} \int_1^\infty t^{-\nu} h(tn, tm) \, dt. \tag{13.33}$$

For $h \in \mathcal{S}_{\text{even}}(\mathbb{R})$, the second expression extends as an entire function of ν. If $\text{Re } \nu < -1$, one has, using Poisson's formula in the middle of the sequence that follows,

$$< \mathcal{F}(\mathfrak{E}_{-\nu}^\natural)_{\text{res}}, h > = \sum_{|n|+|m| \neq 0} \int_1^\infty t^\nu \, (\mathcal{F}h)(tn, tm) \, dt$$

$$= \int_1^\infty t^\nu \Big[\sum_{(n,m) \in \mathbb{Z}^2} (\mathcal{F}h)(tn, tm) - (\mathcal{F}h)(0,0) \Big] \, dt$$

$$= \int_1^\infty t^\nu \Big[\sum_{(n,m) \in \mathbb{Z}^2} t^{-2} h(tn, tm) - (\mathcal{F}h)(0,0) \Big] \, dt$$

$$= \int_0^1 t^{-\nu} \Big[\sum_{(n,m) \in \mathbb{Z}^2} h(tn, tm) - t^{-2} (\mathcal{F}h)(0,0) \Big] \, dt$$

$$= \int_0^1 t^{-\nu} \Big[\sum_{|n|+|m| \neq 0} h(tn, tm) + h(0,0) - t^{-2} (\mathcal{F}h)(0,0) \Big] \, dt \tag{13.34}$$

so that

$$< \mathcal{F}(\mathfrak{E}^{\natural}_{-\nu})_{\text{res}}, h >=< (\mathfrak{E}^{\natural}_{\nu})_{\text{princ}}, h > + \frac{h(0,0)}{1-\nu} + \frac{(\mathcal{F}h)(0,0)}{1+\nu}.$$
$$(13.35)$$

This gives the analytic continuation of $(\mathfrak{E}^{\natural}_{\nu})_{\text{princ}}$, thus that of $\mathfrak{E}^{\natural}_{\nu}$ too; in particular, the residue of either distribution at $\nu = 1$ is the distribution $h \mapsto h(0,0)$, and the residue of either distribution at $\nu = -1$ is the distribution $h \mapsto -\int_{\mathbf{R}^2} h(\xi)\, d\xi$. In the same way as above, one finds

$$< \mathcal{F}(\mathfrak{E}^{\natural}_{-\nu})_{\text{princ}}, h >=< (\mathfrak{E}^{\natural}_{\nu})_{\text{res}}, h > - \frac{h(0,0)}{1-\nu} - \frac{(\mathcal{F}h)(0,0)}{1+\nu}$$
$$(13.36)$$

so that, adding (13.35) and (13.36), the relation $\mathcal{F}\mathfrak{E}^{\natural}_{\nu} = \mathfrak{E}^{\natural}_{-\nu}$ is proven.

Still under the assumption that $\text{Re}\ \nu < -1$, one may write, from (13.28),

$$\widehat{(h^{\flat}_{i\nu})}(\sigma) = \frac{1}{2\pi} \int_0^{\infty} t^{-\nu-1} (\mathcal{F}_1 h)(\frac{\sigma}{t}, t)\, dt : \qquad (13.37)$$

thus

$$\text{res}_0((\widehat{h^{\flat}_{i\nu}})) = \frac{1}{2\pi} \int_0^{\infty} t^{-\nu-1} (\mathcal{F}_1 h)(0,t)\, dt, \qquad (13.38)$$

where the left–hand side was defined in definition 3.3. On the other hand, from (3.16) and (13.29),

$$\text{res}_{-\nu}((\widehat{h^{\flat}_{i\nu}})) = \pi^{\frac{1}{2}-\nu} \frac{\Gamma(\frac{\nu}{2})}{\Gamma(\frac{1-\nu}{2})} \zeta(\nu) \frac{1}{2\pi} \int_0^{\infty} t^{-\nu} h(t,0)\, dt. \qquad (13.39)$$

Equation (13.26) follows from (13.38),(13.39),(13.37) and proposition 3.8.

Finally, let $\mathcal{M}_k$ be the Maass cusp–form characterized by (13.21) and let $\mathfrak{M}_k \in \mathcal{D}'_{i\lambda_k}$ be the associated modular distribution in (13.22). From (13.18) and (13.22),

$$< \mathfrak{M}^{\natural}_k, h > = 4\pi < \mathfrak{M}_k, h^{\flat}_{-\lambda_k} >$$
$$= 2\pi \sum_{n \neq 0} |n|^{\frac{i\lambda_k}{2}} b_n \widehat{(h^{\flat}_{i\nu})}(n), \qquad (13.40)$$

which leads to (13.27), using (13.37) again. The same goes for $\mathfrak{M}^{\natural}_{-k}$, as a consequence of (13.24),(3.32) and (2.15). $\qquad \square$

THEOREM 13.2. *Let* $\mathfrak{S} \in \mathcal{S}'_{\text{even}}(\mathbf{R}^2)$ *be a* Γ*–invariant even tempered distribution in* $\mathbf{R}^2$. *Assume that the two* Γ*–invariant functions on* Π

$$z \mapsto (u_z | \text{Op}(\mathfrak{S}) u_z)$$

and

$$|\Delta - \frac{1}{4}|^{-\frac{1}{2}} (z \mapsto (u^1_z | \text{Op}(\mathfrak{S}) u^1_z))$$

lie in the Petersson space $L^2(\Gamma\backslash\Pi)$. Then, denoting as δ the Dirac measure on $\mathbf{R}^2$, $\mathfrak{S}$ admits a unique decomposition

$$\mathfrak{S} = 2^{\frac{1}{2}}\,c_0 + 2^{-\frac{1}{2}}\,c_0'\,\delta + \frac{1}{8\pi}\int_{-\infty}^{\infty} 2^{-\frac{i\lambda}{2}}\,\Psi(\lambda)\,\mathfrak{E}_{i\lambda}^{\natural}\,d\lambda + \sum_{k\neq 0} c_k\,2^{-\frac{i\lambda_k}{2}}\,\mathfrak{M}_k^{\natural}, \tag{13.41}$$

where the function $\lambda \mapsto \zeta^*(i\lambda)\,\Psi(\lambda)$ lies in $L^2(\mathbf{R})$, $(c_k)_{k\in\mathbf{Z}} \in l^2(\mathbf{Z})$, and the decomposition is valid when tested against any Wigner function $W(u_w, u_z)$ or $W(u_w^1, u_z^1)$. The coefficients are characterized by the validity of the Roelcke–Selberg expansions

$$(u_z|\mathrm{Op}(\mathfrak{S})u_z) = 2^{\frac{1}{2}}\,(c_0 + c_0') + \frac{2^{\frac{1}{2}}}{8\pi}\int_{-\infty}^{\infty}\frac{\Psi(\lambda) + \Psi(-\lambda)}{2}\,E_{\frac{1-i\lambda}{2}}^*(z)\,d\lambda$$
$$+ 2^{\frac{1}{2}}\sum_{k\geq 1}(c_k + c_{-k})\,\mathcal{M}_k(z) \tag{13.42}$$

and

$$(u_z^1|\mathrm{Op}(\mathfrak{S})u_z^1) = 2^{\frac{1}{2}}\,(c_0 - c_0') - \frac{2^{\frac{1}{2}}i}{8\pi}\int_{-\infty}^{\infty}\frac{\lambda(\Psi(\lambda) - \Psi(-\lambda))}{2}\,E_{\frac{1-i\lambda}{2}}^*(z)\,d\lambda$$
$$- 2^{\frac{1}{2}}i\sum_{k\geq 1}\lambda_k(c_k - c_{-k})\,\mathcal{M}_k(z). \tag{13.43}$$

PROOF. First observe that the function $\Phi(\lambda) = \frac{\Psi(\lambda)+\Psi(-\lambda)}{2}\,\zeta^*(1-i\lambda)$ satisfies the property (6.39) or (6.69) that $(\zeta^*(1-i\lambda))^{-1}\Phi(\lambda)$ is even, so that it is indeed the Roelcke–Selberg coefficient of the right-hand side of (13.42). The same goes with the function $\lambda \mapsto \frac{\lambda(\Psi(\lambda)-\Psi(-\lambda))}{2}\,\zeta^*(1-i\lambda)$.

If (13.41) is valid, then, from (13.19), next (3.13) and (13.41):

$$(u_z|\mathrm{Op}(\mathfrak{S})u_z) = 2^{\frac{1}{2}}\,(c_0 + c_0') + \frac{1}{8\pi}\int_{-\infty}^{\infty} 2^{-\frac{i\lambda}{2}}\,\Psi(\lambda)\,2^{\frac{i\lambda+1}{2}}\,(\Theta_{i\lambda}\mathfrak{E}_{i\lambda})(z)\,d\lambda$$
$$+ \sum_{k\neq 0} c_k\,2^{-\frac{i\lambda_k}{2}}\,2^{\frac{i\lambda_k+1}{2}}\,(\Theta_{i\lambda_k}\mathfrak{M}_k)(z)$$
$$= 2^{\frac{1}{2}}\,(c_0 + c_0') + \frac{2^{\frac{1}{2}}}{8\pi}\int_{-\infty}^{\infty}\Psi(\lambda)\,E_{\frac{1-i\lambda}{2}}^*(z)\,d\lambda$$
$$+ 2^{\frac{1}{2}}\sum_{k\neq 0} c_k\,\mathcal{M}_{|k|}(z). \tag{13.44}$$

This is the same as (13.42) thanks to the functional equation of Eisenstein series. In a similar way, still under the assumption (13.41), substituting

(13.20) for (13.19),

$$
(u_z^1 | \mathrm{Op}(\mathfrak{S}) u_z^1) = 2^{\frac{1}{2}} (c_0 - c_0') - \frac{1}{8\pi} \int_{-\infty}^{\infty} 2^{-\frac{i\lambda}{2}} \, \Psi(\lambda) \, 2^{\frac{i\lambda+1}{2}} \, i\lambda \, (\Theta_{i\lambda} \mathfrak{E}_{i\lambda})(z) \, d\lambda
$$

$$
- \sum_{k \neq 0} c_k \, 2^{-\frac{i\lambda_k}{2}} 2^{\frac{i\lambda_k+1}{2}} \, i\lambda_k \, (\Theta_{i\lambda_k} \mathfrak{M}_k)(z)
$$

$$
= 2^{\frac{1}{2}} (c_0 - c_0') - \frac{2^{\frac{1}{2}} i}{8\pi} \int_{-\infty}^{\infty} \lambda \, \Psi(\lambda) \, E_{\frac{1-i\lambda}{2}}^*(z) \, d\lambda
$$

$$
- 2^{\frac{1}{2}} i \sum_{k \neq 0} c_k \lambda_k \, \mathcal{M}_{|k|}(z), \tag{13.45}
$$

which leads to (13.43).

Conversely, under the assumptions in the beginning of the statement of the proposition, the Roelcke–Selberg theorem shows on one hand that the even part of the function $\lambda \mapsto \Psi(\lambda)$, as defined by (13.42), lies in $L^2(\mathbb{R})$ after it has been multiplied by $\zeta^*(i\lambda)$, and that the even part of the sequence $(c_k)_{k \in \mathbb{Z}}$ lies in $l^2(\mathbb{Z})$; on the other hand, defining $|\Delta - \frac{1}{4}|^{-\frac{1}{2}}$ through spectral theory (where the absolute value is meant to convey the fact that $|\Delta - \frac{1}{4}|$ acts on constants as the multiplication by $+\frac{1}{4}$), that the same holds with the odd parts as defined in (13.43). To extend the first of these identities to the case when the Wigner function $W(u_z, u_z)$ is replaced by $W(u_w, u_z)$, first observe that, as a consequence of (13.3), this latter function is $(\mathrm{Im}\,\frac{1}{w} \,\mathrm{Im}\,\frac{1}{z})^{\frac{1}{4}}$ times a sesquiholomorphic function of (w, z) (holomorphic with respect to w). A sesquiholomorphic extension away from the diagonal of the function $(z, z) \mapsto \zeta^*(1 - i\lambda)\, E_{\frac{1-i\lambda}{2}}(z)$ or $(z, z) \mapsto \mathcal{M}_k(z)$ can be obtained from the Fourier expansion (3.29) or (3.27), substituting $\frac{w - \bar{z}}{2i}$ for y and $\frac{w + \bar{z}}{2}$ for x. To account for the extra factor indicated above, one may also note that the constant 1 coincides, on the diagonal, with the function

$$
(w, z) \mapsto (\mathrm{Im}\,\frac{1}{\bar{w}} \,\mathrm{Im}\,\frac{1}{\bar{z}})^{\frac{1}{4}} \times \left(\frac{1}{2i}\left(\frac{1}{\bar{z}} - \frac{1}{w}\right) \right)_{\mathrm{right}}^{-\frac{1}{2}},
$$

in which the subscript in the last factor indicates that the determination of the square–root is the principal one in the right half–plane. This makes it possible to prove that (13.41) is still valid when tested against the function $W(u_w, u_z)$, using estimates of Bessel functions and of the Fourier coefficients of Maass cusp–forms similar to those used in the proof of proposition 7.1; the same holds with $W(u_w^1, u_z^1)$. $\qquad\square$

One should note that, given c_0, c_0', Ψ and $\{c_k\}$ satisfying the conditions of theorem 13.2, the right–hand side of (13.41) does not define a tempered distribution in general, though it does in the case when Ψ has compact support and all but a finite number of the c_k's are zero. Completing the space of all $\mathfrak{S}$'s satisfying the assumptions of the theorem would take us beyond the range of tempered distributions, thus would force us to extend the definition of Op. We plan to do this elsewhere, in connection with the automorphic Weyl calculus.

Remark. With the Fourier transformation $\mathcal{F}$ defined on $L^2(\mathbf{R}^2)$ as

$$(\mathcal{F}h)(\xi) = \int_{\mathbf{R}^2} h(\eta)\, e^{-2i\pi(-\xi_1\eta_2 + \xi_2\eta_1)}\, d\eta \qquad (13.46)$$

(recalling (4.11)), one has

$$(\mathcal{F}h)^\flat_{-\lambda} = \theta_{i\lambda} h^\flat_\lambda \qquad (13.47)$$

as indicated in (4.28). Also, with $\mathcal{G}$ as in (4.12), $\mathcal{G} = 2^{2i\pi\mathcal{E}}\,\mathcal{F}$, so that $\theta_{i\lambda} h^\flat_\lambda = 2^{-i\lambda}\,(\mathcal{G}h)^\flat_{-\lambda}$. These exact formulas extend to a distribution setting: in view of proposition 3.9, one thus gets $\mathfrak{E}^\natural_{-i\lambda} = (\theta_{i\lambda}\,\mathfrak{E}_{i\lambda})^\natural = 2^{-i\lambda}\,\mathcal{G}\mathfrak{E}^\natural_{i\lambda}$, in other words

$$\mathcal{G}\,(2^{-\frac{i\lambda}{2}}\,\mathfrak{E}^\natural_{i\lambda}) = 2^{\frac{i\lambda}{2}}\,\mathfrak{E}^\natural_{-i\lambda}\,. \qquad (13.48)$$

As a consequence of the discussion right after (13.2), it follows that the operator $\mathrm{Op}(\mathfrak{S})$ is even-even (*resp.* odd-odd) if and only if the function Ψ is even, $c_0 = c'_0$ and $c_{-k} = c_k$ for all k (*resp.* Ψ is odd, $c_0 = -c'_0$ and $c_{-k} = -c_k$ for all k).

Definition 13.3. Under the assumptions of theorem 13.2, and with the notations in (13.41), one defines

$$\|\mathfrak{S}\|^2_\Gamma := \frac{2\pi}{3}\,(|c_0|^2 + |c'_0|^2) + \frac{1}{8\pi}\int_{-\infty}^\infty |\Psi(\lambda)|^2\,|\zeta^*(i\lambda)|^2\,d\lambda$$

$$+ 2\sum_{k\neq 0} |c_k|^2\,. \qquad (13.49)$$

Proposition 13.4. *Under the assumptions of theorem 13.2, and with the notations in (13.41),*

$$\|\mathfrak{S}\|^2_\Gamma = \frac{1}{2}\,\|\,z \mapsto (u_z|\mathrm{Op}(\mathfrak{S})u_z)\,\|^2_{L^2(D)}$$

$$+ \frac{1}{8}\,\|\,|\Delta - \frac{1}{4}|^{-\frac{1}{2}}\,(z \mapsto (u^1_z|\mathrm{Op}(\mathfrak{S})u^1_z))\,\|^2_{L^2(D)}\,. \qquad (13.50)$$

Proof. From (13.42), it follows that

$$\|(u_z|\mathrm{Op}(\mathfrak{S})u_z)\|^2_{L^2(D)} = \frac{1}{16\,\pi}\int_{-\infty}^\infty |\Psi(\lambda) + \Psi(-\lambda)|^2\,|\zeta^*(1 - i\lambda)|^2\,d\lambda$$

$$+ \frac{2\pi}{3}\,|c_0 + c'_0|^2 + 2\sum_{k\geq 1} |c_k + c_{-k}|^2\,. \qquad (13.51)$$

On the other hand, from (13.43),

$$\frac{1}{4} \left\| \left| \Delta - \frac{1}{4} \right|^{-\frac{1}{2}} (u_z^1 | \mathrm{Op}(\mathfrak{S}) u_z^1) \right\|_{L^2(D)}^2 = \frac{1}{16\pi} \int_{-\infty}^{\infty} |\Psi(\lambda) - \Psi(-\lambda)|^2 |\zeta^*(1 - i\lambda)|^2 \, d\lambda$$

$$+ \frac{2\pi}{3} |c_0 - c_0'|^2 + 2 \sum_{k \geq 1} |c_k - c_{-k}|^2.$$

$$(13.52)$$

$$\square$$

14. The Hecke decomposition of products or Poisson brackets of two Eisenstein series

The aim of the present section is twofold. First, generalize the theorems in sections 9, 11 and 12 as much as possible so far as the domain of (ν_1, ν_2) is concerned; next, analyze the discrete parts in terms of Hecke's theory, answering as a consequence in the affirmative the question, raised at the end of section 11, whether all Maass forms of a given parity do enter the Roelcke–Selberg decomposition of f_0 (theorem 11.3) or h_0 (theorem 12.2) for generic values of (ν_1, ν_2).

It is necessary to change our notations slightly: in the present section, we shall denote as $(\lambda_k)_{k \geq 1}$ the increasing sequence of positive numbers such that $(\frac{1+\lambda_k^2}{4})_{k \geq 1}$ is the sequence of *distinct even* eigenvalues of Δ; in a similar way, $(\lambda_k^-)_{k \geq 1}$ shall be associated with the increasing sequence of *distinct odd* eigenvalues of Δ. *Even* or *odd* refers to the existence of eigenfunctions invariant or changing to their negatives under the symmetry $z \mapsto -\bar{z}$.

For every k, there is a number $n_k \geq 1$ and a set $(\mathcal{M}_{k,\ell})_{1 \leq \ell \leq n_k}$ of Hecke forms (i.e., Maass forms that are simultaneously eigenfunctions of all Hecke operators) which constitute an orthonormal basis of the space of even eigenfunctions of Δ for the eigenvalue $\frac{1+\lambda_k^2}{4}$. As is customary when dealing with Hecke's theory, it *sometimes* leads to simpler formulas to use instead of $\mathcal{M}_{k,\ell}$ the proportional function $\mathcal{N}_{k,\ell}$ characterized by the condition that its first Fourier coefficient (b_1 in (14.1) *infra*) is 1: *we may assume that* $\mathcal{M}_{k,\ell} = \|\mathcal{N}_{k,\ell}\|^{-1} \mathcal{N}_{k,\ell}$, *so that it has real Fourier coefficients*. A similar convention applies towards the definition of the odd Hecke forms $\mathcal{M}_{k,\ell}^-$ and $\mathcal{N}_{k,\ell}^-$.

If $\mathcal{M}$ is an even Hecke form corresponding to the eigenvalue $\frac{1+\lambda^2}{4}$ with the Fourier expansion

$$\mathcal{M}(x + iy) = y^{\frac{1}{2}} \sum_{n \neq 0} b_n \, K_{\frac{i\lambda}{2}}(2\pi|n|y) \, e^{2i\pi nx}, \qquad (14.1)$$

its associated L-function $L(., \mathcal{M})$ is defined as

$$L(s, \mathcal{M}) = \sum_{n \geq 1} b_n \, n^{-s} \qquad (14.2)$$

(note that we have followed the convention in ([3], p.107), which differs by the factor $\frac{1}{2}$ and by a shift by $\frac{1}{2}$ in the argument from that in ([36], p.242)).

We also set

$$L^*(s, \mathcal{M}) = \pi^{-s} \Gamma(\frac{s}{2} + \frac{i\lambda}{4}) \Gamma(\frac{s}{2} - \frac{i\lambda}{4}) L(s, \mathcal{M}), \qquad (14.3)$$

so that $L^*(s, \mathcal{M})$ extends as an entire function of s satisfying the functional equation $L^*(s, \mathcal{M}) = L^*(1 - s, \mathcal{M})$ (cf. [3], p.107).

We denote as $b_{n;k,\ell}$ the coefficient b_n in (14.1) in the case when $\mathcal{M} = \mathcal{M}_{k,\ell}$. On the other hand, as is well–known, the advantage of using $\mathcal{N}_{k,\ell}$ instead is that the eigenvalues of the Hecke operators can be read directly on its Fourier coefficients. Namely, for every $N \geq 1$,

$$T_N \mathcal{M}_{k,\ell} = \omega_{N;k,\ell} \mathcal{M}_{k,\ell}, \qquad T_N \mathcal{N}_{k,\ell} = \omega_{N;k,\ell} \mathcal{N}_{k,\ell} \qquad (14.4)$$

for some real number $\omega_{N;k,\ell}$, which is just the coefficient b_N in the expansion (14.1) of $\mathcal{N}_{k,\ell}$ (cf. [36], p.242). For every complex number s with Re $s > 1$, one then has

$$\sum_{N \geq 1} N^{-s} T_N \mathcal{N}_{k,\ell} = \sum_{N \geq 1} N^{-s} \omega_{N;k,\ell} \mathcal{N}_{k,\ell}$$
$$= L(s, \mathcal{N}_{k,\ell}) \mathcal{N}_{k,\ell}. \qquad (14.5)$$

Recall from Hecke's theory that the set $\{\omega_{N;k,\ell}\}_{N \geq 1}$ is characterized by its subset with N prime, since

$$\sum_{N \geq 1} N^{-s} T_N = \prod_{p \text{ prime}} (1 - p^{-s} T_p + p^{-2s})^{-1}. \qquad (14.6)$$

Denote as f_{ν_1, ν_2} the function introduced as f in (9.10), and rename $(f_0)_{\nu_1, \nu_2}$ the function f_0 introduced in (9.20), (9.23). Set

$$\Phi(\nu_1, \nu_2; \lambda) =$$
$$\frac{\zeta^*(\frac{1-i\lambda+\nu_1+\nu_2}{2}) \zeta^*(\frac{1-i\lambda-\nu_1-\nu_2}{2}) \zeta^*(\frac{1-i\lambda-\nu_1+\nu_2}{2}) \zeta^*(\frac{1-i\lambda+\nu_1-\nu_2}{2})}{\zeta^*(-i\lambda)} :$$
$$(14.7)$$

this is the function denoted as $\Phi(\lambda)$ in (9.71). Our first aim is to compute the coefficients

$$f_{\nu_1, \nu_2}^{k, \ell} : = (\mathcal{N}_{k,\ell}, f_{\nu_1, \nu_2}) \qquad (14.8)$$

which make the Roelcke–Selberg expansion

$$(f_{\nu_1, \nu_2})(z) = \sum_{k,\ell} f_{\nu_1, \nu_2}^{k, \ell} \|\mathcal{N}_{k,\ell}\|^{-2} \mathcal{N}_{k,\ell}(z) + \frac{1}{8\pi} \int_{-\infty}^{\infty} \Phi(\nu_1, \nu_2; \lambda) E_{\frac{1-i\lambda}{2}}(z) d\lambda \qquad (14.9)$$

valid when (ν_1, ν_2) belongs to the subset Ω_2 of $\mathbb{C}^2$ characterized by the inequalities Re $\nu_1 < -1$, Re $\nu_2 < -1$, $|\text{Re } (\nu_1 - \nu_2)| < 1$ and $\nu_1 \neq \nu_2$ (cf. theorem 9.6).

First note that (14.9) extends to the case when (ν_1, ν_2) lies in the larger set Ω_1 characterized by the inequalities Re $(\nu_1 + \nu_2) < -1$, $|\text{Re } (\nu_1 - \nu_2)| < 1$

together with $\nu_1 \neq -1, \nu_2 \neq -1, \nu_1 \neq \nu_2$. Indeed, a closer look at the argument just before (9.10) shows that f_{ν_1,ν_2} makes sense as an element of $L^2(D)$ if these inequalities are satisfied. On the other hand, $\Phi(\nu_1, \nu_2; \lambda)$ is well-defined for every real λ when Re $(\nu_1 \pm \nu_2) \neq \pm 1$ and the estimates already used in the proof of theorem 11.1 show that the second term on the right-hand side of (14.9) makes sense as a function of z in $L^2(D)$, depending holomorphically on (ν_1, ν_2). Extending (14.9) to our new case thus follows from analytic continuation. One may note, using an argument similar to that in the remark that concludes section 9, that one can dispense with any one of the three conditions $\nu_1 \neq -1$, $\nu_2 \neq -1$, $\nu_1 \neq \nu_2$, though not with the three of them simultaneously: we shall not concern ourselves, in what follows, with these improvements, except in proposition 14.6.

The coefficient $\Phi(\nu_1, \nu_2; \lambda)$ has an interesting structure: for any λ, it is the product of some function of $\nu_1 + \nu_2$ by the same function evaluated at $\nu_1 - \nu_2$. The same holds so far as the coefficients $f_{\nu_1,\nu_2}^{k,\ell}$ are concerned. Though superseded by theorem 14.5 (the proof of which is independent), the following proposition introduces a method which will be used later, in the proof of theorem 14.8.

PROPOSITION 14.1. *There exists a family* $(\alpha^{k,\ell})$ *of coefficients with the property that the Roelcke-Selberg coefficients* $f_{\nu_1,\nu_2}^{k,\ell}$ *characterized by* (14.8) *or* (14.9) *when* (ν_1, ν_2) *lies in the set* Ω_1 *just introduced can be written as*

$$f_{\nu_1,\nu_2}^{k,\ell} = \alpha^{k,\ell} \, L^*(\frac{1 - \nu_1 - \nu_2}{2}, \mathcal{N}_{k,\ell}) \, L^*(\frac{1 + \nu_1 - \nu_2}{2}, \mathcal{N}_{k,\ell}) \, . \tag{14.10}$$

PROOF. Since

$$\overline{(\mathcal{N}_{k,\ell}, E_s)} = \int_D \mathcal{N}_{k,\ell}(z) \, E_{\bar{s}}(z) \, d\mu(z) = 0 \tag{14.11}$$

for all $s \in \mathbb{C}$, $s \neq 1$, one can reduce (14.8) to

$$f_{\nu_1,\nu_2}^{k,\ell} = \int_D \overline{\mathcal{N}_{k,\ell}(z)} \, E_{\frac{1-\nu_1}{2}}^*(z) \, E_{\frac{1-\nu_2}{2}}^*(z) \, d\mu(z) \, , \tag{14.12}$$

which shows that $f_{\nu_1,\nu_2}^{k,\ell}$ extends as a holomorphic function of (ν_1, ν_2) when $\nu_1, \nu_2 \neq -1, 0, 1$.

In the case when (ν_1, ν_2) belongs to the set Ω_2 introduced right after (14.9), one may define also

$$(f_0)_{\nu_1,\nu_2}^{k,\ell} := (\mathcal{N}_{k,\ell}, (f_0)_{\nu_1,\nu_2}) \tag{14.13}$$

thanks to lemma 9.2. From (9.26) and the fact that Hecke operators are self-adjoint with respect to the Petersson scalar product, one may write

$$f^{k,\ell}_{\nu_1,\nu_2} = \sum_{N \geq 1} N^{\frac{\nu_1 + \nu_2 - 1}{2}} \, (T_N \mathcal{N}_{k,\ell}, (f_0)_{\nu_1,\nu_2})$$

$$= L(\frac{1 - \nu_1 - \nu_2}{2}, \mathcal{N}_{k,\ell}) \, (f_0)^{k,\ell}_{\nu_1,\nu_2}, \qquad (14.14)$$

as a consequence of (14.5).

On the other hand, it follows from theorem 11.3 that, when $(\nu_1, \nu_2) \in \Omega_2$, $(f_0)^{k,\ell}_{\nu_1,\nu_2}$ is the product of

$$\pi^{\frac{\nu_1 + \nu_2 - 1}{2}} \, \Gamma(\frac{1 + i\lambda_k - \nu_1 - \nu_2}{4}) \, \Gamma(\frac{1 - i\lambda_k - \nu_1 - \nu_2}{4})$$

by a function of $(\nu_1 - \nu_2; k, \ell)$ independent of $\nu_1 + \nu_2$. Since

$$\pi^{\frac{\nu_1 + \nu_2 - 1}{2}} \, \Gamma(\frac{1 + i\lambda_k - \nu_1 - \nu_2}{4}) \, \Gamma(\frac{1 - i\lambda_k - \nu_1 - \nu_2}{4}) \, L(\frac{1 - \nu_1 - \nu_2}{2}, \mathcal{N}_{k,\ell})$$

$$= L^*(\frac{1 - \nu_1 - \nu_2}{2}, \mathcal{N}_{k,\ell}) \quad (14.15)$$

according to (14.3), one sees that, when $(\nu_1, \nu_2) \in \Omega_2$, one may write

$$f^{k,\ell}_{\nu_1,\nu_2} = L^*(\frac{1 - \nu_1 - \nu_2}{2}, \mathcal{N}_{k,\ell}) \, C(\nu_1 - \nu_2; k, \ell) \qquad (14.16)$$

for some function $C(\nu_1 - \nu_2; k, \ell)$ well-defined and holomorphic as a function of $\nu_1 - \nu_2$ when $|\mathrm{Re}\,(\nu_1 - \nu_2)| < 1$. Finally, (14.12) shows that the extension of $f^{k,\ell}_{\nu_1,\nu_2}$ as a meromorphic function in $\mathbb{C}^2$ (which will turn out at once to be holomorphic) is invariant under the symmetry $(\nu_1, \nu_2) \mapsto (-\nu_1, \nu_2)$, which leads to theorem 14.1.   □

We need to generalize the results of sections 9 and 11 further.

PROPOSITION 14.2. *Let Ω be the subset of $\mathbb{C}^2$ characterized by*

$$(\nu_1, \nu_2) \in \Omega \iff \mathrm{Re}\,(\nu_1 \pm \nu_2) \neq \pm 1. \qquad (14.17)$$

For each $(\nu_1, \nu_2) \in \Omega$, let Σ be the set of pairs $(\varepsilon_1 = \pm 1, \varepsilon_2 = \pm 1)$ such that $\varepsilon_1 \mathrm{Re}\,\nu_1 + \varepsilon_2 \mathrm{Re}\,\nu_2 < 1$: note that Σ depends only on which of the nine components of Ω the point (ν_1, ν_2) lies in. Whenever $(\nu_1, \nu_2) \in \Omega$ and, moreover,

$$\nu_1 \neq -1, 0, 1; \qquad \nu_2 \neq -1, 0, 1; \qquad \nu_1 \pm \nu_2 \neq 0, \qquad (14.18)$$

set

$$f_{\nu_1,\nu_2} := E^*_{\frac{1-\nu_1}{2}} E^*_{\frac{1-\nu_2}{2}} - \sum_{(\varepsilon_1,\varepsilon_2) \in \Sigma} \zeta^*(1 - \varepsilon_1 \nu_1) \, \zeta^*(1 - \varepsilon_2 \nu_2) \, E_{1 - \frac{\varepsilon_1 \nu_1 + \varepsilon_2 \nu_2}{2}}$$

$$(14.19)$$

so that, in particular, our new definition of f_{ν_1,ν_2} coincides with the former one when $(\nu_1, \nu_2) \in \Omega_1$. Then $f_{\nu_1,\nu_2} \in L^2(D)$. Define

$$(f_{\nu_1,\nu_2}^{\mathrm{disc}})(z): = f_{\nu_1,\nu_2}(z) - \frac{1}{8\pi} \int_{-\infty}^{\infty} \Phi(\nu_1,\nu_2;\lambda)\, E_{\frac{1-i\lambda}{2}}(z)\, d\lambda \tag{14.20}$$

whenever $\mathrm{Re}\,(\nu_1 \pm \nu_2) \neq \pm 1$ *and* (14.18) *is satisfied as well: here, the definition of* $\Phi(\nu_1, \nu_2; \lambda)$ *is that given in* (14.7) *and does not depend on which of the components of* Ω *the pair* (ν_1, ν_2) *lies in. Then* $f_{\nu_1,\nu_2}^{\mathrm{disc}}$ *lies in* $L^2(D)$ *too. Moreover, for any fixed* $z \in \Pi$, $(f_{\nu_1,\nu_2}^{\mathrm{disc}})(z)$ *is holomorphic as a function of* (ν_1, ν_2) *in the (connected) subset of* $\mathbf{C}^2$ *characterized by*

$$\nu_1 \neq -1, 0, 1; \qquad \nu_2 \neq -1, 0, 1; \qquad \nu_1 \pm \nu_2 \neq -1, 0, 1. \tag{14.21}$$

PROOF. That f_{ν_1,ν_2} lies in $L^2(D)$ follows the same argument as that used just before (9.10). Next, from its definition (14.7), it follows that the function $\Phi(\nu_1, \nu_2; \lambda)$ is well–defined for every real λ when $(\nu_1, \nu_2) \in \Omega$ and the same estimates as those used in the proof of theorem 11.1 show that the function $z \mapsto \int_{-\infty}^{\infty} \Phi(\nu_1, \nu_2; \lambda)\, E_{\frac{1-i\lambda}{2}}(z)\, d\lambda$ lies in $L^2(D)$; also that, as a function of (ν_1, ν_2), it is holomorphic in Ω.

The origin of the discontinuities, as the pair (ν_1, ν_2) crosses one of the hyperplanes $\mathrm{Re}\,(\nu_1 \pm \nu_2) = \pm 1$, of the second (integral) term on the right–hand side of (14.9), lies in the fact that the function $\lambda \mapsto \Phi(\nu_1, \nu_2; \lambda)$ then gets poles on the real line, corresponding to the poles $0, 1$ of the function ζ^*. However, as is easily seen, no two factors occurring in the definition (14.7) of this function can be singular simultaneously, in view of the condition $\nu_1 \pm \nu_2 \neq -1, 0, 1$: since we may analyze the discontinuities one at a time, we may be satisfied with a discussion of what happens when one crosses one, and only one, of the hyperplanes. In view of the symmetries $\nu_1 \mapsto -\nu_1,\ \nu_2 \mapsto -\nu_2$, it finally suffices to discuss what happens across the hyperplane $\mathrm{Re}\,(\nu_1 + \nu_2) = -1$.

To prove the last point, it thus suffices to show that, as a function of (ν_1, ν_2), $(f_{\nu_1,\nu_2}^{\mathrm{disc}})(z)$ extends continuously across the hyperplane $\mathrm{Re}\,(\nu_1 + \nu_2) = -1$, away from any other hyperplane $\mathrm{Re}\,(\nu_1 \pm \nu_2) = \pm 1$. To that effect we introduce the new integrand

$$F(\nu_1, \nu_2; \lambda; z): = \Phi(\nu_1, \nu_2; \lambda)\, E_{\frac{1-i\lambda}{2}}(z)$$
$$+ \left[\frac{2i}{\lambda + i(1 + \nu_1 + \nu_2)} - \frac{2i}{\lambda - i(1 + \nu_1 + \nu_2)} \right] \zeta^*(1+\nu_1)\, \zeta^*(1+\nu_2)\, E_{1+\frac{\nu_1+\nu_2}{2}}(z) \tag{14.22}$$

and observe that, as a function of (ν_1, ν_2), it remains holomorphic in the region under consideration. Indeed, on one hand,

$$\zeta^*(\frac{1 - i\lambda + \nu_1 + \nu_2}{2}) \zeta^*(\frac{1 - i\lambda - \nu_1 - \nu_2}{2})$$
$$+ \left[\frac{2i}{\lambda + i(1 + \nu_1 + \nu_2)} - \frac{2i}{\lambda - i(1 + \nu_1 + \nu_2)}\right] \zeta^*(1 + \nu_1 + \nu_2)$$

is a continuous function of $(\nu_1 + \nu_2, \lambda)$ when Re $(\nu_1 + \nu_2) < 1$, $\lambda \in \mathbb{R}$ and $\nu_1 + \nu_2 \neq 0, -1$, holomorphic with respect to $\nu_1 + \nu_2$. On the other hand, the function

$$\frac{\zeta^*(\frac{1-i\lambda-\nu_1+\nu_2}{2}) \zeta^*(\frac{1-i\lambda+\nu_1-\nu_2}{2})}{\zeta^*(-i\lambda)} E_{\frac{1-i\lambda}{2}}(z)$$

coincides with

$$\frac{\zeta^*(-\nu_1) \zeta^*(-\nu_2)}{\zeta^*(-1 - \nu_1 - \nu_2)} E_{-\frac{\nu_1+\nu_2}{2}}(z) = \frac{\zeta^*(1 + \nu_1) \zeta^*(1 + \nu_2)}{\zeta^*(1 + \nu_1 + \nu_2)} E_{1 + \frac{\nu_1 + \nu_2}{2}}(z)$$

when either $i\lambda = 1 + \nu_1 + \nu_2$ or $i\lambda = -1 - \nu_1 - \nu_2$, where the equality of the two sides arises from the functional equation of the zeta function together with that of Eisenstein series.

Thus the function

$$z \mapsto \int_{-\infty}^{\infty} F(\nu_1, \nu_2; \lambda; z) \, d\lambda, \tag{14.24}$$

for any fixed z, is a holomorphic function of (ν_1, ν_2) when Re $(\nu_1 + \nu_2) < 1$, $\nu_1, \nu_2 \neq -1, 0, 1$, $\nu_1 \pm \nu_2 \neq 0$, $\nu_1 + \nu_2 \neq -1$. Now one has

$$\frac{1}{8\pi} \int_{-\infty}^{\infty} \left[\frac{2i}{\lambda + i(1 + \nu_1 + \nu_2)} - \frac{2i}{\lambda - i(1 + \nu_1 + \nu_2)}\right] d\lambda =$$
$$\begin{cases} \frac{1}{2} & \text{if Re } (\nu_1 + \nu_2) > -1 \\ -\frac{1}{2} & \text{if Re } (\nu_1 + \nu_2) < -1 \end{cases} \tag{14.25}$$

so that

$$\frac{1}{8\pi} \int_{-\infty}^{\infty} \left(\Phi(\nu_1, \nu_2; \lambda) E_{\frac{1-i\lambda}{2}}(z) - F(\nu_1, \nu_2; \lambda; z)\right) d\lambda$$
$$= \pm \frac{1}{2} \zeta^*(1 + \nu_1) \zeta^*(1 + \nu_2) E_{1 + \frac{\nu_1 + \nu_2}{2}}(z) \tag{14.26}$$

where the sign is $+$ or $-$ according to whether Re $(\nu_1 + \nu_2) < -1$ or Re $(\nu_1 + \nu_2) > -1$.

From (14.26) and (14.19), one sees that, when moving from Re $(\nu_1 + \nu_2) < -1$, $|\text{Re } (\nu_1 - \nu_2)| < 1$ to $-1 < \text{Re } (\nu_1 + \nu_2) < -1$, $|\text{Re } (\nu_1 - \nu_2)| < 1$, the jumps of the two terms on the right–hand side of (14.20) exactly cancel out. This proves the last assertion in proposition 14.2.

$\square$

Remark. The point of proposition 14.2 is to show that, contrary to f_{ν_1,ν_2}, the function $f_{\nu_1,\nu_2}^{\text{disc}}$ does not suffer from the discontinuities implied by the definition (14.19) of f_{ν_1,ν_2}. Justifying the terminology, i.e., showing that $f_{\nu_1,\nu_2}^{\text{disc}}$ is, indeed, the discrete part of the Roelcke–Selberg decomposition of f_{ν_1,ν_2}, does not seem to follow easily from considerations of analytic continuation (it does not seem to be be possible, with this kind of arguments, to cross the hyperplanes $\text{Re}\,(\nu_1 \pm \nu_2) = \pm 1$) and will be deferred somewhat.

PROPOSITION 14.3. *Assume that* $\text{Re}\,(\nu_1 + \nu_2) < -1$, $\text{Re}\,(\nu_1 - \nu_2) < -1$, *and that the conditions* (14.18) *are satisfied too, in which case* $f := f_{\nu_1,\nu_2}$ *is given as*

$$(f_{\nu_1,\nu_2})(z) = E^*_{\frac{1-\nu_1}{2}} E^*_{\frac{1-\nu_2}{2}} - \zeta^*(1-\nu_1)\zeta^*(1-\nu_2) E_{1-\frac{\nu_1+\nu_2}{2}}$$

$$- \zeta^*(1-\nu_1)\zeta^*(1+\nu_2) E_{1+\frac{\nu_2-\nu_1}{2}} \quad (14.27)$$

according to proposition 14.2. *Then theorem* 9.6 *extends.*

PROOF. First assume, with $\ell_1 = \text{Re}\,\nu_1$, $\ell_2 = \text{Re}\,\nu_2$, that $\ell_2 < -1$ and $\ell_1 - \ell_2 > -2$ (together with $\ell_1 - \ell_2 < -1$). Set

$$f_0(z) = (f_0)_{\text{main}}(z) - \pi^{\frac{\nu_1+\nu_2-1}{2}} \Gamma(\frac{1-\nu_1}{2}) \Gamma(-\frac{\nu_2}{2}) E_{1+\frac{\nu_2-\nu_1}{2}}(z), \quad (14.28)$$

with $(f_0)_{\text{main}}$ defined just as in (9.20). Still defining f_{main} by (9.15), we can extend (9.22) and (9.26) (which express that f_{main} is the image of $(f_0)_{\text{main}}$, and f that of f_0, under the operator $\sum_{N\geq 1} N^{\frac{\nu_1+\nu_2-1}{2}} T_N$) in a straightforward way. Lemmas 9.2 and 9.3 extend, with the modification that the second term on the right–hand side of (9.28), not compensated any longer by any extra term in the new definition of $f_0(z)$, must be taken care of on its own, writing

$$\sum_{n_1 \in \mathbb{Z}} y^{\frac{1-\ell_2}{2}} (\frac{|z-n_1|^2}{y})^{\frac{\ell_1-1}{2}} = y^{\frac{\ell_1-\ell_2}{2}} \sum_{n_1 \in \mathbb{Z}} (1+\frac{(x-n_1)^2}{y^2})^{\frac{\ell_1-1}{2}}$$

$$\leq C\, y^{1+\frac{\ell_1-\ell_2}{2}}. \quad (14.29)$$

The proof of theorem 9.5 then generalizes, noting that in our new case $\ell := \max(\ell_1, \ell_2) = \ell_2$ and that $-1 - \ell_1 - \ell_2 < 1 - 2\ell$ since $\ell_1 - \ell_2 > -2$, so that (9.63) still defines a non–void strip. Theorem 9.6 follows from theorem 9.5 under the conditions stated at the beginning of the present proof, and analytic continuation to the case covered by the statement of the theorem follows, since it does not involve crossing any of the hyperplanes $\text{Re}\,(\nu_1 \pm \nu_2) = \pm 1$. $\square$

PROPOSITION 14.4. *Assume* $-\frac{3}{2} < \ell_2 = \text{Re}\,\nu_2 < -1$, $\ell_2 - 2 < \ell_1 = \text{Re}\,\nu_1 < \ell_2 - 1$, *and that the conditions* (14.18) *are satisfied. A priori, the* *function $b_n(\mu)$ defined in (11.21) is holomorphic for $1 - \ell_1 + \ell_2 < \operatorname{Im} \mu < 1 - \ell_1 - \ell_2$. It extends as a meromorphic function in the strip*

$$-2 - \ell_1 + \ell_2 < \operatorname{Im} \mu < 1 - \ell_1 - \ell_2, \qquad (14.30)$$

holomorphic except at the points $\mu_1 = i(1 - \nu_1 + \nu_2)$, $\mu_2 = i(-1 - \nu_1 + \nu_2)$, at the points $-i\omega$ with $\zeta^*(\omega) = 0$ and at all points λ_k, where $\frac{1+\lambda_k^2}{4}$ is an even eigenvalue of Δ. With d_n defined by (11.22), the formula (11.23) then extends.

The function $\zeta_n(s,t)$ extends as a meromorphic function for $\operatorname{Re} t > -\frac{1}{2}$, $1 < \operatorname{Re}(s-t) < 2$, $t \neq 1$, $s + t \neq 1$, holomorphic outside the set of points (s,t) with $s + t = 1 - i\lambda_k, \frac{1+\lambda_k^2}{4}$ in the even part of the spectrum of Δ, or $s + t = \omega$, a non–trivial zero of the zeta function..

PROOF. The first point comes from a simple examination of the Gamma factors in the definition of $b_n(\mu)$, together with the fact that $\zeta_n(s,t)$ is defined by a convergent series when $\operatorname{Re} s > 1$, $\operatorname{Re} t > 1$. The analysis between (11.30) and (11.32) shows that, in our new case, the only singularities of $(\tilde{c}_n)^+_{\min}(\mu)$ within the half–plane $\operatorname{Im} \mu > -2 - \ell_1 + \ell_2$ are μ_1 and μ_2: note that the first one corresponds to a singularity of $\zeta_n(s,t)$, in the domain under consideration, when $t = 1$, whereas the second one simply appears as a pole in the second of the extra Gamma factors apparent on the right-hand side of (11.21).

Following the same analysis as in the proof of theorem 11.3, what has to be shown for the main part is that $(\tilde{c}_n)^+_{\text{maj}}(\mu)$, defined by (11.37) in the strip $-1 - \ell_1 - \ell_2 < \operatorname{Im} \mu < 1 - 2\ell_2$, differs from $b_n(\mu)$, holomorphic in the same strip, by a function that can be analytically continued to the strip

$$-2 - \ell_1 + \ell_2 < \operatorname{Im} \mu < 1 - 2\ell_2. \qquad (14.31)$$

The major novelty, from our previous situation, arises from the consideration of the second term $C \sigma^{-1}$ in (11.40), which contributes to $(\tilde{c}_n)^+_{\text{maj}}(\mu)$ the product of $\zeta_n^-(\frac{-\nu_1 + \nu_2 - i\mu + 3}{2}, \frac{\nu_1 - \nu_2 - i\mu + 3}{2})$ by some coefficient holomorphic when $\operatorname{Im} \mu < 1 - 2\ell_2$. Now, it follows from theorem 12.3 that this function analytically extends when $\operatorname{Im} \mu > \max(-5 - \ell_1 + \ell_2, -2)$, which covers the case defined by (14.30).

The proof of the last assertion follows the same lines as that of corollary 11.4.

□

THEOREM 14.5. *The coefficients which enter the statement of proposition 14.1 are all given as $\alpha^{k,\ell} = \frac{1}{2}$, i.e.,*

$$f^{k,\ell}_{\nu_1,\nu_2} := \int_D \overline{\mathcal{N}_{k,\ell}(z)} \, E^*_{\frac{1-\nu_1}{2}}(z) \, E^*_{\frac{1-\nu_2}{2}}(z) \, d\mu(z)$$

$$= \frac{1}{2} L^*(\frac{1 - \nu_1 - \nu_2}{2}, \mathcal{N}_{k,\ell}) \, L^*(\frac{1 + \nu_1 - \nu_2}{2}, \mathcal{N}_{k,\ell}) .$$

$$(14.32)$$

PROOF. We may assume that $\text{Re}\,(\nu_1 + \nu_2) < -1$, $\text{Re}\,(\nu_1 - \nu_2) < -1$ as in the statement of proposition 14.3 (and that of 14.4). Using (3.29) and (14.1), we get the constant term of $\overline{N}_{k,\ell}(z)\, E^*_{\frac{1-\nu_2}{2}}(z)$ as

$$a_0(y) = 2\,y \sum_{n \neq 0} |n|^{-\frac{\nu_2}{2}}\, \sigma_{\nu_2}(|n|)\, \omega_{|n|;k,\ell}\, K_{\frac{i\lambda_k}{2}}\,(2\pi|n|y)\, K_{\frac{\nu_2}{2}}\,(2\pi|n|y) \tag{14.33}$$

so that, by theorem 7.3 or, in this case, by the usual Rankin–Selberg "unfolding" method (cf. remark right after theorem 7.3), one has

$$\int_D \overline{N}_{k,\ell}(z)\, E^*_{\frac{1-\nu_2}{2}}(z)\, E^*_{\frac{1-i\lambda}{2}}(z)\, d\mu(z) = \zeta^*(1 - i\lambda)\, \Phi(-\lambda), \tag{14.34}$$

where $\Phi(-\lambda)$ is the value, continued at $\mu = \lambda$ from above, of

$$\int_0^\infty a_0(y)\, y^{-\frac{3}{2} - \frac{i\mu}{2}}\, dy.$$

Using the Weber–Schafheitlin integral (cf. [27], p.101) and making the substitution $i\mu \mapsto \nu_1$, we get

$$f^{k,\ell}_{\nu_1,\nu_2} = \frac{1}{2}\, \pi^{\frac{-1+\nu_1}{2}}\, (\Gamma(\frac{1-\nu_1}{2}))^{-1}\, \zeta^*(1-\nu_1)$$
$$\Gamma(\frac{1-\nu_1+\nu_2+i\lambda_k}{4})\,\Gamma(\frac{1-\nu_1+\nu_2-i\lambda_k}{4})\,\Gamma(\frac{1-\nu_1-\nu_2+i\lambda_k}{4})\,\Gamma(\frac{1-\nu_1-\nu_2-i\lambda_k}{4})$$
$$\sum_{n \geq 1} n^{\frac{-1+\nu_1-\nu_2}{2}}\, \sigma_{\nu_2}(n)\, \omega_{|n|;k,\ell}. \tag{14.35}$$

Now

$$\sum_{n\geq 1} n^{\frac{-1+\nu_1-\nu_2}{2}} \sigma_{\nu_2}(n)\, T_n$$

$$= \prod_{p \text{ prime}} \sum_{j\geq 0} p^{j(\frac{-1+\nu_1-\nu_2}{2})}(1+p^{\nu_2}+\cdots+p^{j\nu_2})\, T_{p^j}$$

$$= \prod_p \sum_{j\geq 0} p^{j(\frac{-1+\nu_1-\nu_2}{2})} \frac{1-p^{(j+1)\nu_2}}{1-p^{\nu_2}}\, T_{p^j}$$

$$= \zeta(-\nu_2) \prod_p \sum_{j\geq 0} [\, p^{j(\frac{-1+\nu_1-\nu_2}{2})} - p^{\nu_2+j(\frac{-1+\nu_1+\nu_2}{2})}\,]\, T_{p^j}$$

$$= \zeta(-\nu_2) \prod_p [\,(1-T_p\, p^{\frac{-1+\nu_1-\nu_2}{2}} + p^{-1+\nu_1-\nu_2})^{-1}$$

$$\qquad\qquad - p^{\nu_2}\,(1-T_p\, p^{\frac{-1+\nu_1+\nu_2}{2}} + p^{-1+\nu_1+\nu_2})^{-1}\,]$$

$$= \zeta(-\nu_2) \prod_p (1-T_p\, p^{\frac{-1+\nu_1-\nu_2}{2}} + p^{-1+\nu_1-\nu_2})^{-1}\ \times$$

$$\prod_p (1-T_p\, p^{\frac{-1+\nu_1+\nu_2}{2}} + p^{-1+\nu_1+\nu_2})^{-1}\ \times$$

$$\prod_p (1+p^{-1+\nu_1+\nu_2} - p^{\nu_2} - p^{-1+\nu_1}) \tag{14.36}$$

so that, by (14.4) and (14.6),

$$\sum_{n\geq 1} n^{\frac{-1+\nu_1-\nu_2}{2}} \sigma_{\nu_2}(n)\, \omega_{|n|;k,\ell}$$

$$= \zeta(-\nu_2)\, L(\frac{1-\nu_1+\nu_2}{2}, \mathcal{N}_{k,\ell})\, L(\frac{1-\nu_1-\nu_2}{2}, \mathcal{N}_{k,\ell}) \prod_p (1-p^{-1+\nu_1})(1-p^{\nu_2})$$

$$= (\zeta(1-\nu_1))^{-1}\, L(\frac{1-\nu_1+\nu_2}{2}, \mathcal{N}_{k,\ell})\, L(\frac{1-\nu_1-\nu_2}{2}, \mathcal{N}_{k,\ell})\,, \tag{14.37}$$

which permits to conclude the proof of theorem 14.5, starting from (14.35). $\square$

PROPOSITION 14.6. *Let $\nu \in \mathbb{C}$ satisfy $|\mathrm{Re}\,\nu| < 1$ so that, by corollary 11.4, the function*

$$\mu \mapsto \zeta_n(\frac{1-i\mu-\nu}{2}, \frac{1-i\mu+\nu}{2}) \tag{14.38}$$

extends as a meromorphic function for $\mathrm{Im}\ \mu > |\mathrm{Re}\ \nu| - 1$. *Its residue at* $\mu = \lambda_k$ *is given as*

$$\text{residue of } \zeta_n\left(\frac{1 - i\mu - \nu}{2}, \frac{1 - i\mu + \nu}{2}\right) \quad \text{at } \mu = \lambda_k$$

$$= i\pi^{-\frac{1+i\lambda_k}{2}} |4n|^{\frac{i\lambda_k}{2}} \frac{\Gamma(-i\lambda_k)}{\Gamma(\frac{1-i\lambda_k+\nu}{4})\Gamma(\frac{1-i\lambda_k-\nu}{4})} \times$$

$$\sum_\ell b_{n;k,\ell}\, L^*\left(\frac{1+\nu}{2}, \mathcal{M}_{k,\ell}\right) \quad (14.39)$$

(recall that the coefficients $b_{n;k,\ell}$ are the Fourier coefficients b_n in the expansion (14.1) of the $L^2(D)$-normalized function $\mathcal{M}_{k,\ell}$). In the case when $-2 < \mathrm{Re}\ \nu < -1$, the function $\zeta_n(\frac{1-i\mu-\nu}{2}, \frac{1-i\mu+\nu}{2})$ extends as a meromorphic function for $\mathrm{Im}\ \mu > -2 - \mathrm{Re}\ \nu$ by proposition 14.4, and the same formula holds.

PROOF. Set $\nu = \nu_1 - \nu_2$. In the case when (ν_1, ν_2) belongs to the set Ω_2 introduced right after (14.9) (this assumption can be dispensed with by analytic continuation), (14.14) and theorem 14.5 yield

$$(f_0)^{k,\ell}_{\nu_1,\nu_2} = \frac{1}{2} \frac{L^*(\frac{1-\nu_1-\nu_2}{2}, \mathcal{N}_{k,\ell})}{L(\frac{1-\nu_1-\nu_2}{2}, \mathcal{N}_{k,\ell})} L^*\left(\frac{1+\nu_1-\nu_2}{2}, \mathcal{N}_{k,\ell}\right). \tag{14.40}$$

On one hand, the projection of f_0 on the even eigenspace of Δ for the eigenvalue $\frac{1+\lambda_k^2}{4}$ is given as

$$(\mathrm{Pr}_{\lambda_k} f_0)(z) = \sum_\ell (f_0)^{k,\ell}_{\nu_1,\nu_2} \|\mathcal{N}_{k,\ell}\|^{-2} \mathcal{N}_{k,\ell}(z). \tag{14.41}$$

On the other hand, using theorem 11.3, one has in the case when $|\mathrm{Re}\ \nu| < 1$

$$(\mathrm{Pr}_{\lambda_k} f_0)(z) = y^{\frac{1}{2}} \sum_{n \neq 0} d_n\, K_{\frac{i\lambda_k}{2}}(2\pi|n|y)\, e^{2i\pi nx}, \tag{14.42}$$

with

$$d_n = -8i\pi\, |n|^{-\frac{i\lambda_k}{2}} \times \text{residue of } b_n(\mu) \text{ at } \mu = \lambda_k, \tag{14.43}$$

where $b_n(\mu)$ has been defined in (11.21). Comparing (14.41) and (14.42), we are done. Trading theorem 11.3 for proposition 14.4, we get the same result when $-2 < \mathrm{Re}\ \nu < -1$.

The careful reader may have noticed that the special case when $\nu = 0$ has not been excluded from the statement of proposition 14.6: this would correspond to $\nu_1 = \nu_2$ in theorem 11.3. But then, we noted in the remark preceding corollary 11.4 that, using the trick expounded at the end of section 9, one could indeed extend theorem 11.3 so as to include the case $\nu_1 = \nu_2$. $\square$

Remark. With $s = \frac{1-i\mu-\nu}{2}$, $t = \frac{1-i\mu+\nu}{2}$, the two conditions Re $(s+t) >$ $\frac{3}{2}$, Re $t < 0$ are compatible only if Re $\nu < -\frac{3}{2}$, in which case the analytic continuation of $\zeta_n(s,t)$ can be related to that of Kloosterman–Selberg series by theorem 10.5: this does not hold in the first case considered in proposition 14.6, and arguments based on analytic continuation do not permit to cross the line Re $\nu = -1$.

Some of our readers may wish to check that the residue computed in (14.39) agrees with the one that could be derived from known results relative to the Kloosterman–Selberg series themselves. We briefly show that this is the case, starting from the exposition by Iwaniec (cf. [20]). With

$$KS_s(j,n): = \sum_{m \geq 1} m^{-2s} S(j,n;m), \qquad (14.44)$$

Iwaniec introduces ([20], (9.1) and (9.2)) a certain series $Z_s(j,n)$ which satisfies the property that $KS_s(j,n)$ has the same poles and polar parts, on the line Re $s = \frac{1}{2}$, as the function

$$2^{2-2s}\, \pi^{1-2s}\, j^{1-s}\, |n|^{1-s}\, \Gamma(2s)\, Z_s(j,n).$$

His theorem 9.2 states that $Z_s(j,n)$ has a simple pole at $s = s_k: = \frac{1-i\lambda_k}{2}$ with residue

$$\frac{1}{2\,s_k - 1} \sum_\ell \bar\rho_{k,\ell}(j)\, \rho_{k,\ell}(n),$$

where the coefficients $2\,\rho_{k,\ell}(n)$ (k fixed) are the Fourier coefficients b_n (cf. (14.1)) relative to some orthonormal basis (of course we choose $\{\mathcal{M}_{k,\ell}\}$ as such) of the eigenspace of Δ for the eigenvalue $\frac{1+\lambda_k^2}{4}$. Actually, we made a global change of sign in this quotation: for the use of the resolvant equation right before theorem 7.5 in [20] indeed demands that the right inverse $R_s = (-\Delta + s(1-s))^{-1}$ (with our Δ) be used, and (cf. [20], (5.4)) the integral kernel of R_s is $-G_s(z/z')$, not $G_s(z/z')$.

With our notations, we get

$$\rho_{k,\ell}(n) = \frac{1}{2}\, |n|^{-\frac{1}{2}}\, b_{n;k,\ell} \,, \qquad (14.45)$$

so that, after some easy computations, the residue of the function $\mu \mapsto KS_{\frac{1-i\mu}{2}}(j,n)$ at $\mu = \lambda_k$ is given as

$$i\,(2\pi)^{i\lambda_k}\, \Gamma(-i\lambda_k)\, |jn|^{\frac{i\lambda_k}{2}} \sum_\ell b_{n;k,\ell}\, b_{j;k,\ell} \,. \qquad (14.46)$$

Now, theorem 10.5 can be rephrased, in the case when Im $\mu > \frac{1}{2}$ and $1+$ Im $\mu + $ Re $\nu < 0$, using $\zeta_n = \zeta_{-n}$ together with $S(j,-n;m) = S(-j,n;m)$, as

$$\zeta_n\left(\frac{1-i\mu-\nu}{2},\frac{1-i\mu+\nu}{2}\right) = \frac{1}{2}\,\pi^{\frac{\nu-i\mu}{2}}\frac{\Gamma\left(\frac{1+i\mu-\nu}{4}\right)}{\Gamma\left(\frac{1+i\mu+\nu}{4}\right)}$$

$$\sum_{j\geq 1} j^{\frac{-1-i\mu+\nu}{2}}\left[KS_{\frac{1-i\mu}{2}}(j,n) + KS_{\frac{1-i\mu}{2}}(-j,n)\right].\quad(14.47)$$

Using the expression (14.46) of the residues of $KS_{\frac{1-i\mu}{2}}(\pm j,n)$ at $\mu = \lambda_k$ and

$$\sum_{j\geq 1} j^{\frac{\nu-1}{2}}\, b_{j;k,\ell} = L\left(\frac{1-\nu}{2},\mathcal{M}_{k,\ell}\right)$$

$$= \pi^{\frac{1-\nu}{2}}\left[\Gamma\left(\frac{1+i\lambda_k-\nu}{4}\right)\Gamma\left(\frac{1-i\lambda_k-\nu}{4}\right)\right]^{-1} L^*\left(\frac{1+\nu}{2},\mathcal{M}_{k,\ell}\right),$$
$$(14.48)$$

we would just find (14.39) as a consequence, provided we checked (we have not done so, but there is a similar argument at the end of the proof of theorem 15.6) that series of residues at $\mu = \lambda_k$, in our present circumstances, can be performed.

PROPOSITION 14.7. *Whenever (ν_1,ν_2) belongs to the set Ω introduced in proposition 14.2 and (14.18) is satisfied, the function $f_{\nu_1,\nu_2}^{\mathrm{disc}}$, as defined in (14.20), is the discrete part, with respect to the Roelcke–Selberg decomposition, of f_{ν_1,ν_2}.*

PROOF. When $|\mathrm{Re}\,(\nu_1-\nu_2)| < 1$ and $\mathrm{Re}\,(\nu_1+\nu_2) < -1$, this is a consequence of theorem 9.6 and analytic continuation; when $\mathrm{Re}\,(\nu_1-\nu_2) < -1$ and $\mathrm{Re}\,(\nu_1+\nu_2) < -1$, this is proposition 14.3. Using also the symmetries $(\nu_1,\nu_2) \mapsto (\pm\nu_1,\pm\nu_2)$ implied by the functional equation of Eisenstein series and that of the zeta function, one can see that all cases (Ω has nine components) are covered, except the "central" one, when $|\mathrm{Re}\,(\nu_1 \pm \nu_2)| < 1$. In this last case, one cannot use the original definition (3.14) of Eisenstein series, and one must work instead with their Fourier expansions (3.29); also, it will be just as well (even if less illuminating) to work with f in a direct way, not through a preliminary study of f_0.

Under our new conditions, $f: = f_{\nu_1,\nu_2}$ is defined (proposition 14.2) as

$$f(z) = E^*_{\frac{1-\nu_1}{2}}(z)\,E^*_{\frac{1-\nu_2}{2}}(z) - \sum_{\varepsilon_1^2=\varepsilon_2^2=1} \zeta^*(1-\varepsilon_1\nu_1)\,\zeta^*(1-\varepsilon_2\nu_2)\,E_{1-\frac{\varepsilon_1\nu_1+\varepsilon_2\nu_2}{2}}(z)$$

$$(14.49)$$

so that, using (3.29), one sees that the "constant term" $a_0(y)$ in the Fourier expansion of f may be written, just like in (9.43), as

$$a_0(y) = (a_0)_{\mathrm{maj}}(y) + (a_0)_{\mathrm{min}}(y),\quad(14.50)$$

where the second term on the right–hand side is a linear combination, with coefficients expressed in terms of the zeta function, of powers of y, and

$$(a_0)_{\mathrm{maj}}(y) = 4 \sum_{n \neq 0} |n|^{-\frac{\nu_1+\nu_2}{2}} y \, \sigma_{\nu_1}(|n|) \, \sigma_{\nu_2}(|n|) \, K_{\frac{\nu_1}{2}}(2\pi|n|y) \, K_{\frac{\nu_2}{2}}(2\pi|n|y) \,.$$
$$(14.51)$$

Just like in (9.51), one computes $(C_0)^{\pm}_{\mathrm{min}}(\mu)$, and one finds that this expression is analytic except at $\mu = \pm i \, (1 - \varepsilon_1 \nu_1 - \varepsilon_2 \nu_2)$, one point of a pair disjoint from $\mathbb{R}$; moreover, the analytic continuations of $(C_0)^{+}_{\mathrm{min}}(\mu)$ and $(C_0)^{-}_{\mathrm{min}}(\mu)$ agree on the real line. Then, from theorem 7.3, $C_0(\lambda)$ is the value at $\mu = \lambda$, computed from above, of

$$\frac{1}{8\pi} \int_0^{\infty} \frac{\pi^{-\frac{i\mu}{2}}}{\Gamma(-\frac{i\mu}{2})} y^{-\frac{3}{2} - \frac{i\mu}{2}} \, dy \times$$

$$8 \sum_{n \geq 1} n^{-\frac{\nu_1+\nu_2}{2}} \sigma_{\nu_1}(n) \, \sigma_{\nu_2}(n) \, y \, K_{\frac{\nu_1}{2}}(2\pi ny) \, K_{\frac{\nu_2}{2}}(2\pi ny)$$

$$= \frac{1}{8} \frac{\pi^{-\frac{3}{2}}}{\Gamma(-\frac{i\mu}{2}) \Gamma(\frac{1-i\mu}{2})} \Gamma(\tfrac{1-i\mu+\nu_1+\nu_2}{4}) \Gamma(\tfrac{1-i\mu+\nu_1-\nu_2}{4}) \Gamma(\tfrac{1-i\mu-\nu_1+\nu_2}{4}) \Gamma(\tfrac{1-i\mu-\nu_1-\nu_2}{4})$$

$$\sum_{n \geq 1} n^{\frac{-1+i\mu-\nu_1-\nu_2}{2}} \sigma_{\nu_1}(n) \, \sigma_{\nu_2}(n) \,, \qquad\qquad (14.52)$$

where we have assumed $\mathrm{Im}\,\mu > 1 + |\mathrm{Re}\,\nu_1| + |\mathrm{Re}\,\nu_2|$. Working with Eulerian products (as is classically done) just like in (14.36), one gets

$$\sum_{n \geq 1} n^{\frac{-1+i\mu-\nu_1-\nu_2}{2}} \sigma_{\nu_1}(n) \, \sigma_{\nu_2}(n) = \zeta(1 - i\mu)^{-1} \times$$

$$\zeta(\frac{1 - i\mu + \nu_1 + \nu_2}{2}) \zeta(\frac{1 - i\mu + \nu_1 - \nu_2}{2}) \zeta(\frac{1 - i\mu - \nu_1 + \nu_2}{2}) \zeta(\frac{1 - i\mu - \nu_1 - \nu_2}{2})$$
$$(14.$$

a formula which goes back to Ramanujan, as indicated in ([37], p.163 or [21], p.232). It follows, using (7.24), that the continuous part of f, under the Roelcke–Selberg decomposition, is again given by the same formula as in theorem 9.6. □

We now turn to the study of the Roelcke–Selberg decomposition, coupled with Hecke's theory, of h or h_0 (as introduced in (12.1) or (12.3)). This is simpler than the corresponding problem relative to f and f_0, discussed so far in the present section, since no continuous part is present; also, the sole condition $\nu_1 \neq -1, 0, 1, \nu_2 \neq -1, 0, 1$ ensures that h, or h_0, is well-defined and lies in $L^2(\Gamma \backslash \Pi)$, so that no crossing of hyperplanes is involved in the discussion.

If $\mathcal{M}^-$ is an odd Hecke form for the eigenvalue $\frac{1+(\lambda^-)^2}{4}$ with the Fourier expansion

$$\mathcal{M}^-(x+iy) = y^{\frac{1}{2}} \sum_{n \neq 0} b_n^- K_{\frac{i\lambda^-}{2}}(2\pi|n|y) \, e^{2i\pi nx} , \qquad (14.54)$$

one still sets

$$L(s, \mathcal{M}^-) = \sum_{n \geq 1} b_n^- \, n^{-s} \qquad (14.55)$$

but

$$L^*(s, \mathcal{M}^-) = \pi^{-s} \Gamma(\frac{s+1}{2} + \frac{i\lambda^-}{4}) \Gamma(\frac{s+1}{2} - \frac{i\lambda^-}{4}) L(s, \mathcal{M}^-), \qquad (14.56)$$

so as to get

$$L^*(s, \mathcal{M}^-) = -L^*(1-s, \mathcal{M}^-) \qquad (14.57)$$

(*cf.* [3], p.107, noting that Bump's ϵ should be 1).

If $\mathcal{M}^- = \mathcal{N}^-_{k,\ell}$, set

$$b_n^- = (\text{sign } n) \, \omega^-_{|n|;k,\ell} \qquad (14.58)$$

so that

$$T_N \mathcal{N}^-_{k,\ell} = \omega^-_{N;k,\ell} \mathcal{N}^-_{k,\ell}, \qquad T_N \mathcal{M}^-_{k,\ell} = \omega^-_{N;k,\ell} \mathcal{M}^-_{k,\ell} \qquad (14.59)$$

for all $N \geq 1$ in view of the normalization chosen. Under the assumptions of theorem 12.2,

$$(h_0)^{k,\ell}_{\nu_1,\nu_2} := (\mathcal{N}^-_{k,\ell}, (h_0)_{\nu_1,\nu_2}) \qquad (14.60)$$

is

$$\pi^{\frac{\nu_1+\nu_2-1}{2}} \Gamma(\frac{3-\nu_1-\nu_2+i\lambda^-_k}{4}) \Gamma(\frac{3-\nu_1-\nu_2-i\lambda^-_k}{4})$$

times a function of $(\nu_1 - \nu_2; k, \ell)$, and

$$h^{k,\ell}_{\nu_1,\nu_2} = (\mathcal{N}^-_{k,\ell}, h_{\nu_1,\nu_2})$$
$$= \sum_{N \geq 1} N^{\frac{\nu_1+\nu_2-1}{2}} (T_N \mathcal{N}^-_{k,\ell}, (h_0)_{\nu_1,\nu_2})$$
$$= L(\frac{1-\nu_1-\nu_2}{2}, \mathcal{N}^-_{k,\ell})(h_0)^{k,\ell}_{\nu_1,\nu_2}, \qquad (14.61)$$

so that, just like in proposition 14.1, one gets

$$h^{k,\ell}_{\nu_1,\nu_2} = \beta^{k,\ell} L^*(\frac{1-\nu_1-\nu_2}{2}, \mathcal{N}^-_{k,\ell}) L^*(\frac{1+\nu_1-\nu_2}{2}, \mathcal{N}^-_{k,\ell}) \qquad (14.62)$$

for some family $(\beta^{k,\ell})$ which we now proceed to compute.

THEOREM 14.8.

$$h_{\nu_1,\nu_2}^{k,\ell} := \int_D \overline{\mathcal{N}_{k,\ell}^-}(z) \{E_{\frac{1-\nu_1}{2}}^*, E_{\frac{1-\nu_2}{2}}^*\}(z)\, d\mu(z)$$

$$= \frac{1}{i}\, L^*(\frac{1-\nu_1-\nu_2}{2}, \mathcal{N}_{k,\ell}^-)\, L^*(\frac{1+\nu_1-\nu_2}{2}, \mathcal{N}_{k,\ell}^-).$$

$$(14.63)$$

PROOF. By an argument of analytic continuation, one may assume that $\mathrm{Re}\,(\nu_2 - \nu_1) > \frac{3}{2}$, as will be needed later. It saves computations to use theorem 12.2 and Iwaniec's results [20] instead of Eulerian products, in the way developed in the alternative proof of (14.39) (cf. remark just after (14.43)). As a consequence of (14.61), (14.62) and (14.56), one has

$$(h_0)_{\nu_1,\nu_2}^{k,\ell} = \beta^{k,\ell}\, \pi^{\frac{\nu_1+\nu_2-1}{2}}$$

$$\Gamma(\frac{3-\nu_1-\nu_2+i\lambda_k^-}{4})\, \Gamma(\frac{3-\nu_1-\nu_2-i\lambda_k^-}{4})\, L^*(\frac{1+\nu_1-\nu_2}{2}, \mathcal{N}_{k,\ell}^-).\quad (14.64)$$

Consider the orthogonal projection of $(h_0)_{\nu_1,\nu_2}$ onto the odd eigenspace of Δ for the eigenvalue $\frac{1+(\lambda_k^-)^2}{4}$, i.e.,

$$(\mathrm{Pr}_{\lambda_k^-}\, h_0)(z) = \sum_\ell (h_0)_{\nu_1,\nu_2}^{k,\ell}\, \|\mathcal{N}_{k,\ell}^-\|^{-2}\, \mathcal{N}_{k,\ell}^-(z),\qquad (14.65)$$

together with its Fourier expansion

$$(\mathrm{Pr}_{\lambda_k^-}\, h_0)(z) = y^{\frac{1}{2}} \sum_{n\neq 0} (-d_n^-)\, K_{\frac{i\lambda_k^-}{2}}(2\pi|n|y)\, e^{2i\pi nx},$$

$$(14.66)$$

where the notation is compatible with that in (12.7). In view of (14.64) and (14.58), one has

$$-d_n^- = (\mathrm{sign}\, n)\, \pi^{\frac{\nu_1+\nu_2-1}{2}}\, \Gamma(\frac{3-\nu_1-\nu_2+i\lambda_k^-}{4})\, \Gamma(\frac{3-\nu_1-\nu_2-i\lambda_k^-}{4})$$

$$\sum_\ell \beta^{k,\ell}\, \|\mathcal{N}_{k,\ell}^-\|^{-2}\, \omega_{|n|;k,\ell}^-\, L^*(\frac{1+\nu_1-\nu_2}{2}, \mathcal{N}_{k,\ell}^-).\quad (14.67)$$

On the other hand, using theorem 12.2 and the duplication formula for the Gamma function, one gets

$$-d_n^- = -2^{-i\lambda_k^-}\, i\, \pi^{\frac{-2+\nu_1+\nu_2-i\lambda_k^-}{2}}\, |n|^{-\frac{i\lambda_k^-}{2}} \times$$

$$\frac{\Gamma(\frac{3-\nu_1-\nu_2+i\lambda_k^-}{4})\, \Gamma(\frac{3+\nu_1-\nu_2-i\lambda_k^-}{4})\, \Gamma(\frac{3-\nu_1+\nu_2-i\lambda_k^-}{4})\, \Gamma(\frac{3-\nu_1-\nu_2-i\lambda_k^-}{4})}{\Gamma(-i\lambda_k^-)} \times$$

$$\text{residue of } \zeta_n^-(\frac{1-i\mu-\nu_1+\nu_2}{2}, \frac{1-i\mu+\nu_1-\nu_2}{2}) \text{ at } \mu = \lambda_k^-.$$

$$(14.68)$$

From (10.30), (14.44) and the equation $\zeta_{-n}^{-}(s,t) = -\zeta_{-n}^{-}(s,t)$, we get when $\frac{1}{2} < \mathrm{Im}\,\mu < -1 + \mathrm{Re}\,(\nu_2 - \nu_1)$ the relation

$$\zeta_n^-\left(\frac{1 - i\mu - \nu_1 + \nu_2}{2}, \frac{1 - i\mu + \nu_1 - \nu_2}{2}\right) = \frac{1}{2i}\,\pi^{\frac{\nu_1 - \nu_2 - i\mu}{2}}\,\frac{\Gamma\left(\frac{3 + i\mu - \nu_1 + \nu_2}{4}\right)}{\Gamma\left(\frac{3 - i\mu + \nu_1 - \nu_2}{4}\right)}$$

$$\sum_{j \geq 1} j^{\frac{-1 - i\mu + \nu_1 - \nu_2}{2}}\left[KS_{\frac{1-i\mu}{2}}(j,n) - KS_{\frac{1-i\mu}{2}}(-j,n)\right]. \quad (14.69)$$

Using Iwaniec's results [20] in just the same way as the one used to prove (14.46), and not forgetting that, this time, the Fourier coefficients $n \mapsto \rho_{k,\ell}(n)$ (cf. (14.45)) are odd functions, we get the residue of the function $\mu \mapsto KS_{\frac{1-i\mu}{2}}(j,n)$ at $\mu = \lambda_k^-$ as

$$i\,(2\pi)^{i\lambda_k^-}\,\Gamma(-i\lambda_k^-)\,\langle jn\rangle^{\frac{i\lambda_k^-}{2}}\sum_\ell b_{|n|;k,\ell}^-\,b_{|j|;k,\ell}^-$$

where, again, the numbers $b_{n;k,\ell}^-$ are the Fourier coefficients of the $L^2(D)$-normalized function $\mathcal{M}_{k,\ell}^-$; next, using (14.69), we may express the residue of $\zeta_n^-\left(\frac{1-i\mu-\nu_1+\nu_2}{2}, \frac{1-i\mu+\nu_1-\nu_2}{2}\right)$ at $\mu = \lambda_k^-$ as

$$2^{i\lambda_k^-}\,\pi^{\frac{1+i\lambda_k^-}{2}}\,\Gamma(-i\lambda_k^-)\left[\Gamma\left(\frac{3 + \nu_1 - \nu_2 - i\lambda_k^-}{4}\right)\Gamma\left(\frac{3 - \nu_1 + \nu_2 - i\lambda_k^-}{4}\right)\right]^{-1}$$

$$|n|^{\frac{i\lambda_k^-}{2}}\sum_\ell b_{n;k,\ell}^-\,L^*\left(\frac{1 - \nu_1 + \nu_2}{2}, \mathcal{M}_{k,\ell}^-\right). \quad (14.70)$$

Substituting this last result for the residue on the right–hand side of (14.68) and comparing to (14.67), we get theorem 14.8.

$$\square$$

From (10.30), (14.14) and the equation $G_n'(s,\tau) = -(-1)... $ we get

where $\gamma = \frac{1}{2} + it$, $\nu = Re(z + \tau)$, the relation

15. A generating series of sorts for Maass cusp-forms

One should not place too much hope in the availability of a representation of Maass cusp–forms as series of functions as explicit, say, as Eisenstein series. Also, one might expect that such a representation could not be the result of investigations from which algebraic number theory, for one, would be totally absent, even though experience in this direction is tied to congruence groups rather than Γ itself. In sections 19 and 20, we shall try some other approach also based on quadratic fields.

In the present section, we introduce families (F_μ) of functions defined as series resembling Eisenstein series in a way. Each F_μ, though $\mathbf{Z}$–periodic and an eigenfunction of the Laplacian, fails to be a non–holomorphic modular form because the equation $F_\mu(z) = F_\mu(-\frac{1}{z})$ fails to hold. However, it is approximately correct in the sense that the difference $F_\mu(z) - F_\mu(-\frac{1}{z})$ extends, as a function of μ, to a larger domain that $F_\mu(z)$: this constitutes an *a priori* explanation of the fact that the polar parts of our family of (analytically continued) functions do bring to light modular forms, actually the sought–after Maass forms and certain Eisenstein series.

We prepare for this with some elementary arithmetic considerations. Let m be an integer ≥ 1. In the ring $\mathbf{Z}/m\mathbf{Z}$, the solutions of the equation $r(1-r) = 0$ are exactly the powers $x^{\phi(m)}$, $x \in \mathbf{Z}/m\mathbf{Z}$, where ϕ is Euler's function: indeed, $r = r^2 \Rightarrow r = r^{\phi(m)}$, and in the other direction the validity of the equation $x^{\phi(m)}(x^{\phi(m)} - 1) \equiv 0 \mod m$ for every x is essentially the Fermat–Euler theorem ([16], p.63). If $r(1-r) \equiv 0 \mod m$, there exists a unique pair m_1, m_2 of positive integers with $m_1 | r-1$, $m_2 | r$ and $m_1 m_2 = m$: conversely, given a pair m_1, m_2 with $(m_1, m_2) = 1$ and $m_1 m_2 = m$, there exists a unique $r \mod m$ such that $r \equiv 1 \mod m_1$ and $r \equiv 0 \mod m_2$. Consequently,

$$\#\{r \bmod m: r(1-r) = 0\} = 2^{\#\{p \text{ prime}: \, p | m\}} \tag{15.1}$$

and

$$\log(\#\{r \bmod m: r(1-r) = 0\}) = O(\frac{\log m}{\log\log m}). \tag{15.2}$$

It makes sense, for every complex parameter ν and every integer $m \geq 1$, to define

$$A_{\nu,n}(m) = \sum_{\substack{r \bmod m \\ r(1-r) \equiv 0}} \left(\frac{m_1}{m_2}\right)^{\frac{\nu}{2}} \cos 2\pi n \frac{r}{m}, \qquad A_{\nu,n}^-(m) = \sum_{\substack{r \bmod m \\ r(1-r) \equiv 0}} \left(\frac{m_1}{m_2}\right)^{\frac{\nu}{2}} \sin 2\pi n \frac{r}{m},$$

$$(15.3)$$

where the integers m_1, m_2 are associated with the pair m, r in the way indicated above.

One application of proposition 14.6 occurs when $\nu = 0$ (only in the even case): we display this special situation in view of the simple arithmetic nature of the coefficients involved. We abbreviate $A_{0,n}(m) + i\, A_{0,n}^-(m)$ as

$$a_n(m): = \sum_{\substack{r \bmod m \\ r(1-r) \equiv 0}} e^{2i\pi n \frac{r}{m}}. \qquad (15.4)$$

PROPOSITION 15.1. *Let* $m = p_1^{\alpha_1} \ldots p_k^{\alpha_k}$, *with* $p_1 < \cdots < p_k$, *be the decomposition of* m *into prime factors. For each* $j = 1, \ldots, k$, *let* x_j *be the solution (mod* m*) of the Chinese problem* $x_j \equiv 1 \bmod p_j^{\alpha_j}$, $x_j \equiv 0 \bmod p_\ell^{\alpha_\ell}$ *for* $\ell \neq j$. *Then*

$$a_n(m) = \prod_{j=1}^{k} (1 + e^{2i\pi n \frac{x_j}{m}}). \qquad (15.5)$$

One may also write

$$a_n(m) = e^{i\pi \frac{n}{m}} \times \sum_{\substack{q \bmod 2m \\ q^2 \equiv 1 \bmod 4m}} e^{i\pi n \frac{q}{m}}. \qquad (15.6)$$

PROOF. With x_j as defined in the statement of proposition 15.1, (15.5) will be proven if we show that

$$r = \varepsilon_1 x_1 + \cdots + \varepsilon_k x_k, \qquad (15.7)$$

with $\varepsilon_j = 0$ or 1 for all j, is the general solution of the equation $r(1-r) \equiv 0$. Since all these numbers are pairwise distinct mod m, it will follow from (15.1) that, indeed, this is the general solution if it is a solution at all: now, this comes from the decomposition

$$(r_1 + r_2)(1 - r_1 - r_2) = r_1(1 - r_1) + r_2(1 - r_2) - 2\, r_1 r_2 \qquad (15.8)$$

applied in the case when r_1 and r_2 are two solutions of the equation under discussion satisfying the property that $r_1 \equiv 0 \bmod n_1$ and $r_2 \equiv 0 \bmod n_2$ with $m \mid \mathrm{l.c.m.}(n_1, n_2)$.

The expression (15.6) is equivalent to (15.4) under the change of variable $q = 2r - 1$. $\qquad\square$

Remark. More generally,

$$A_{\nu,n}(m) + i\, A_{\nu,n}^{-}(m) = \prod_{j=1}^{k}(p_j^{-\frac{\nu a_j}{2}} + p_j^{\frac{\nu a_j}{2}}\, e^{2i\pi n\frac{x_j}{m}}). \qquad (15.9)$$

THEOREM 15.2. *For every $n \neq 0$, set*

$$Z_n(s) = \sum_{m \geq 1} a_n(m)\, m^{-s} \qquad (15.10)$$

where $a_n(m)$ was defined in (15.4). This Dirichlet series is absolutely convergent for $\mathrm{Re}\ s > 1$ and extends as a meromorphic function in the half-plane $\mathrm{Re}\ s > 0$. The points of this half-plane (with the exception of the point $s = 1$) which are poles of at least one of the functions Z_n are precisely the points $\frac{\omega}{2}$ with $\zeta^*(\omega) = 0$ and the points $s_k = \frac{1 - i\lambda_k}{2}$, $s_k(1 - s_k) = \frac{1 + \lambda_k^2}{4}$ in the even part of the discrete spectrum of the modular Laplacian, which satisfy the additional property that for at least one value of ℓ, the L-function $L(\sigma, \mathcal{M}_{k,\ell})$ does not vanish at $\sigma = \frac{1}{2}$.

PROOF. Before giving the proof, let us mention that we do not know whether the additional condition is satisfied for all k. Starting from (10.5), we may write

$$\zeta_n(s,s) = \sum_{\substack{m_1 \geq 1, m_2 \geq 1 \\ (m_1, m_2) = 1}} (m_1 m_2)^{-s} \cos 2\pi n \frac{\overline{m}_2}{m_1},$$

$$(\overline{m}_2 m_2 \equiv 1 \bmod m_1). \quad (15.11)$$

This can be written as $\sum_{m \geq 1} b_n(m)\, m^{-s}$, provided we set

$$b_n(m) = \sum_{\substack{m_1 \geq 1 \\ m_1 m_2 = m \\ (m_1, m_2) = 1}} \cos 2\pi n \frac{\overline{m}_2}{m_1}. \qquad (15.12)$$

Now, setting $r = m_2 \overline{m}_2$, one immediately gets $b_n(m) = A_{0,n}(m)$, with $A_{0,n}$ as defined in (15.3): indeed, it suffices to recall the argument immediately before (15.1). Using the symmetry $r \mapsto 1 - r$ of the equation $r(1 - r) \equiv 0$ mod m together with (15.2) and

$$e^{2i\pi n \frac{1-r}{m}} = e^{-2i\pi n \frac{r}{m}}(1 + O(\frac{1}{m})), \qquad (15.13)$$

one sees that the difference $Z_n(s) - \zeta_n(s,s)$ extends as a holomorphic function in the half-plane $\mathrm{Re}\ s > 0$, so that our statement is equivalent to the same one in which $Z_n(s)$ is to be replaced by $\zeta_n(s,s)$. Applying proposition

14.6, we get the value of the residue of $\zeta_n(s, s)$ at s_k (whether this is an actual pole or not) as

$$\frac{1}{2}\,(4\pi|n|)^{\frac{i\lambda_k}{2}}\,\frac{\Gamma(-i\lambda_k)\,\Gamma(\frac{1+i\lambda_k}{4})}{\Gamma(\frac{1-i\lambda_k}{4})}\,\sum_\ell b_{n;k,\ell}\,L(\tfrac{1}{2}, \mathcal{M}_{k,\ell}). \tag{15.14}$$

If some point s_k fails to be a pole of any of the functions $s \mapsto \zeta_n(s, s)$, it follows from the last formula that all the Fourier coefficients of the linear combination

$$\sum_\ell L(\tfrac{1}{2}, \mathcal{M}_{k,\ell})\,\mathcal{M}_{k,\ell}$$

are zero, so that $L(\tfrac{1}{2}, \mathcal{M}_{k,\ell}) = 0$ for all ℓ.

The proof is not completely finished since the application of corollary 11.4 to the study of the singularities of the function $\zeta_n(s, s)$ still leaves the point $s = \tfrac{1}{2}$ unsettled: however, this will be removed in proposition 15.8. $\square$

One may note that, near the diagonal $s = t$, the pair (s, t) lies outside the range of values for which theorem 10.5, expressing $\zeta_n(s, t)$ as a series of Kloosterman–Selberg series, could be applicable. To our knowledge, the analytic continuation of $\zeta_n(s, s)$ could not be obtained in any simple way from results already known about the Kloosterman–Selberg series themselves. The same remark holds for the next theorem.

THEOREM 15.3. *Assume* $|\mathrm{Re}\,\nu| < 1$. *For* $\mathrm{Re}\,s > 1 + \tfrac{1}{2}|\mathrm{Re}\,\nu|$, *one has*

$$\zeta_n\big(s - \frac{\nu}{2}, s + \frac{\nu}{2}\big) = \sum_{m \geq 1} A_{n,\nu}(m)\,m^{-s} \tag{15.15}$$

and

$$\zeta_n^-\big(s - \frac{\nu}{2}, s + \frac{\nu}{2}\big) = i \sum_{m \geq 1} A_{n,\nu}^-(m)\,m^{-s}, \tag{15.16}$$

where the coefficients were defined in (15.3). These two functions extend as meromorphic functions in the half–plane $\mathrm{Re}\,s > \tfrac{1}{2}|\mathrm{Re}\,\nu|$. *The points of this half–plane (with the exception of the points* $s = 1 \pm \tfrac{\nu}{2}$*) which are poles of at least one of the functions* $s \mapsto \zeta_n(s - \tfrac{\nu}{2}, s + \tfrac{\nu}{2})$ *are precisely the points* $\tfrac{\omega}{2}$ *with* $\zeta^*(\omega) = 0$ *in the half–plane and the points* $s_k = \tfrac{1 - i\lambda_k}{2}$, $s_k(1 - s_k) = \tfrac{1 + \lambda_k^2}{4}$ *in the even part of the discrete spectrum of the modular Laplacian, which satisfy the additional property that for at least one value of* ℓ, *the* L–function $L(\sigma, \mathcal{M}_{k,\ell})$ *does not vanish at* $\sigma = \tfrac{1 + \nu}{2}$. *The same holds with the functions* $s \mapsto \zeta_n^-(s - \tfrac{\nu}{2}, s + \tfrac{\nu}{2})$, *substituting* λ_k^- *for* λ_k *and the set* $\{\mathcal{M}_{k,\ell}^-\}$ *for* $\{\mathcal{M}_{k,\ell}\}$, *with the difference, however, that the half–zeroes of the zeta function do not contribute in this case.*

PROOF. First observe that the non–vanishing condition is certainly satisfied for generic values of ν. The formulas (15.15) and (15.16) are proved in just the same way as the identity for $\zeta_n(s,s)$ in the proof of theorem 15.2. Also, from proposition 14.6, the residue of the function $s \mapsto \zeta_n(s - \frac{\nu}{2}, s + \frac{\nu}{2})$ at $s = s_k$ $(s_k = \frac{1-i\lambda_k}{2})$ is

$$\frac{1}{2} \pi^{\frac{i\lambda_k - \nu}{2}} (4|n|)^{\frac{i\lambda_k}{2}} \frac{\Gamma(-i\lambda_k)\,\Gamma(\frac{1+\nu+i\lambda_k}{4})}{\Gamma(\frac{1-\nu-i\lambda_k}{4})} \sum_\ell b_{n;k,\ell}\, L(\frac{1+\nu}{2}, \mathcal{M}_{k,\ell}), \tag{15.17}$$

which proves theorem 15.3 in the even case. The odd case is covered with the help of the computation at the very end of the proof of theorem 14.8. That the functions under consideration are regular at $s = \frac{1}{2}$ will have to wait until proposition 15.8 is proven (*cf.* the very end of the proof of theorem 15.2). □

Recall the definition

$$\zeta(1 - i\mu)\, E_{\frac{1-i\mu}{2}}(z) = \frac{1}{2} \sum_{|m|+|n|\neq 0} \left(\frac{|mz-n|^2}{\operatorname{Im} z}\right)^{\frac{i\mu-1}{2}}, \tag{15.18}$$

valid for $\operatorname{Im} \mu > 1$, of Eisenstein series. One may consider that there are some elements of similarity between this function and the *non-modular* function

$$F_\mu(z) = \frac{1}{2} \sum_{\substack{n\in\mathbf{Z},\, m\in\mathbf{Z}^\times \\ m|n(n-1)}} |m|^{\frac{1-i\mu}{2}} \left(\frac{|mz-n|^2}{\operatorname{Im} z}\right)^{\frac{i\mu-1}{2}}. \tag{15.19}$$

More generally, ν being another complex parameter satisfying $|\operatorname{Re} \nu| < 1$, one may consider the function

$$F_{\mu,\nu}(z) = \frac{1}{2} \sum_{\substack{n\in\mathbf{Z},\, m\in\mathbf{Z}^\times \\ m|n(n-1)}} (\frac{m_1}{m_2})^{\frac{\nu}{2}} |m|^{\frac{1-i\mu}{2}} \left(\frac{|mz-n|^2}{\operatorname{Im} z}\right)^{\frac{i\mu-1}{2}}, \tag{15.20}$$

in which (just as in (15.3)) the pair m_1, m_2 is the pair of positive integers characterized by the conditions $|m| = m_1 m_2$, $m_1|n-1$, $m_2|n$).

THEOREM 15.4. *Assume* $|\operatorname{Re} \nu| < 1$. *The series* (15.20) *defining* $F_{\mu,\nu}(z)$ *converges (uniformly for z in a compact subset of Π) when* $\operatorname{Im} \mu > 1 + |\operatorname{Re} \nu|$. *It satisfies the equation* $\Delta F_{\mu,\nu} = \frac{1+\mu^2}{4} F_{\mu,\nu}$. *It is $\mathbf{Z}$-periodic. The function*

$$F_{\mu,\nu}(z) - F_{\mu,\nu}(-\frac{1}{z})$$

extends as a holomorphic function for $\operatorname{Im} \mu > -1 + |\operatorname{Re} \nu|$, $\mu \neq i(1 \pm \nu)$.

PROOF. The first point is a consequence of the inequality (when $n \neq 0$)

$$|m|^{\frac{|\text{Re } \nu| + 1 + \text{Im } \mu}{2}} (m^2 + n^2)^{\frac{-1 - \text{Im } \mu}{2}} \leq C \, |n|^{\frac{|\text{Re } \nu| - 1 - \text{Im } \mu}{2}}$$

(15.21)

together with a trivial bound for the number of divisors of $n(n-1)$. The differential equation and the periodicity are trivial. The terms with $n = 0$ in (15.20) add up to $\zeta(\frac{1+\nu-i\mu}{2}) \left(\frac{|z|^2}{\text{Im } z} \right)^{\frac{i\mu-1}{2}}$. We also isolate the terms with $n = 1$, for which $m_2 = 1$: they add up to $\frac{1}{2} \left(\frac{|z|^2}{\text{Im } z} \right)^{\frac{i\mu-1}{2}} \times \sum_{m \neq 0} |m|^{\frac{-1+i\mu+\nu}{2}} |1 - \frac{1}{zm}|^{i\mu-}$ where the coefficient is the sum of $\zeta(\frac{1-i\mu-\nu}{2})$ and of a function holomorphic for Im $\mu > -1 + \text{Re } \nu$. With

$$G_{\mu,\nu}(z) := \frac{1}{2} \sum_{\substack{n \neq 0,1 \\ m|n(n-1)}} (\frac{m_1}{m_2})^{\frac{\nu}{2}} |m|^{\frac{1-i\mu}{2}} \left(\frac{|mz - n|^2}{\text{Im } z} \right)^{\frac{i\mu-1}{2}},$$

(15.22)

where m_1 and m_2 depend on the pair m, n as indicated right after (15.20), we presently show that $G_{\mu,\nu}(z) - G_{\mu,\nu}(-\frac{1}{z})$ extends as a holomorphic function in the half-plane Im $\mu > -1 + |\text{Re } \nu|$.

Given an irreducible fraction $\frac{a}{d} \neq 0$, the set of pairs m, n with $mn \neq 0$, $m|n(n-1)$ and $\frac{n}{m} = \frac{a}{d}$ can be parametrized, setting $n = ka$, $m = kd$ so that $d|ka - 1$, as $m = (\bar{a} + jd)d$, $n = (\bar{a} + jd)a$, with $j \in \mathbf{Z}$, where $\bar{a} \in \mathbf{Z}$ is any fixed integer chosen so that $\bar{a}a \equiv 1 \bmod d$. Moreover, for such a pair m, n, the pair m_1, m_2 characterized by $|m| = m_1 m_2$, $m_1|n-1$, $m_2|n$ is given as $m_1 = |d|$, $m_2 = |\bar{a} + jd|$; finally, the condition $n \neq 1$ becomes $a\bar{a} + jad \neq 1$. This permits to write, if Im $\mu > 1 + |\text{Re } \nu|$,

$$G_{\mu,\nu}(z) = \frac{1}{4} \sum_{\substack{ad \neq 0 \\ (a,d)=1}} |ad|^{\frac{1-i\mu+\nu}{2}} \left(\frac{|dz - a|^2}{\text{Im } z} \right)^{\frac{i\mu-1}{2}} \times$$

$$\sum_{\substack{j \in \mathbf{Z} \\ \bar{a}+jd \neq 0 \\ a\bar{a}+jad \neq 1}} |a\bar{a} + jad|^{\frac{-1+i\mu-\nu}{2}}, \qquad a\bar{a} \equiv 1 \bmod d \quad (15.23)$$

or

$$G_{\mu,\nu}(z) = \frac{1}{4} \sum_{\substack{ad \neq 0 \\ (a,d)=1}} |ad|^{\frac{1-i\mu+\nu}{2}} \left(\frac{|dz - a|^2}{\text{Im } z} \right)^{\frac{i\mu-1}{2}} \sum_{\substack{q \neq 0,1 \\ q \equiv 0 \bmod a \\ q \equiv 1 \bmod d}} |q|^{\frac{-1+i\mu-\nu}{2}}.$$

(15.24)

One can express $G_{\mu,\nu}(-\frac{1}{z})$ in the same way: trading the pair (a,d) for the pair $(d,-a)$, one gets just the same expression, except that in the pair of constraints $(q \equiv 0 \bmod a, q \equiv 1 \bmod d)$, a and d must be exchanged. Since

$$\left| |q|^{\frac{-1+i\mu-\nu}{2}} - |1-q|^{\frac{-1+i\mu-\nu}{2}} \right| \le C\,|q|^{\frac{-3-\mathrm{Im}\,\mu-\mathrm{Re}\,\nu}{2}},$$

$$(15.25)$$

one has

$$|G_{\mu,\nu}(z) - G_{\mu,\nu}(-\tfrac{1}{z})| \le C \sum_{\substack{ad\neq 0 \\ (a,d)=1}} |ad|^{\frac{1+\mathrm{Im}\,\mu+\mathrm{Re}\,\nu}{2}} \times$$

$$\left(\frac{|dz - a|^2}{\mathrm{Im}\,z}\right)^{\frac{-\mathrm{Im}\,\mu-1}{2}} \sum_{\substack{q\neq 0,1 \\ q\equiv 0 \bmod a \\ q\equiv 1 \bmod d}} |q|^{\frac{-3-\mathrm{Im}\,\mu-\mathrm{Re}\,\nu}{2}}. \quad (15.26)$$

Doing in reverse the same operations as those which led from (15.22) to the expression (15.24) of $G_{\mu,\nu}(z)$, we get

$$|G_{\mu,\nu}(z) - G_{\mu,\nu}(-\tfrac{1}{z})| \le C \sum_{\substack{n\neq -1,0,1 \\ m|n(n-1)}} \left(\frac{|mz - n|^2}{\mathrm{Im}\,z}\right)^{\frac{-1-\mathrm{Im}\,\mu}{2}}$$

$$|m|^{\frac{1+\mathrm{Im}\,\mu}{2}} \left(\frac{m_1}{m_2}\right)^{\frac{\mathrm{Re}\,\nu}{2}} |n|^{-1}. \quad (15.27)$$

Since

$$\sum_{\substack{mn\neq 0 \\ m|n(n-1)}} (m^2 + n^2)^{\frac{-1-\mathrm{Im}\,\mu}{2}} |m|^{\frac{1+\mathrm{Im}\,\mu}{2}} \left(\frac{m_1}{m_2}\right)^{\frac{\mathrm{Re}\,\nu}{2}} |n|^{-1}$$

$$\le C \sum_{mn\neq 0} (m^2 + n^2)^{\frac{-1-\mathrm{Im}\,\mu}{2}} |m|^{\frac{1+\mathrm{Im}\,\mu+|\mathrm{Re}\,\nu|}{2}} |n|^{-1}, \quad (15.28)$$

we are done thanks to (15.21).

$\square$

Remark. An immediate consequence is that if we can prove that $F_{\mu,\nu}(z)$ extends as a meromorphic function of μ in some open subset of the half-plane $\mathrm{Im}\,\mu > -1 + |\mathrm{Re}\,\nu|$ (this will be done presently), with a certain pole μ_0, then the coefficients of negative orders in the Laurent expansion of $F_{\mu,\nu}(z)$ at $\mu = \mu_0$ furnish non-holomorphic modular forms, since $F_{\mu,\nu}(z) - F_{\mu,\nu}(-\tfrac{1}{z})$ is regular at that point.

PROPOSITION 15.5. *The Fourier series expansion of $F_{\mu,\nu}(z)$ is given, under the assumptions $|\mathrm{Re}\,\nu| < 1$, $\mathrm{Im}\,\mu > 1 + |\mathrm{Re}\,\nu|$, as*

$$F_{\mu,\nu}(z) = \pi^{\frac{1}{2}} \frac{\Gamma(-\frac{i\mu}{2})}{\Gamma(\frac{1-i\mu}{2})} \frac{\zeta(\frac{1-i\mu-\nu}{2})\zeta(\frac{1-i\mu+\nu}{2})}{\zeta(1-i\mu)} y^{\frac{1+i\mu}{2}} + \frac{2\pi^{\frac{1-i\mu}{2}}}{\Gamma(\frac{1-i\mu}{2})} \times$$

$$\sum_{n\neq 0} \zeta_n(\frac{1-i\mu-\nu}{2}, \frac{1-i\mu+\nu}{2}) y^{\frac{1}{2}} e^{2i\pi nx} |n|^{-\frac{i\mu}{2}} K_{\frac{i\mu}{2}}(2\pi|n|y).$$

$$(15.29)$$

PROOF. Start from (15.20), rewritten as

$$F_{\mu,\nu}(z) = \frac{1}{2} \sum_{m\neq 0} |m|^{\frac{1-i\mu}{2}} \sum_{\substack{r \bmod m \\ r(r-1)\equiv 0}} (\frac{m_1}{m_2})^{\frac{\nu}{2}} \sum_{k\in\mathbb{Z}} \left[\frac{(m(x+k)+r)^2 + m^2 y^2}{y}\right]^{\frac{i\mu-1}{2}},$$

$$(|m| = m_1 m_2, \ r \equiv 1 \bmod m_1, \ r \equiv 0 \bmod m_2). \quad (15.30)$$

Recall that, for $\mathrm{Im}\,\mu > 0$, the "true value" of the function $\sigma \mapsto |\sigma|^{-\frac{i\mu}{2}} K_{\frac{i\mu}{2}}(2\pi|\sigma|)$ at $\sigma = 0$ is (*cf.* (2.12) or [27], p.66) $\frac{1}{2}\pi^{\frac{i\mu}{2}}\Gamma(-\frac{i\mu}{2})$. Using (2.12) again,

$$\int_{-\infty}^{\infty} e^{-2i\pi s\sigma} \left[\frac{(m(x+s)+r)^2 + m^2 y^2}{y}\right]^{\frac{i\mu-1}{2}} ds$$

$$= \frac{2\pi^{\frac{1-i\mu}{2}}}{\Gamma(\frac{1-i\mu}{2})} |m|^{i\mu-1} y^{\frac{1}{2}} e^{2i\pi(x+\frac{r}{m})\sigma} |\sigma|^{-\frac{i\mu}{2}} K_{\frac{i\mu}{2}}(2\pi y|\sigma|) \quad (15.31)$$

so that, by Poisson's formula,

$$F_{\mu,\nu}(z) = \frac{2\pi^{\frac{1-i\mu}{2}}}{\Gamma(\frac{1-i\mu}{2})} \sum_{m\geq 1} m^{\frac{i\mu-1}{2}} \sum_{\substack{r \bmod m \\ r(r-1)\equiv 0}} (\frac{m_1}{m_2})^{\frac{\nu}{2}} \times$$

$$\left\{\frac{1}{2}\pi^{\frac{i\mu}{2}}\Gamma(-\frac{i\mu}{2}) y^{\frac{1+i\mu}{2}} + y^{\frac{1}{2}} \sum_{n\neq 0} e^{2i\pi nx}(\cos 2\pi n\frac{r}{m})|n|^{-\frac{i\mu}{2}} K_{\frac{i\mu}{2}}(2\pi|n|y)\right\}.$$

$$(15.32)$$

On one hand, from (10.5), (10.6),

$$\sum_{m\geq 1} m^{\frac{i\mu-1}{2}} \sum_{\substack{r \bmod m \\ r(r-1)\equiv 0}} (\frac{m_1}{m_2})^{\frac{\nu}{2}} = \sum_{\substack{m_1\geq 1, m_2\geq 1 \\ (m_1,m_2)=1}} m_1^{\frac{-1+i\mu+\nu}{2}} m_2^{\frac{-1+i\mu-\nu}{2}}$$

$$= \frac{\zeta(\frac{1-i\mu-\nu}{2})\zeta(\frac{1-i\mu+\nu}{2})}{\zeta(1-i\mu)}. \quad (15.33)$$

On the other hand, from (15.3), (15.15),

$$\sum_{m \geq 1} m^{\frac{i\mu-1}{2}} \sum_{\substack{r \bmod m \\ r(r-1) \equiv 0}} (\frac{m_1}{m_2})^{\frac{\nu}{2}} \cos 2\pi n \frac{r}{m} = \sum_{m \geq 1} m^{\frac{i\mu-1}{2}} A_{\nu,n}(m)$$

$$= \zeta_n(\frac{1-i\mu-\nu}{2}, \frac{1-i\mu+\nu}{2}).$$

(15.34)

$\square$

THEOREM 15.6. *Assume* $|\text{Re } \nu| < 1$. *The function*

$$\mu \mapsto F_{\mu,\nu}(z) = \frac{1}{2} \sum_{\substack{n \in \mathbf{Z}, m \in \mathbf{Z}^{\times} \\ m|n(n-1)}} (\frac{m_1}{m_2})^{\frac{\nu}{2}} |m|^{\frac{1-i\mu}{2}} \left(\frac{|mz-n|^2}{\text{Im } z} \right)^{\frac{i\mu-1}{2}},$$

introduced in (15.20) *extends as a holomorphic function for* Im $\mu > -1 + |\text{Re } \nu|$, $\mu \neq i(1 \pm \nu)$, *except for the following poles: the points* $-i\omega$ *with* $\zeta^*(\omega) = 0$ *contained in this half-plane, and the points* λ_k, $\frac{1+\lambda_k^2}{4}$ *in the even part of the discrete spectrum of* Δ. *Near a point* $-i\omega$, *the function* $\mu \mapsto \text{Err}_{-i\omega}(\mu; z)$, *defined as*

$$\text{Err}_{-i\omega}(\mu; z) = F_{\mu,\nu}(z) - \pi^{\frac{1}{2}} \frac{\Gamma(-\frac{i\mu}{2})}{\Gamma(\frac{1-i\mu}{2})} \frac{\zeta(\frac{1-i\mu-\nu}{2}) \zeta(\frac{1-i\mu+\nu}{2})}{\zeta(1-i\mu)} E_{\frac{1+i\mu}{2}}(z),$$

(15.35)

remains holomorphic. A point λ_k *can only be a simple pole: it is if and only if one has* $L(\frac{1+\nu}{2}, \mathcal{M}) \neq 0$ *for at least one even cusp-form corresponding to the eigenvalue* $\frac{1+\lambda_k^2}{4}$, *and one has*

$$\text{Res}_{\mu=\lambda_k} F_{\mu,\nu}(z) = i \pi^{-\frac{\nu}{2}} \frac{\Gamma(-\frac{i\lambda_k}{2}) \Gamma(\frac{1+i\lambda_k+\nu}{4})}{\Gamma(\frac{1-i\lambda_k-\nu}{4})} \mathcal{M}_{\nu;k}(z),$$

(15.36)

with

$$\mathcal{M}_{\nu;k} := \sum_{\ell} L(\frac{1+\nu}{2}, \mathcal{M}_{k,\ell}) \mathcal{M}_{k,\ell}.$$

(15.37)

PROOF. We first examine separately the analytic continuation of the individual Fourier coefficients. When Im $\mu > -1 + |\text{Re } \nu|$, it follows from corollary 11.4 and from a direct examination of the coefficient of $y^{\frac{1+i\mu}{2}}$ on the right-hand side of (15.29) that these coefficients can be singular only when $\mu = i(1 \pm \nu)$, or $\mu = -i\omega$ for some ω with $\zeta^*(\omega) = 0$, or $\mu = \lambda_k$ for some k.

Looking back at the Fourier expansion (3.29) of Eisenstein series, one sees that the coefficient of $y^{\frac{1+i\mu}{2}}$ in $\mathrm{Err}_{-i\omega}(\mu; z)$ is zero. The other cuspidal coefficient (that of $y^{\frac{1-i\mu}{2}}$) comes solely from the second term in $\mathrm{Err}_{-i\omega}(\mu; z)$: from (3.29), it is

$$-\pi^{\frac{1}{2}} \frac{\Gamma(-\frac{i\mu}{2})}{\Gamma(\frac{1-i\mu}{2})} \frac{\zeta(\frac{1-i\mu-\nu}{2})\zeta(\frac{1-i\mu+\nu}{2})}{\zeta(1-i\mu)} \times \frac{\zeta^*(1-i\mu)}{\zeta^*(1+i\mu)}$$

and thus remains holomorphic near $-i\omega$.

Still near $-i\omega$, we show that the n-th Fourier coefficient of

$$\pi^{\frac{1}{2}} \frac{\Gamma(-\frac{i\mu}{2})}{\Gamma(\frac{1-i\mu}{2})} \frac{\zeta(\frac{1-i\mu-\nu}{2})\zeta(\frac{1-i\mu+\nu}{2})}{\zeta(1-i\mu)} E_{\frac{1+i\mu}{2}}(z),$$

i.e., (cf. (3.29), using $E_{\frac{1+i\mu}{2}}(z) = (\zeta^*(-i\mu))^{-1} E^*_{\frac{1-i\mu}{2}}(z))$,

$$c_n(\mu) = 2\pi^{\frac{1}{2}} \frac{\Gamma(-\frac{i\mu}{2})}{\Gamma(\frac{1-i\mu}{2})} \frac{\zeta(\frac{1-i\mu-\nu}{2})\zeta(\frac{1-i\mu+\nu}{2})}{\zeta(1-i\mu)\zeta^*(-i\mu)} |n|^{-\frac{i\mu}{2}} \sigma_{i\mu}(|n|), \tag{15.38}$$

has a polar part, near $\mu = -i\omega$, which is just the same as that of

$$\frac{2\pi^{\frac{1-i\mu}{2}}}{\Gamma(\frac{1-i\mu}{2})} \zeta_n\left(\frac{1-i\mu-\nu}{2}, \frac{1-i\mu+\nu}{2}\right) |n|^{-\frac{i\mu}{2}}. \tag{15.39}$$

From theorem 11.3, the function

$$\mu \mapsto \zeta_n\left(\frac{1-i\mu-\nu}{2}, \frac{1-i\mu+\nu}{2}\right)$$

agrees near $-i\omega$, up to a holomorphic error term, with

$$8\pi^{\frac{3+i\mu-\nu_1-\nu_2}{2}} \times \frac{1}{8\pi} \sigma_{i\mu}(|n|) \frac{\Phi_0(\mu)}{\zeta^*(1-i\mu)}$$

$$\times \frac{\Gamma(-\frac{i\mu}{2})\Gamma(\frac{1-i\mu}{2})}{\Gamma(\frac{1+i\mu-\nu_1-\nu_2}{4})\Gamma(\frac{1-i\mu+\nu_1-\nu_2}{4})\Gamma(\frac{1-i\mu-\nu_1+\nu_2}{4})\Gamma(\frac{1-i\mu-\nu_1-\nu_2}{4})}, \tag{15.40}$$

i.e., using the expression (11.15) of $\frac{\Phi_0(\mu)}{\zeta^*(1-i\mu)}$, with

$$\sigma_{i\mu}(|n|) \frac{\zeta(\frac{1-i\mu-\nu_1+\nu_2}{2})\zeta(\frac{1-i\mu+\nu_1-\nu_2}{2})}{\zeta(1-i\mu)\zeta(-i\mu)}.$$

Comparing $c_n(\mu)$, as expressed in (15.38), with $\frac{2\pi^{\frac{1-i\mu}{2}}}{\Gamma(\frac{1-i\mu}{2})}$ times this last expression, in which ν has been set for $\nu_1 - \nu_2$, we are done for this coefficient, i.e., we have shown that all the Fourier coefficients of $\mathrm{Err}_{-i\omega}(\mu; z)$ remain holomorphic near $\mu = -i\omega$.

Finally, let us examine the (non–cuspidal only, in this case) Fourier coefficients $c_n(\mu)$ near $\mu = \lambda_k$. This is a simple pole, so that we can satisfy ourselves with computing residues. The residue at $\mu = \lambda_k$ of the coefficient $c_n(\mu)$ arising from $F_{\mu,\nu}(z)$ is

$$
\frac{2\pi^{\frac{1-i\lambda_k}{2}}}{\Gamma(\frac{1-i\lambda_k}{2})} |n|^{-\frac{i\lambda_k}{2}} \operatorname{Res}_{\mu=\lambda_k} \zeta_n(\frac{1-i\mu-\nu}{2}, \frac{1-i\mu+\nu}{2}),
$$

and we wish (looking back at the notation $b_{n;k,\ell}$ from section 14, just before (14.4)) to equate it with

$$
i\,\pi^{-\frac{\nu}{2}} \frac{\Gamma(-\frac{i\lambda_k}{2})\Gamma(\frac{1+i\lambda_k+\nu}{4})}{\Gamma(\frac{1-i\lambda_k-\nu}{4})} \sum_\ell L(\frac{1+\nu}{2}, \mathcal{M}_{k,\ell})\, b_{n;k,\ell}.
$$

This is just the result of proposition 14.6, using again the duplication formula for the Gamma product $\Gamma(-\frac{i\lambda_k}{2})\Gamma(\frac{1-i\lambda_k}{2})$.

We have finished the examination of the analytic continuation of the Fourier coefficients of the function $F_{\mu,\nu}(z)$ and of the nature of their singularities: we must now perform a summation with respect to n to get the continuation of the function $F_{\mu,\nu}(z)$ itself, remembering that we are quite satisfied with any fixed z (z in any fixed compact subset of Π would do just the same). Now, if one gets back to what was done in sections 9 and 11 concerning the analytic continuation of the functions $\zeta_n(\frac{1-i\mu-\nu}{2}, \frac{1-i\mu+\nu}{2})$, in a way that was not uniform with respect to n, one can see that (when Im μ stays within any two given bounds) all the estimates we got there would involve only possible losses of the order of a power of n: these are quite compensated (for fixed z) by the presence of the factor $K_{\frac{i\mu}{2}}(2\pi|n|y)$.

Concerning the extra terms $\mathcal{M}_{\nu;k}(z)$ which had to be added to $F_{\mu,\nu}(z)$ so as to get $\operatorname{Err}_{\lambda_k}(\mu; z)$, it should be emphasized that, since these were not added simultaneously, but only one at a time so as to get the nature of the singularity of $F_{\mu,\nu}(z)$ near some given λ_k, no uniformity with respect to k whatsoever is needed: in particular, Smith's result [35] as recalled (and made use of) in (7.3) is not needed at that point. The same remark (to wit, again, the fact that no uniformity with respect to μ is required), applies to the Eisenstein series $E_{\frac{1+i\mu}{2}}(z)$ used as an extra term in the definition of $\operatorname{Err}_{-i\omega}(\mu; z)$, since in this case we are only interested in what happens near $\mu = -i\omega$.

$\square$

The odd case is much simpler since the zeros of the zeta function do not contribute to the poles. We shall not restate the odd case version of theorem 15.4, for which there is no difference, only that of theorem 15.6.

THEOREM 15.7. *Assume* $|\operatorname{Re} \nu| < 1$. *The function*

$$\mu \mapsto F_{\mu,\nu}^-(z) = \frac{1}{2} \sum_{\substack{n \in \mathbb{Z}, m \in \mathbb{Z}^\times \\ m \mid n(n-1)}} (\frac{m_1}{m_2})^{\frac{\nu}{2}} |m|^{\frac{1-i\mu}{2}} \operatorname{sign}(m) \left(\frac{|mz - n|^2}{\operatorname{Im} z} \right)^{\frac{i\mu-1}{2}},$$

extends as a holomorphic function for $\operatorname{Im} \mu > -1 + |\operatorname{Re} \nu|$, $\mu \neq i(1 \pm \nu)$, *except for the (at most simple) poles* λ_k^-, $\frac{1+(\lambda_k^-)^2}{4}$ *in the odd part of the discrete spectrum of* Δ. *One has*

$$\operatorname{Res}_{\mu=\lambda_k^-} F_{\mu,\nu}^-(z) = i \, \pi^{-\frac{\nu}{2}} \frac{\Gamma(-\frac{i\lambda_k^-}{2}) \Gamma(\frac{3+i\lambda_k^-+\nu}{4})}{\Gamma(\frac{3-i\lambda_k^--\nu}{4})} M_{\nu;k}^-(z),$$

$$(15.41)$$

with

$$M_{\nu;k}^- := \sum_\ell L(\frac{1+\nu}{2}, M_{k,\ell}^-) M_{k,\ell}^-. \qquad (15.42)$$

PROOF. Equation (15.29) becomes in the odd case

$$F_{\mu,\nu}^-(z) = \frac{2 \, i \, \pi^{\frac{1-i\mu}{2}}}{\Gamma(\frac{1-i\mu}{2})} \sum_{n \neq 0} \zeta_n^-(\frac{1-i\mu-\nu}{2}, \frac{1-i\mu+\nu}{2})$$

$$y^{\frac{1}{2}} e^{2i\pi n x} |n|^{-\frac{i\mu}{2}} K_{\frac{i\mu}{2}}(2\pi|n|y). \qquad (15.43)$$

Then (14.70) gives the residue of the function $\zeta_n^-(\frac{1-i\mu-\nu}{2}, \frac{1-i\mu+\nu}{2})$ at $\mu = \lambda_k^-$; the rest of the proof of theorem 15.6 goes without change. $\qquad\square$

PROPOSITION 15.8. *In corollary* 11.4, *the assumption* $s + t \neq 1$ *can be removed, i.e., the function* $\zeta_n(s,t)$, *extended as a meromorphic function for* $\operatorname{Re} s > 0$, $\operatorname{Re} t > 0$, $|\operatorname{Re}(s-t)| < 1$, $s \neq 1$, $t \neq 1$, *is regular near the points where* $s + t = 1$. *Under the same assumptions, so is the function* $\zeta_n^-(s,t)$, *which extends part of theorem* 12.2.

PROOF. Applying Cauchy's formula to $\zeta_n(s,t)$ regarded as a function of $s + t$ when $s - t$ is fixed, it suffices to show, looking back at the proof of corollary 11.4, that the function $\mu \mapsto \zeta_n(\frac{1-i\mu-\nu_1+\nu_2}{2}, \frac{1-i\mu+\nu_1-\nu_2}{2})$ is regular through $\mu = 0$ if $|\operatorname{Re}(\nu_1 - \nu_2)| < 1$. The difficulty with the point $\mu = 0$ originated from the presence, in the denominator of the right–hand side of (11.21), of the factor $\Gamma(-\frac{i\mu}{2})$: it prevented us from deriving the behaviour of $\zeta_n(\frac{1-i\mu-\nu_1+\nu_2}{2}, \frac{1-i\mu+\nu_1-\nu_2}{2})$ near $\mu = 0$ from that of $b_n(\mu)$. However, the former function can only have a simple pole at $\mu = 0$, and so does the function $\mu \mapsto F_{\mu,\nu}(z)$ (with $\nu = \nu_1 - \nu_2$), as a consequence of (15.29). Now, from proposition 15.4, the function $\mu \mapsto \operatorname{Res}_{\mu=0} F_{\mu,\nu}(z)$ has to be a (non–holomorphic) modular form, and (15.29) indicates (since $\frac{\Gamma(-\frac{i\mu}{2})}{\zeta(1-i\mu)}$ is regular at $\mu = 0$), that it can only be a cusp–form. It has to be zero since it corresponds to the eigenvalue $\frac{1}{4}$ of Δ.

The same proof applies to the function $\zeta_n^-(s,t)$, starting from theorem 12.2 and (15.43). $\qquad\qquad\qquad\qquad\qquad\qquad\qquad\qquad\qquad\qquad$ $\square$

16. Some arithmetic distributions

As opposed to (non–holomorphic) modular forms, automorphic distributions (section 13) live on $\mathbf{R}^2$ rather than Π. This does not have the sole advantage that Δ gives way to the simpler Euler operator $\mathcal{E}$. Another one, from an arithmetic point of view, is of course that $\mathbf{R}^2$ is the natural habitat of lattices: we shall concentrate, here, on distributions supported in $\mathbf{Z}^2$.

Consider Dirac's comb

$$\mathfrak{D}(\xi): = 2\pi \sum_{|m|+|n|\neq 0} \delta(\xi_1 - n, \xi_2 - m). \tag{16.1}$$

As shown in (16.2), it contains the whole information about Eisenstein series. But Γ–invariant measures supported in $\mathbf{Z}^2\backslash\{0\}$ will never exhibit any "cusp–distribution" $\mathfrak{M}_k^{\sharp}$ in their expansion (13.41).

The present section is not systematic: we are fully aware, for instance, that substituting $\mathbf{Q}^2\backslash\{0\}$ for $\mathbf{Z}^2\backslash\{0\}$ in what follows could be carried up to some point. One of its purposes is to familiarize ourselves with *some* automorphic distributions in the sense of section 13. Also, since one of the reasons for this work is an attempt to connect modular form theory to a rather wide range of classical analysis, we found it useful to show that a certain number of calculations of a style more familiar in an adelic environment, based on localisation, Eulerian products and Hecke operators, could actually be carried in a classical distribution setting, where it has an almost, but not quite, identical meaning.

We first give the decompositions of $\mathfrak{D}$ into homogeneous parts.

PROPOSITION 16.1. *One has, in a weak sense in* $S'_{\mathrm{even}}(\mathbf{R}^2)$,

$$\mathfrak{D} = 2\pi + \int_{-\infty}^{\infty} \mathfrak{E}_{i\lambda}^{\sharp}\, d\lambda. \tag{16.2}$$

PROOF. Let $h \in S(\mathbf{R}^2)$. Extending (4.20) with a complex μ substituted for λ,

$$h_\mu(\xi) = \frac{1}{2\pi} \int_0^{\infty} t^{i\mu} h(t\xi)\, dt, \qquad \xi \neq 0, \tag{16.3}$$

one gets, after a contour deformation in (4.19),

$$h(\xi) = \int_{ia-\infty}^{ia+\infty} h_{-\mu}(\xi)\, d\mu, \tag{16.4}$$

for any real $a > 0$, thus

$$< \mathfrak{D}, h >= \int_{ia-\infty}^{ia+\infty} < \mathfrak{D}, h_{-\mu} > d\mu. \qquad (16.5)$$

On the other hand, from (16.3) and (16.1), if Im $\mu > 1$,

$$< \mathfrak{D}, h_{-\mu} > = \sum_{|m|+|n|\neq 0} \int_0^\infty t^{-i\mu} h(tn, tm) \, dt$$

$$=< \mathfrak{C}_{i\mu}^\sharp, h > \qquad (16.6)$$

according to (13.25), so that (16.5) reads

$$< \mathfrak{D}, h >= \int_{ia-\infty}^{ia+\infty} < \mathfrak{C}_{i\mu}^\sharp, h > d\mu \qquad (16.7)$$

if $a > 1$. We move back the contour to the line Re $\mu = 0$, remembering from proposition 13.1 that the residue of the function $\mu \mapsto \mathfrak{C}_{i\mu}^\sharp$ at $\mu = i$ is the constant i: thus

$$< \mathfrak{D}, h >= 2\pi < 1, h > + \int_{-\infty}^\infty < \mathfrak{C}_{i\lambda}^\sharp, h > d\lambda. \qquad (16.8)$$

$\square$

Remark. Just for the peace of the mind, let us observe that the integral term on the right–hand side of (16.8) is invariant under the symplectic Fourier transformation $\mathcal{F}$ in view of proposition 13.1: this is just right since Poisson's formula means that $\mathfrak{D} + 2\pi\delta$ is invariant under $\mathcal{F}$.

The Γ–invariant distribution $\mathfrak{D}$ does not belong to the pre–Hilbert space introduced in theorem 13.2. The shortest way to see it is to compare (16.2) and (13.41): the function $\lambda \mapsto 2^{\frac{i\lambda}{2}} \zeta^*(i\lambda)$ is not square–integrable on $\mathbb{R}$ because of its singularity at $\lambda = 0$. This is why the decomposition (16.2) has been stated as a weak decomposition in $\mathcal{S}'_{even}(\mathbb{R}^2)$ rather than a spectral decomposition.

On the other hand, since

$$(t^{2i\pi\mathcal{E}}h)(\xi) = t \, h(t\xi), \qquad t > 0 \qquad (16.9)$$

if $h \in \mathcal{S}(\mathbb{R}^2)$, it is natural to set

$$< t^{2i\pi\mathcal{E}} \mathfrak{S}, h >=< \mathfrak{S}, t^{-2i\pi\mathcal{E}} h > \qquad (16.10)$$

if $h \in \mathcal{S}(\mathbb{R}^2)$ and $\mathfrak{S} \in \mathcal{S}'(\mathbb{R}^2)$, or

$$< t^{-1-2i\pi\mathcal{E}} \mathfrak{S}, h >=< \mathfrak{S}, \xi \mapsto h(t\xi) > . \qquad (16.11)$$

Functions, in some spectral–theoretic sense, of $\mathcal{E}$, cannot be applied to tempered distributions in general. However, given *any* measure supported in $\mathbf{Z}^2\backslash\{0\}$,

$$\mathfrak{S}_a = \sum_{|m|+|n|\neq 0} a(n,m)\,\delta(\xi_1 - n, \xi_2 - m) \tag{16.12}$$

and an *arbitrary* Dirichlet series $\sum_{k\geq 1} c_k\, k^{-1-2i\pi\mathcal{E}}$, one may define, for any $h \in \mathcal{S}(\mathbf{R}^2)$,

$$< (\sum_{k\geq 1} c_k\, k^{-1-2i\pi\mathcal{E}})\,\mathfrak{S}_a,\, h > = \sum_{k\geq 1} c_k \sum_{|m|+|n|\neq 0} a(n,m)\,h(kn, km). \tag{16.13}$$

DEFINITION 16.2. We let arbitrary Dirichlet series in $2i\pi\mathcal{E}$ as an argument act on arbitrary measures supported in $\mathbf{Z}^2\backslash\{0\}$ through

$$(\sum_{k\geq 1} c_k\, k^{-1-2i\pi\mathcal{E}}) \sum_{|m|+|n|\neq 0} a(n,m)\,\delta(\xi_1 - n, \xi_2 - m)$$

$$= \sum_{|m|+|n|\neq 0} b(n,m)\,\delta(\xi_1 - n, \xi_2 - m), \tag{16.14}$$

where

$$b(n,m) = \sum_{\substack{k\geq 1 \\ k|(n,m)}} c_k\, a(\frac{n}{k}, \frac{m}{k}). \tag{16.15}$$

Given a pair n, m with $|n| + |m| \neq 0$ and a prime number p, we set

$$v_p(n,m) := \max\{\alpha \in \mathbf{N}: p^\alpha | (n,m)\} \tag{16.16}$$

and define, as operators acting on measures supported in $\mathbf{Z}^2\backslash\{0\}$, the operators X_p and Y_p as follows: X_p acts as the multiplication of the coefficient $a(n,m)$ in (16.12) by $v_p(n,m)$, and

$$Y_p := \sum_{j\geq 0}(p^{j+1})^{-1-2i\pi\mathcal{E}}$$

$$= \frac{p^{-1-2i\pi\mathcal{E}}}{1 - p^{-1-2i\pi\mathcal{E}}}, \tag{16.17}$$

where the second expression is really a compact version of the first one, the meaning of which is explained in definition 16.2.

PROPOSITION 16.3. *One has*

$$[X_p, Y_p] = Y_p(Y_p + 1) \tag{16.18}$$

and any X_p commutes with any X_q or Y_q with $q \neq p$. If $a(n,m) = a(pn, pm)$ for all pairs n, m, one has $X_p\,\mathfrak{S}_a = Y_p\,\mathfrak{S}_a$. In particular,

$$X_p\,\mathfrak{D} = Y_p\,\mathfrak{D}. \tag{16.19}$$

PROOF. Given a function $a \colon \mathbf{Z}^2 \backslash \{0\} \to \mathbf{C}$, consider the measure $\mathfrak{S}_a$ introduced in (16.12). Dropping the subscript p which is kept fixed in the present proof, set $X \mathfrak{S}_a = \mathfrak{S}_{\tilde{X}a}$ and $Y \mathfrak{S}_a = \mathfrak{S}_{\tilde{Y}a}$. One has $(\tilde{X}a)(n, m) = v_p(n, m) \, a(n, m)$ and (a consequence of (16.12) and (16.17))

$$(\tilde{Y}a)(n, m) = \sum_{\substack{j \geq 0 \\ p^{j+1} \mid (n,m)}} a\left(\frac{n}{p^{j+1}}, \frac{m}{p^{j+1}}\right). \tag{16.20}$$

Thus $XY \mathfrak{S}_a = \mathfrak{S}_b$ and $YX \mathfrak{S}_a = \mathfrak{S}_c$ with

$$b(n, m) = v_p(n, m) \sum_{\substack{j \geq 0 \\ p^{j+1} \mid (n,m)}} a\left(\frac{n}{p^{j+1}}, \frac{m}{p^{j+1}}\right) \tag{16.21}$$

and

$$c(n, m) = \sum_{\substack{j \geq 0 \\ p^{j+1} \mid (n,m)}} [\, v_p(n, m) - j - 1 \,] \, a\left(\frac{n}{p^{j+1}}, \frac{m}{p^{j+1}}\right). \tag{16.22}$$

On the other hand, with $P = p^{-1-2i\pi\mathcal{E}}$, one has in the sense of formal power series $Y = \frac{P}{1-P}$, thus $Y(Y+1) = \frac{P}{(1-P)^2}$, in other words

$$Y(Y+1) = \sum_{j \geq 0} (j+1) \, (p^{j+1})^{-1-2i\pi\mathcal{E}}. \tag{16.23}$$

Using (16.21), (16.22) and (16.23), one gets (16.18).

Finally, if $a(n, m) = a(pn, pm)$ for all pairs n, m, one has

$$(\tilde{Y}a)(n, m) = \sum_{\substack{j \geq 0 \\ p^{j+1} \mid (n,m)}} a(n, m)$$

$$= v_p(n, m) \, a(n, m)$$

$$= (\tilde{X}a)(n, m). \tag{16.24}$$

$\square$

PROPOSITION 16.4. *Fix a prime p and drop the subscript p. Given a polynomial $f(X)$, there is a unique polynomial $(\Lambda f)(Y)$ with*

$$f(X) \, \mathfrak{D} = (\Lambda f)(Y) \, \mathfrak{D}. \tag{16.25}$$

The map Λ does not depend on p, and preserves the degree. Explicitly,

$$\Lambda(X^n) = n! \sum_{j=0}^{n} c_{j,n} \, Y^j, \tag{16.26}$$

where the coefficients are borrowed from the generating series

$$(e^t - 1)^j = \sum_{n \geq j} c_{j,n} t^n, \quad j \geq 0.$$ (16.27)

PROOF. Existence follows from the equations $X\mathfrak{D} = Y\mathfrak{D}$ and $[X, Y] = Y(Y+1)$.

With $P = p^{-1-2i\pi\mathcal{E}}$, one has for all $j \geq 0$

$$(Y + 1)^j = (1 - P)^{-j}$$

$$= \sum_{\ell \geq 0} \frac{(j)_\ell}{\ell!} (p^\ell)^{-1-2i\pi\mathcal{E}},$$ (16.28)

with $(j)_\ell = j(j+1)\dots(j+\ell-1)$. Thus $(Y+1)^j \mathfrak{D} = 2\pi\, \mathfrak{S}_a$ with

$$a(n, m) = \sum_{\substack{\ell \geq 0 \\ p^\ell | (n,m)}} \frac{(j)_\ell}{\ell!}.$$ (16.29)

It is clear (comparing $(j)_\ell$ and $(j')_\ell$ for large ℓ) that there can be no linear dependence between the measures $(Y+1)^j\mathfrak{D}$, $j \geq 0$, whence the uniqueness.

With $c_{j,n}$ defined by (16.27), set $c_{j,n} = 0$ if $j > n$ or $j < 0$, and assume that $X^n\mathfrak{D} = f_n(Y)\mathfrak{D}$ for some polynomial $f_n(Y)$ and some $n \geq 0$. Then

$$X^{n+1}\mathfrak{D} = f_n(Y)X\mathfrak{D} + [X, f_n(Y)]\mathfrak{D}$$
$$= Y f_n(Y)\mathfrak{D} + [X, Y] f_n'(Y)\mathfrak{D}$$
$$= Y f_n(Y)\mathfrak{D} + Y(Y+1) f_n'(Y)\mathfrak{D},$$ (16.30)

thus

$$X^{n+1}\mathfrak{D} = f_{n+1}(Y)\mathfrak{D}$$ (16.31)

with

$$f_{n+1}(Y) = Y\frac{d}{dY} f_n(Y) + Y\frac{d}{dY}(Y f_n(Y)).$$ (16.32)

On the other hand, with $e_j(t) = (e^t - 1)^j$,

$$\frac{d}{dt} e_j(t) = j(e_j(t) + e_{j-1}(t)).$$ (16.33)

Thus, as $e_j(t) = \sum_n c_{j,n} t^n$, one has

$$(n + 1) c_{j,n+1} = j(c_{j,n} + c_{j-1,n}).$$ (16.34)

From (16.32), it follows that if, for some n,

$$f_n(Y) = n! \sum_j c_{j,n} Y^j,$$ (16.35)

one then has

$$f_{n+1}(Y) = n! \sum_j j\, c_{j,n}\, Y^j + n! \sum_j (j+1)\, c_{j,n}\, Y^{j+1}$$

$$= n! \sum_j j\, (c_{j,n} + c_{j-1,n})\, Y^j$$

$$= (n+1)! \sum_j c_{j,n+1}\, Y^j . \tag{16.36}$$

$\square$

PROPOSITION 16.5. *With* $X : = X_p$ *and* $Y : = Y_p$, *one has for every* $t \in \mathbb{C}$

$$e^{tX}\, \mathfrak{D} = (1 + (1 - e^t)Y)^{-1}\, \mathfrak{D}, \tag{16.37}$$

where the operator on the right-hand side makes sense as a power series in Y, *i.e., as a Dirichlet series in* $2i\pi\mathcal{E}$, *since there is no constant term in the expansion* (16.17) *of* Y.

PROOF.

$$\frac{d}{dt}(1 + (1 - e^t)Y)^{-1} = e^t Y\, (1 + (1 - e^t)Y)^{-2} \tag{16.38}$$

thus

$$X\,(1 + (1 - e^t)Y)^{-1}\, \mathfrak{D}$$

$$= (1 + (1 - e^t)Y)^{-1}\, X\, \mathfrak{D} + [X, (1 + (1 - e^t)Y)^{-1}]\, \mathfrak{D}$$

$$= (1 + (1 - e^t)Y)^{-1}\, X\, \mathfrak{D} + (e^t - 1)(1 + (1 - e^t)Y)^{-2}\, [X, Y]\, \mathfrak{D}$$

$$= (1 + (1 - e^t)Y)^{-1}\, Y\, \mathfrak{D} + (e^t - 1)(1 + (1 - e^t)Y)^{-2}\, Y(Y + 1)\, \mathfrak{D}$$

$$= (1 + (1 - e^t)Y)^{-2}\, [(1 + Y - e^t Y)Y + e^t(Y^2 + Y) - Y^2 - Y]\, \mathfrak{D}$$

$$= e^t Y\, (1 + (1 - e^t)Y)^{-2}\, \mathfrak{D} \tag{16.39}$$

so that

$$(\frac{d}{dt} - X)(1 + (1 - e^t)Y)^{-1}\, \mathfrak{D} = 0 \tag{16.40}$$

and

$$(1 + (1 - e^t)Y)^{-1}\, \mathfrak{D} = e^{tX}\, \mathfrak{D} . \tag{16.41}$$

$\square$

PROPOSITION 16.6. *Given a function* $r \mapsto \psi(r)$, $r = 1, 2, \ldots$, *denote as* $\mathrm{Mult}(\psi(r))$ *the operator, acting on measures supported in* $\mathbb{Z}^2 \backslash \{0\}$, *which consists in multiplying the coefficient* $a(n, m)$ *in* (16.12) *by* $\psi((n, m))$, *with* $(n, m) = g.c.d.(n, m)$. *Given a prime* p, *denote as* $|\ |_p$ *the standard* p–*adic absolute value on* $\mathbb{Q}$. *Then, for every* $\nu \in \mathbb{C}$,

$$\mathrm{Mult}\,(|r|_p^{-\nu})\, \mathfrak{D} = \frac{1 - p^{-1-2i\pi\mathcal{E}}}{1 - p^{-1+\nu-2i\pi\mathcal{E}}}\, \mathfrak{D} \tag{16.42}$$

and

$$\text{Mult}\,(\frac{1 - p^\nu |r|_p^{-\nu}}{1 - p^\nu})\,\mathfrak{D} = \frac{1}{1 - p^{-1+\nu-2i\pi\mathcal{E}}}\,\mathfrak{D}\,, \qquad (16.43)$$

where the right–hand side makes sense according to definition 16.2.

PROOF. If $\nu = \frac{t}{\log p}$, one has $p^\nu = e^t$ so that (16.37) reads $p^{\nu X}\mathfrak{D} = (1+(1-p^\nu)Y)^{-1}\,\mathfrak{D}$. On one hand, $p^{\nu X}\mathfrak{D}$ is the image of $\mathfrak{D}$ by the operator $\text{Mult}\,(p^{\nu v_p(r)}) = \text{Mult}\,(|r|_p^{-\nu})$. On the other hand, with $P = p^{-1-2i\pi\mathcal{E}}$ again so that $Y = \frac{P}{1-P}$, one has

$$(1 + (1 - p^\nu)Y)^{-1} = \frac{1 - P}{1 - p^\nu P} : \qquad (16.44)$$

this proves (16.42). Starting from this equation, one also gets

$$\text{Mult}\,(\frac{1 - p^\nu |r|_p^{-\nu}}{1 - p^\nu})\,\mathfrak{D} = (\frac{1}{1 - p^\nu} - \frac{p^\nu}{1 - p^\nu}\frac{1 - P}{1 - p^\nu P})\,\mathfrak{D}$$

$$= \frac{1}{1 - p^\nu P}\,\mathfrak{D}\,. \qquad (16.45)$$

$\square$

We now perform Eulerian products, in the usual way. What makes this possible is the full force of proposition 16.3, *i.e.*, the fact that not only $X_p\,\mathfrak{D} = Y_p\,\mathfrak{D}$ but, also, $X_p\,\mathfrak{S}_a = Y_p\,\mathfrak{S}_a$ whenever $a(n,m)$ only depends on the collection of pairs $(|n|_q, |m|_q)$ for $q \neq p$. The same extension could be carried into propositions 16.5 and 16.6.

Given any integer $r \geq 1$, $|r|_p^{-\nu}$ or $\frac{1-p^\nu|r|_p^{-\nu}}{1-p^\nu}$ is equal to 1 for almost every prime p and, if $r = p_1^{\alpha_1}\ldots p_k^{\alpha_k}$, one has

$$\prod_p |r|_p^{-\nu} = \prod_j (p_j^{-\alpha_j})^{-\nu}$$

$$= r^\nu \qquad (16.46)$$

and

$$\prod_p \frac{1 - p^\nu |r|_p^{-\nu}}{1 - p^\nu} = \prod_j \frac{1 - p_j^{\nu(1+\alpha_j)}}{1 - p_j^\nu}$$

$$= \prod_j (1 + p_j^\nu + \cdots + p_j^{\alpha_j \nu})$$

$$= \sum_{\substack{n \geq 1 \\ n|r}} n^\nu$$

$$= \sigma_\nu(r)\,. \qquad (16.47)$$

Thus, from (16.42) and (16.43),

$$\text{Mult}\,(r^{\nu})\,\mathfrak{D} = (\prod_{p} \frac{1 - p^{-1-2i\pi\mathcal{E}}}{1 - p^{-1+\nu-2i\pi\mathcal{E}}})\,\mathfrak{D} \tag{16.48}$$

and

$$\text{Mult}\,(\sigma_{\nu}(r))\,\mathfrak{D} = (\prod_{p} \frac{1}{1 - p^{-1+\nu-2i\pi\mathcal{E}}})\,\mathfrak{D}: \tag{16.49}$$

here, the operator on the right–hand side of (16.48) or (16.49) is, for all $\nu \in \mathbb{C}$, well-defined, according to definition 16.2, as a Dirichlet series in $2i\pi\mathcal{E}$.

One would expect that formulas like

$$\text{Mult}\,(r^{\nu})\,\mathfrak{D} = \frac{\zeta(1 - \nu + 2i\pi\mathcal{E})}{\zeta(1 + 2i\pi\mathcal{E})}\,\mathfrak{D} \tag{16.50}$$

or

$$\text{Mult}\,(\sigma_{\nu}(r))\,\mathfrak{D} = \zeta(1 - \nu + 2i\pi\mathcal{E})\,\mathfrak{D} \tag{16.51}$$

should be valid too, in some sense closer to a spectral–theoretic one and in a certain range of values of ν. We show that this is, indeed, the case. Recall (16.2) and the equation

$$2i\pi\mathcal{E}\,\mathfrak{E}_{i\lambda}^{\natural} = -i\lambda\,\mathfrak{E}_{i\lambda}^{\natural}. \tag{16.52}$$

THEOREM 16.7. *Let the function* $\psi = \psi(r)$, $r = 1, 2, \ldots$, *satisfy* $\sum_{r \geq 1} \frac{|\psi(r)|}{r} < \infty$. *Then, in a weak sense in* $S'_{\text{even}}(\mathbb{R}^2)$,

$$\text{Mult}\,(\psi(r))\,\mathfrak{D} = \frac{12}{\pi} \sum_{r \geq 1} \frac{\psi(r)}{r^2} + \int_{-\infty}^{\infty} \frac{\sum_{r \geq 1} \psi(r)\,r^{-1+i\lambda}}{\zeta(1 - i\lambda)}\,\mathfrak{E}_{i\lambda}^{\natural}\,d\lambda. \tag{16.53}$$

PROOF. Set, for $h \in S_{\text{even}}(\mathbb{R}^2)$,

$$< \mathfrak{D}^{\text{prime}}, h >: = 2\pi \sum_{\substack{m \in \mathbb{Z},\, n \in \mathbb{Z} \\ (m,n)=1}} h(n, m). \tag{16.54}$$

Following the proof of proposition 16.1, note that

$$< \mathfrak{D}^{\text{prime}}, h_{-\mu} > = \sum_{(n_1, m_1)=1} \int_{0}^{\infty} t^{-i\mu}\,h(tn_1, tm_1)\,dt \tag{16.55}$$

if $\text{Im}\,\mu > 1$. Since any pair $(n, m) \in \mathbb{Z} \times \mathbb{Z} \setminus \{(0,0)\}$ can be uniquely written as (an_1, am_1) with $a \geq 1$, $(n_1, m_1) = 1$, one gets, comparing (16.55) and (16.6),

$$\zeta(1 - i\mu) < \mathfrak{D}^{\text{prime}}, h_{-\mu} > = < \mathfrak{D}, h_{-\mu} > . \tag{16.56}$$

This leads to the formula

$$\mathfrak{D}^{\text{prime}} = \frac{12}{\pi} + \int_{-\infty}^{\infty} (\zeta(1 - i\lambda))^{-1}\, \mathfrak{E}_{i\lambda}^{\natural}\, d\lambda \tag{16.57}$$

after a contour deformation. Then

$$< \text{Mult}\,(\psi(r))\,\mathfrak{D}\,,\, h > = 2\pi \sum_{|n|+|m|\neq 0} \psi(\,(n,m)\,)\,h(n,m)$$

$$= 2\pi \sum_{r\geq 1} \psi(r) \sum_{(n,m)=1} h(rn, rm)$$

$$= \sum_{r\geq 1} \psi(r) < r^{-1-2i\pi\mathcal{E}}\,\mathfrak{D}^{\text{prime}}\,,\, h > \tag{16.58}$$

according to (16.11): this leads to (16.53). $\qquad\qquad\square$

New terms may appear when the condition $\sum_{r\geq 1} \frac{|\psi(r)|}{r} < \infty$ is not satisfied. For instance, one has

PROPOSITION 16.8. *If* Re $\nu < 0$ *or* $\nu = 0$, *one has in a weak sense in* $S'_{\text{even}}(\mathbb{R}^2)$

$$\text{Mult}\,(r^{\nu})\,\mathfrak{D} = \frac{12}{\pi}\,\zeta(2-\nu) + \int_{-\infty}^{\infty} \frac{\zeta(1-\nu-i\lambda)}{\zeta(1-i\lambda)}\,\mathfrak{E}_{i\lambda}^{\natural}\, d\lambda\,. \tag{16.59}$$

If Re $\nu > 0$, $\nu \neq 1$,

$$\text{Mult}\,(r^{\nu})\,\mathfrak{D} = \frac{12}{\pi}\,\zeta(2-\nu) + \frac{2\pi}{\zeta(1+\nu)}\,\mathfrak{E}_{-\nu}^{\natural} + \int_{-\infty}^{\infty} \frac{\zeta(1-\nu-i\lambda)}{\zeta(1-i\lambda)}\,\mathfrak{E}_{i\lambda}^{\natural}\, d\lambda\,. \tag{16.60}$$

If Re $\nu = 0$, $\nu \neq 0$,

$$\text{Mult}\,(r^{\nu})\,\mathfrak{D} = \frac{12}{\pi}\,\zeta(2-\nu) + \frac{\pi}{\zeta(1+\nu)}\,\mathfrak{E}_{-\nu}^{\natural} + \text{FP} \int_{-\infty}^{\infty} \frac{\zeta(1-\nu-i\lambda)}{\zeta(1-i\lambda)}\,\mathfrak{E}_{i\lambda}^{\natural}\, d\lambda\,, \tag{16.61}$$

where the symbol FP *stands for "finite part" (or Cauchy's principal value in this case): it means that one should take the limit as ε goes to zero of the integral on the union* $]-\infty, -\varepsilon[\cup]\varepsilon, \infty[$.

PROOF. Let us first remark that the first of the three formulas is the correct meaning of the formula (16.50), guessed from an "adelic" argument: this can be seen from (16.2) as a consequence of (16.52) and of the equation $2i\pi\mathcal{E}.1 = 1$.

When Im $\mu > \max(\text{Re}\,\nu, 1)$, one has

$$< \text{Mult}\,(r^{\nu})\,\mathfrak{D}\,,\, h_{-\mu} > = \sum_{a\geq 1} a^{\nu-1+i\mu} \sum_{(n_1,m_1)=1} \int_0^{\infty} t^{-i\mu}\, h(tn_1, tm_1)\, dt\,, \tag{16.62}$$

i.e.,

$$< \text{Mult} \, (r^\nu) \, \mathfrak{D} \, , h_{-\mu} > = \frac{\zeta(1 - \nu - i\mu)}{\zeta(1 - i\mu)} < \mathfrak{D}, h_{-\mu} >$$

$$= \frac{\zeta(1 - \nu - i\mu)}{\zeta(1 - i\mu)} < \mathfrak{E}^\sharp_{i\mu}, h > . \tag{16.63}$$

To move the contour to the real line, starting from the integral

$$\int_{ia-\infty}^{ia+\infty} \frac{\zeta(1 - \nu - i\mu)}{\zeta(1 - i\mu)} < \mathfrak{E}^\sharp_{i\mu}, h > d\mu$$

with $a > \max(\text{Re} \, \nu, 1)$, we have to take into account the poles at $\mu = i$ like in (16.7), and at $\mu = i\nu$ if $\text{Re} \, \nu \geq 0$ (a double pole in the case when $\nu = 1$). This easily leads to the proof of proposition 16.8, at the price of a small half–circular detour in the upper half–plane around $i\nu$ in the case when $\text{Re} \, \nu = 0$, $\nu \neq 0$.   □

Remark. In the case when $\nu = 1$, we have not completed the calculation, which can be performed with the help of (3.28) together with proposition 3.6, the latter one in order to compute $\text{res}_{-i\mu}(\widehat{h^\flat_\mu})$.

PROPOSITION 16.9. *If* $\text{Re} \, \nu < 0$ *or* $\nu = 0$,

$$\text{Mult} \, (\sigma_\nu(r)) \, \mathfrak{D} = 2\pi \, \zeta(2 - \nu) + \int_{-\infty}^{\infty} \zeta(1 - \nu - i\lambda) \, \mathfrak{E}^\sharp_{i\lambda} \, d\lambda \, . \tag{16.64}$$

If $\text{Re} \, \nu > 0$, $\nu \neq 1$,

$$\text{Mult} \, (\sigma_\nu(r)) \, \mathfrak{D} = 2\pi \, \zeta(2 - \nu) + 2\pi \, \mathfrak{E}^\sharp_{-\nu} + \int_{-\infty}^{\infty} \zeta(1 - \nu - i\lambda) \, \mathfrak{E}^\sharp_{i\lambda} \, d\lambda \, . \tag{16.65}$$

PROOF. It is just the same proof, using [**16**, p. 250]

$$\sum_{a \geq 1} \sigma_\nu(a) \, a^{-1+i\mu} = \zeta(1 - i\mu) \, \zeta(1 - \nu - i\mu) \tag{16.66}$$

again, so that

$$< \text{Mult} \, (\sigma_\nu(r)) \, \mathfrak{D} \, , h_{-\mu} >= \zeta(1 - \nu - i\mu) < \mathfrak{E}^\sharp_{i\mu}, h > \, ; \tag{16.67}$$

the third formula (16.61) extends too.

□

We close this section with a few remarks concerning the way Hecke operators can be defined on the level of distributions on $\mathbb{R}^2$.

Recall that, in section 8, we lifted the Hecke operators T_N, acting on modular forms, to operators T^λ_N acting on $\mathbb{Z}$–periodic distributions in the space $\mathcal{D}'_{i\lambda}$, so that (*cf.* (8.8)), the equation

$$\Theta_{i\lambda} T^\lambda_N \mathfrak{T} = T_N \Theta_{i\lambda} \mathfrak{T} \tag{16.68}$$

should be satisfied for every $\mathfrak{T} \in \mathcal{D}'_{i\lambda}$. Again, we go one step further and define a version T_N^{dist} acting on $\mathcal{S}'^{\text{per}}_{\text{even}}(\mathbb{R}^2)$: this is by definition the subspace of $\mathcal{S}'_{\text{even}}(\mathbb{R}^2)$ consisting of distributions invariant under the *linear* action of the matrix $\left(\begin{smallmatrix} 1 & 1 \\ 0 & 1 \end{smallmatrix} \right)$, *i.e.*, under the transformation $\xi \mapsto (\xi_1 + \xi_2, \xi_2)$. Of course, it contains the space of Γ–invariant tempered distributions.

DEFINITION 16.10. Given $\mathfrak{S} \in \mathcal{S}'^{\text{per}}_{\text{even}}(\mathbb{R}^2)$ and an integer $N \geq 1$, we define the distribution $T_N^{\text{dist}}\mathfrak{S}$ through

$$< T_N^{\text{dist}}\mathfrak{S}, h >: \; = N^{-\frac{1}{2}} \sum_{\substack{ad=N, d>0 \\ b \bmod d}} < \mathfrak{S}, \xi \mapsto h\left(\frac{d\xi_1 - b\xi_2}{\sqrt{N}}, \frac{a\xi_2}{\sqrt{N}} \right) > . \tag{16.69}$$

Remark. If $\mathfrak{S}$ is a measure supported in $\mathbb{Z}^2 \backslash \{0\}$, $T_N^{\text{dist}}\mathfrak{S}$ is not, in general, but $N^{-i\pi\mathcal{E}} T_N^{\text{dist}}\mathfrak{S}$ is, since

$$< N^{-i\pi\mathcal{E}} T_N^{\text{dist}}\mathfrak{S}, h >: \; = \sum_{\substack{ad=N, d>0 \\ b \bmod d}} < \mathfrak{S}, \xi \mapsto h(d\xi_1 - b\xi_2, a\xi_2) > \tag{16.70}$$

(the remark 3.8 "Why $GL(2)$ and not $SL(2)$" in [9, p. 249] may be quoted at that point). On homogeneous distributions, of course, $N^{-i\pi\mathcal{E}}$ acts as a scalar.

Note that if $\mathfrak{S}$ is the distribution associated with an (even) function h_1 on $\mathbb{R}^2$, *i.e.*, $< \mathfrak{S}, h >= \int_{\mathbb{R}^2} h_1(\xi)\, h(\xi)\, d\xi$, the distribution $T_N^{\text{dist}}\mathfrak{S}$ is associated, as shown by a simple change of variables, with the function

$$(T_N^{\text{dist}}h_1)(\xi) = N^{-\frac{1}{2}} \sum_{\substack{ad=N, d>0 \\ b \bmod d}} h_1\left(\frac{a\xi_1 + b\xi_2}{\sqrt{N}}, \frac{d\xi_2}{\sqrt{N}} \right) . \tag{16.71}$$

If h is a function living on Ξ (*i.e.*, an even function on $\mathbb{R}^2$), recall that its decomposition into homogeneous parts, followed by some obvious restriction, led us to defining a certain function $h_\lambda^\flat$ on the real line for almost all real λ (*cf.* (4.19)–(4.21)). Performing the change of variables $\xi_1 = st$, $\xi_2 = t$ in the integral

$$< h_\lambda^\flat, u >= \frac{1}{2\pi} \int_{-\infty}^{\infty} \left(\int_0^\infty t^{i\lambda} h(ts, t)\, dt \right) u(s)\, ds , \tag{16.72}$$

one may define, more generally,

$$< \mathfrak{S}_\lambda^\flat, u >= \frac{1}{4\pi} < \mathfrak{S}, \xi \mapsto |\xi_2|^{-1+i\lambda} u(\frac{\xi_1}{\xi_2}) > , \tag{16.73}$$

in other words (*cf.* (4.35))

$$< \mathfrak{S}_\lambda^\flat, u >= \frac{1}{4\pi} < \mathfrak{S}, u_{-\lambda}^\natural > , \tag{16.74}$$

whenever $\mathfrak{S}$ is an even distribution on $\mathbf{R}^2$, provided, of course, that it does make sense.

Using (16.73) when meaningful, one may check with this definition, using also $\frac{a}{\sqrt{N}} = (\frac{a}{d})^{\frac{1}{2}}$, that

$$< (T_N^{\text{dist}}\mathfrak{S})_\lambda^b, u > = \frac{1}{4\pi} < T_N^{\text{dist}}\mathfrak{S}, \xi \mapsto |\xi_2|^{-1+i\lambda} u(\frac{\xi_1}{\xi_2}) >$$

$$= \frac{1}{4\pi} N^{-\frac{1}{2}} \sum_{\substack{ad=N, d>0 \\ b \bmod d}} (\frac{a}{d})^{\frac{-1+i\lambda}{2}} < \mathfrak{S}, \xi \mapsto |\xi_2|^{-1+i\lambda} u(\frac{d\xi_1 - b\xi_2}{a\xi_2}) >,$$

$$(16.75)$$

whereas, from (8.6),

$$< T_N^\lambda \mathfrak{S}_\lambda^b, u > = N^{-\frac{1}{2}} \sum_{\substack{ad=N, d>0 \\ b \bmod d}} (\frac{a}{d})^{\frac{-1+i\lambda}{2}} < \mathfrak{S}_\lambda^b, s \mapsto u(\frac{ds - b}{a}) >,$$

$$(16.76)$$

an expression which is readily seen to be the same as the one in (16.75) if one applies (16.74) again. Thus

$$(T_N^{\text{dist}}\mathfrak{S})_\lambda^b = T_N^\lambda \mathfrak{S}_\lambda^b,$$

$$(16.77)$$

an identity which, coupled with (16.68), shows the compatibility of the definition of T_N^{dist} with that of Hecke operators on the modular form level.

PROPOSITION 16.11. *The operator* T_N^{dist} *commutes with the Euler operator* $\mathcal{E}$ *defined in* (4.8) *as well as with the symplectic Fourier transformation* $\mathcal{F}$ *defined in* (4.11).

PROOF. In such formal computations, it is no loss of generality to use (16.71), *i.e.*, to apply the various operators involved to a function h_1 rather than a distribution. That $\mathcal{E}T_N^{\text{dist}}h_1 = T_N^{\text{dist}}\mathcal{E}h_1$ is obvious. Starting from

$$(\mathcal{F}T_N^{\text{dist}}h_1)(\xi) = N^{-\frac{1}{2}} \sum_{\substack{ad=N, d>0 \\ b \bmod d}} \int h_1 \left(\frac{a\eta_1 + b\eta_2}{\sqrt{N}}, \frac{d\eta_2}{\sqrt{N}} \right) e^{2i\pi(\xi_1\eta_2 - \xi_2\eta_1)} \, d\eta$$

$$(16.78)$$

and performing the change of variables (with determinant 1) $(\eta_1, \eta_2) \mapsto \left(\frac{d\eta_1 - b\eta_2}{\sqrt{N}}, \frac{a\eta_2}{\sqrt{N}} \right)$, we get

$$(\mathcal{F} T_N^{\text{dist}} h_1)(\xi) = N^{-\frac{1}{2}} \sum_{\substack{ad=N, d>0 \\ b \bmod d}} \int h_1(\eta) \exp 2i\pi \left(\frac{a\xi_1 + b\xi_2}{\sqrt{N}} \eta_2 - \frac{d\xi_2}{\sqrt{N}} \eta_1 \right) d\eta$$

$$= N^{-\frac{1}{2}} \sum_{\substack{ad=N, d>0 \\ b \bmod d}} (\mathcal{F} h_1) \left(\frac{a\xi_1 + b\xi_2}{\sqrt{N}}, \frac{d\xi_2}{\sqrt{N}} \right)$$

$$= (T_N^{\text{dist}} \mathcal{F} h_1)(\xi). \tag{16.79}$$

The proof is over, but before we leave it we note that the last property is equivalent to the fact that the Hecke operator T_N^λ as defined in (8.6) satisfies the equation

$$T_N^{-\lambda} \theta_{i\lambda} = \theta_{i\lambda} T_N^\lambda, \tag{16.80}$$

where the intertwining operator $\theta_{i\lambda}$ was defined in (2.15). Indeed, using (4.28) then (16.76), we get

$$(\mathcal{F} T_N^{\text{dist}} h_1)_{-\lambda}^b = \theta_{i\lambda} (T_N^{\text{dist}} h_1)_\lambda^b = \theta_{i\lambda} T_N^\lambda (h_1)_\lambda^b; \tag{16.81}$$

on the other hand, starting from the result of last proposition, and using again (16.76) and (4.28), this is also

$$(T_N^{\text{dist}} \mathcal{F} h_1)_{-\lambda}^b = T_N^{-\lambda} (\mathcal{F} h_1)_{-\lambda}^b = T_N^{-\lambda} \theta_{i\lambda} (h_1)_\lambda^b. \tag{16.82}$$

$\square$

The equation $\theta_\nu \phi_z^\nu = \phi_z^{-\nu}$, already observed in the remark following (3.33), explains why an equation like (16.80) has no analogue on the modular form level, in connection with the usual Hecke operators: it is for the same reason that the functions $E_{\frac{1-\nu}{2}}^*$ and $E_{\frac{1+\nu}{2}}^*$ are identical, while the distributions $\mathfrak{E}_\nu$ and $\mathfrak{E}_{-\nu}$ are only the images of each other under the canonical intertwining operator.

One can handle all Hecke operators simultaneously through the consideration of the one–parameter family of operators

$$\mathcal{L}(s) := \sum_{N \geq 1} N^{-s} T_N^{\text{dist}}$$

$$= \prod_p (1 - p^{-s} T_N^{\text{dist}} + p^{-2s})^{-1}. \tag{16.83}$$

PROPOSITION 16.12. *For* Re $\nu < -1$ *and* Re $s > 1 - \frac{\text{Re } \nu}{2}$, *one has*

$$\mathcal{L}(s) \mathfrak{E}_\nu^\sharp = \zeta\left(s - \frac{\nu}{2}\right) \zeta\left(s + \frac{\nu}{2}\right) \mathfrak{E}_\nu^\sharp, \tag{16.84}$$

where $\mathfrak{E}_\nu^\natural$ has been defined in (13.25). *For any Maass cusp–form $\mathcal{M}_k$, and with $\mathfrak{M}_{\pm k}^\natural$ as defined in (13.27), one has for* Re s *large enough*

$$\mathcal{L}(s)\,\mathfrak{M}_{\pm k}^\natural = L(s,\mathcal{M}_k)\,\mathfrak{M}_{\pm k}^\natural\,. \qquad (16.85)$$

PROOF. With $h \in \mathcal{S}_{\text{even}}(\mathbb{R})$, we may rephrase (13.25) as

$$< \mathfrak{E}_\nu^\natural, h > = \frac{1}{2}\,\zeta(1-\nu) \sum_{g \in \Gamma \backslash \Gamma_\infty} \int_{-\infty}^{\infty} |t|^{-\nu}\,h(t(g.\varepsilon))\,dt\,, \qquad (16.86)$$

where $\varepsilon = \left(\begin{smallmatrix}1\\0\end{smallmatrix}\right) \in \mathbb{R}^2$. Then, using (16.69) together with the change of variable $t \mapsto N^{\frac{1}{2}}t$, and remembering (8.12) (which obviously works also with the two factors in the reverse order), we get

$$< T_N^{\text{dist}}\,\mathfrak{E}_\nu^\natural, h > = \frac{1}{2}\,\zeta(1-\nu)\,N^{-\frac{\nu}{2}} \sum_{g \in M_N(\mathbb{Z}) \backslash \Gamma_\infty} \int_{-\infty}^{\infty} |t|^{-\nu}\,h(t(g.\varepsilon))\,dt\,. \qquad (16.87)$$

In [36, p. 246], it is explained why, given a pair a, c with $(a,c) = k \geq 1$, there are k ways to complete $\left(\begin{smallmatrix}a\\c\end{smallmatrix}\right)$ into a class in $M_N(\mathbb{Z})\backslash\Gamma_\infty$: since $k^{-1}(g.\varepsilon) = k^{-1}\left(\begin{smallmatrix}a\\c\end{smallmatrix}\right)$ if $g = \left(\begin{smallmatrix}a & b\\c & d\end{smallmatrix}\right)$, we get after the change of variable $t \mapsto k^{-1}t$ the relation

$$T_N^{\text{dist}}\,\mathfrak{E}_\nu^\natural = N^{-\frac{\nu}{2}}\,\sigma_\nu(N)\,\mathfrak{E}_\nu^\natural\,, \qquad (16.88)$$

from which one concludes thanks to [16, p. 290].

The relation concerning $\mathfrak{M}_{\pm k}^\natural$ can be traced back to $\mathcal{L}(s)\,\mathcal{M}_k = L(s,\mathcal{M}_k)\,\mathcal{M}_k$ which is after all what Hecke's theory is made for.

□

17. Quantization, products and Poisson brackets

The aim of the present section is to explain, as promised in the introduction, why the existence of formulas such as (8.16) and (8.26), linking the
bilinear operations $L^{\pm}_{i\lambda_1,i\lambda_2;i\lambda}$ to the seemingly simpler ones of extracting the
components, with respect to the spectral decomposition of the Laplacian on
II, of the product or Poisson bracket of the two functions involved, had to
be expected. The link between the two sides is quantization, more precisely
the composition formula for symbols, which, in the case of the familiar Weyl
calculus, was the object of section 5. In our discussion, in the next section,
of the equivalence between the Lax–Phillips scattering theory for the automorphic wave equation and that of automorphic distributions on $\mathbb{R}^2$ (cf.
section 13), we shall also need a few formulas from the present section.

The holomorphic discrete series of representations of G

We first present the holomorphic discrete series of G: this is of course
well-known, but we must fix our notations.

PROPOSITION 17.1. Let τ be a real number > -1, and let $H_{\tau+1}$ be the
Hilbert space of all (classes of) measurable functions on the half-line $(0, \infty)$
such that

$$\|v\|^2_{\tau+1} := \int_0^\infty |v(s)|^2 s^{-\tau}\, ds < \infty. \qquad (17.1)$$

There exists a unitary projective representation $\pi = \mathcal{D}_{\tau+1}$ of G in $H_{\tau+1}$ with
the following properties, in the statement of which $g = \left(\begin{smallmatrix} a & b \\ c & d \end{smallmatrix}\right)$:

1. for every pair (g, g_1) of elements of G, the complex number
   $\pi(gg_1)^{-1}\pi(g)\pi(g_1)$ belongs to the group $\exp(2i\pi\tau\mathbb{Z})$;
2. for every g with $b < 0$, $\pi(g) = e^{i\pi(\tau+1)}\pi(-g)$;
3. if $b = 0$, $a > 0$, $(\pi(g)v)(s) = a^{\tau-1}v(a^{-2}s)e^{2i\pi\frac{c}{a}s}$;
4. if $b > 0$,

$$(\pi(g)v)(s) = e^{-i\pi\frac{\tau+1}{2}}\frac{2\pi}{b}\int_0^\infty v(t)\left(\frac{s}{t}\right)^{\frac{\tau}{2}}\exp\left(2i\pi\frac{ds+at}{b}\right)J_\tau\left(\frac{4\pi}{b}\sqrt{st}\right)dt. \qquad (17.2)$$

PROOF. That one can give a real-type realization of the holomorphic
discrete series in which the integral kernels of the unitary operators $\pi(g)$ are
realized with the help of Bessel–type functions is well-known in a much more
general context [14, 15]. An elementary reference is ([39], proposition 1.5)
(note that in the latter reference, the representation was denoted as $g \mapsto$
$M_{g^{-1}}$ and that our present $\mathcal{D}_{\tau+1}(g)$ is just M_{g_1} with $g_1 = \left(\begin{smallmatrix} 0 & 1 \\ 1 & 0 \end{smallmatrix}\right)g^{-1}\left(\begin{smallmatrix} 0 & 1 \\ 1 & 0 \end{smallmatrix}\right)$,
and τ substituted for λ).

Of course, if τ is an integer, $\mathcal{D}_{\tau+1}$ can be chosen as a genuine representation in a unique way; in any case, $\mathcal{D}_{\tau+1}(g)$ is uniquely defined up to the multiplication by some number in $\exp(2i\pi\tau\mathbf{Z})$. $\qquad\square$

PROPOSITION 17.2. *Under the assumption that* $\tau > 0$, *consider the Hilbert space* $\tilde{H}_{\tau+1}$ *of all holomorphic functions* f *in* Π *with*

$$\|f\|^2 := \int_\Pi |f(z)|^2 (\operatorname{Im} z)^{\tau+1}\, d\mu(z) < \infty \qquad (17.3)$$

together with the map $v \mapsto f$,

$$f(z) = (4\pi)^{\frac{\tau}{2}} (\Gamma(\tau))^{-\frac{1}{2}} z^{-\tau-1} \int_0^\infty v(s)\, e^{-2i\pi s z^{-1}}\, ds. \qquad (17.4)$$

The map just defined is an isometry from $H_{\tau+1}$ *onto* $\tilde{H}_{\tau+1}$: *it intertwines the representation* $\mathcal{D}_{\tau+1}$ *of* G *in* $H_{\tau+1}$ *and a representation* $\tilde{\mathcal{D}}_{\tau+1}$ *of* G *in* $\tilde{H}_{\tau+1}$, *taken from the holomorphic (projective) discrete series, characterized up to scalar factors in the group* $\exp(2i\pi\tau\mathbf{Z})$ *by the fact that*

$$(\tilde{\mathcal{D}}_{\tau+1}(g)f)(z) = (-cz+a)^{-\tau-1} f\left(\frac{dz-b}{-cz+a}\right) \qquad (17.5)$$

if $c < 0$. *In all this the fractional powers which occur are those associated with the principal determination of the logarithm in* Π.

PROOF. This is just proposition 1.5 in [39] if the intertwining operator (from a real to a complex realization) used there is followed by the map $\bar{f}_1 \mapsto f$ from antiholomorphic functions on the right half-plane $-i\Pi$ to holomorphic functions on Π defined by

$$f(z) = z^{-\tau-1} \bar{f}_1\left(\frac{-i}{\bar{z}}\right). \qquad (17.6)$$

Note that if $k = \tau + 1$ is a positive integer, $\tilde{\mathcal{D}}_{\tau+1}(g^{-1})f = f|_k g$ if one switches to a notation familiar in the theory of holomorphic modular forms. $\qquad\square$

Our reason for changing our conventions from preceding works ([39] to [42]) is twofold. First, the shift by one unit $\tau \mapsto \tau + 1$ and the choice of the upper half-plane Π instead of the right–half plane $-i\Pi$ were made, of course, to please the majority: still, note that the choice of the right–half plane, a very reasonable one in quantization theory (there is some justification of this remark between (13.5) and (13.6)), cannot be maintained when dealing with modular form theory. On the other hand, the conjugation by the matrix $\left(\begin{smallmatrix}0&1\\1&0\end{smallmatrix}\right)$ in the very definition of $\mathcal{D}_{\tau+1}$ (*cf.* proof of proposition 17.1) permits to get a perfect agreement with the metaplectic representation. Recall that this is a projective unitary representation Met of G in $L^2(\mathbf{R})$ with the following properties:

1. the scalar factors $\mathrm{Met}(gg_1)^{-1}\mathrm{Met}(g)\mathrm{Met}(g_1)$ can only be ± 1;
2. if $b = c = 0$, $a > 0$ and $u \in L^2(\mathbf{R})$, one has $(\mathrm{Met}(g)u)(x) = a^{-\frac{1}{2}}u(a^{-1}x)$;

3. if $g = \left(\begin{smallmatrix} 1 & 0 \\ c & 1 \end{smallmatrix}\right)$, then $(\mathrm{Met}(g)u)(x) = e^{i\pi c x^2} u(x)$;

4. If $g = \left(\begin{smallmatrix} 0 & -1 \\ 1 & 0 \end{smallmatrix}\right)$, then $\mathrm{Met}(g) = e^{\frac{i\pi}{4}}\mathcal{F}^{-1}$, where the one-dimensional inverse Fourier transformation is normalized as

$$(\mathcal{F}^{-1}u)(x) = \int_{\mathbb{R}} u(y)\, e^{2i\pi xy}\, dy. \tag{17.7}$$

The metaplectic representation is not irreducible, but splits into its even and odd parts, corresponding to even or odd functions on the real line. The following proposition is easy with the help of proposition 17.1 and played a basic role in [40].

PROPOSITION 17.3. *The map* $Sq_{\mathrm{even}}: v \mapsto u$, *with* $u(x) = 2^{-\frac{3}{4}}\, |x|\, v(\frac{x^2}{2})$, *is an isometry from* $H_{\frac{1}{2}}$ *onto* $L^2_{\mathrm{even}}(\mathbb{R})$, *which intertwines the representation* $\mathcal{D}_{\frac{1}{2}}$ *with the even part of the metaplectic representation. Similarly, the map* $Sq_{\mathrm{odd}}: v \mapsto u$, *with* $u(x) = 2^{-\frac{1}{4}}\, v(\frac{x^2}{2})\, \mathrm{sign}\, x$, *is an isometry from* $H_{\frac{3}{2}}$ *onto* $L^2_{\mathrm{odd}}(\mathbb{R})$, *which intertwines the representation* $\mathcal{D}_{\frac{3}{2}}$ *with the odd part of the metaplectic representation.*

Symbolic calculi with Π as a phase space

We now turn to the definition of several symbolic calculi of operators acting on $H_{\tau+1}$, under the standing assumption that $\tau > -1$. The most important species of symbol in what follows is the so–called *passive* symbol, but we shall first recall the definition of the *contravariant* and *covariant* symbols from Berezin's theory [1, 2], which some readers may already be familiar with. Given $z \in \Pi$, the function $\psi_z^{\tau+1}$, with

$$\psi_z^{\tau+1}(s) = (\Gamma(\tau+1))^{-\frac{1}{2}} (4\pi)^{\frac{\tau+1}{2}} \left(\mathrm{Im}\left(-\frac{1}{z}\right)\right)^{\frac{\tau+1}{2}} s^{\tau}\, e^{2i\pi s\bar{z}^{-1}}, \tag{17.8}$$

has norm 1 in $H_{\tau+1}$ as soon as $\tau > -1$: also ([39], proposition 4.3),

$$\mathcal{D}_{\tau+1}(g)\psi_z^{\tau+1} = \omega\, \psi_{g.z}^{\tau+1} \tag{17.9}$$

for all $g \in G$, $z \in \Pi$ and some ω, $|\omega| = 1$ depending on (g, z). If $\tau > 0$, the function f introduced in proposition 17.2 can be written as

$$f(z) = (4\pi)^{-\frac{1}{2}}\, \tau^{\frac{1}{2}}\, z^{-\tau-1}(\mathrm{Im}\left(-\frac{1}{z}\right))^{-\frac{\tau+1}{2}} (\psi_z^{\tau+1}, u), \tag{17.10}$$

where the scalar product in $H_{\tau+1}$, linear with respect to the *second* variable, occurs on the right-hand side: hence

$$\|v\|^2_{\tau+1} = \frac{\tau}{4\pi} \int |(\psi_z^{\tau+1}, v)|^2\, d\mu(z). \tag{17.11}$$

Thus the family $(\psi_z^{\tau+1})$ constitutes a family of coherent states for the given (square–integrable) representation in a sense already alluded to, in the particular case when $\tau + 1 = \frac{3}{2}$ (cf. proposition 17.3) in (13.6): the slight discrepancy of notation with this former reference is of no consequence.

Given an integrable function a on Π (a bounded function would do in the case when $\tau > 0$), one may define the bounded operator $\mathrm{Op}_{\mathrm{Ber}}(a)$ with *contravariant* symbol a by the equation

$$\mathrm{Op}_{\mathrm{Ber}}(a)v = \int_{\Pi} a(z)\,(\psi_z^{\tau+1}, v)\,\psi_z^{\tau+1}\,d\mu(z). \tag{17.12}$$

For instance, from (17.11), it follows that, if $\tau > 0$, the identity operator has a contravariant symbol, given as the constant $\frac{\tau}{4\pi}$. Given a bounded linear operator A on $H_{\tau+1}$, one defines its *covariant* symbol by the equation

$$(\mathrm{Symb}_{\mathrm{Ber}}(A))(z) = (\psi_z^{\tau+1},\, A\psi_z^{\tau+1}): \tag{17.13}$$

it is a bounded function on Π.

Let us observe the basic role played in definitions (17.12) and (17.13) by the orthogonal projections P_z on the functions $\psi_z^{\tau+1}$: indeed, (17.12) expresses $\mathrm{Op}_{\mathrm{Ber}}(a)$ as an integral linear combination of the operators P_z's, while (17.13) may also be written as $(\mathrm{Symb}_{\mathrm{Ber}}(A))(z) = \mathrm{Tr}(AP_z)$. Substituting for P_z the operator $2\,e^{-i\pi\frac{\tau+1}{2}}\,\mathcal{D}_{\tau+1}(\sigma_z)$, where $\sigma_z = y^{-1}\begin{pmatrix} x & -x^2-y^2 \\ 1 & -x \end{pmatrix} \in G$ is associated with the geodesic symmetry around z (observe that the extra factor $e^{-i\pi\frac{\tau+1}{2}}$ has been chosen so as to get a positive symmetric operator as a consequence of the definition of $\mathcal{D}_{\tau+1}(\sigma_z)$ according to proposition 17.1), one would get the *active* and *passive* symbols in the place of the contravariant and covariant ones: the calculus so obtained was studied in [39] and [42] and plays the major role in the present section.

The covariant symbol satisfies the *covariance* property (a possibly unfortunate terminology: all four species of symbols mentioned do satisfy this property!) that, given a bounded operator A on $H_{\tau+1}$ and $g \in G$, the covariant symbol of $\mathcal{D}_{\tau+1}(g)A\mathcal{D}_{\tau+1}(g)^{-1}$ is $b \circ g^{-1}$ in terms of the covariant symbol b of A. A fully similar rule holds for contravariant symbols, or for active and passive symbols as well. As a consequence, if a is an integrable function on Π, the covariant symbol b of $\mathrm{Op}_{\mathrm{Ber}}(a)$ is linked to a by an equation $b = \Lambda a$, where the operator Λ on $L^2(\Pi)$ commutes with the quasi-regular action of G on Π. This makes it possible to compute Λ explicitly, and one finds, for instance as a consequence of ([40], theorem 3.3), that

$$\Lambda = 4\pi\,\frac{\Gamma(\tau + \frac{1}{2} + i\sqrt{\Delta - 1/4})\,\Gamma(\tau + \frac{1}{2} - i\sqrt{\Delta - 1/4})}{(\Gamma(\tau + 1))^2}. \tag{17.14}$$

One should note here that a computation of the analogue of the operator Λ in the case when a general hermitian symmetric space is used in the place of II has been done in [2], where the result was announced for the classical series of such spaces, and in [46], where it was stated and proved in the most general case.

Since the Gamma function is rapidly decreasing at infinity on vertical lines on the complex plane, it is clear that one should not expect in general to be able to recover any contravariant symbol from the covariant one: this has been explained in [42], where it was also shown that the calculus based on the active and passive symbols instead does not suffer from this defect, at least for τ large enough. Indeed, as shown in ([39], theorem 5.4), the operator F_τ which transforms the active symbol of some operator on $H_{\tau+1}$ into its passive one is given, in spectral-theoretic terms, as

$$F_\tau = 2\pi \frac{\Gamma(\frac{\tau}{2} + \frac{1}{4} + \frac{i}{2}\sqrt{\Delta - 1/4})\,\Gamma(\frac{\tau}{2} + \frac{1}{4} - \frac{i}{2}\sqrt{\Delta - 1/4})}{\Gamma(\frac{\tau}{2} + \frac{3}{4} + \frac{i}{2}\sqrt{\Delta - 1/4})\,\Gamma(\frac{\tau}{2} + \frac{3}{4} - \frac{i}{2}\sqrt{\Delta - 1/4})} \quad (17.15)$$

and F_τ is a bounded operator in $L^2(\Pi)$ when $\tau > -\frac{1}{2}$. Also, since ([39], theorem 5.2)

$$F_\tau F_{\tau+1} = 4\pi^2(\Delta + \tau(\tau + 1))^{-1}, \qquad (17.16)$$

one can see that, if $\tau > \frac{1}{2}$, F_τ^{-1} is the product of a bounded operator by $\Delta + \tau(\tau - 1)$.

However, if one concentrates on symbols in $L^2(\Pi)$ which lie in $L_0^2(\Pi)$, the set of images under the spectral measure for Δ of compact intervals in $]\frac{1}{4}, \infty[$, one may freely circulate between the covariant and contravariant symbols, as well as between those and the active and passive ones. It will be enough for our present purpose to recall ([40], theorems 3.3 and 3.4) that the active symbol f of an operator $\mathrm{Op_{Ber}}(a)$ is linked to a by the equation

$$f = \frac{2^\tau \pi^{-\frac{1}{2}}}{\Gamma(\tau + 1)} \Gamma(\frac{1}{2}(\tau + \frac{3}{2} + i\sqrt{\Delta - 1/4}))\,\Gamma(\frac{1}{2}(\tau + \frac{3}{2} - i\sqrt{\Delta - 1/4}))\,a. \tag{17.17}$$

From (17.12) and (17.13) it easily follows that if $a \in L^1(\Pi) \cap L^2(\Pi)$ is such that $\mathrm{Op_{Ber}}(a)$ is a trace-class operator on $H_{\tau+1}$, then its Hilbert-Schmidt norm is given by

$$\|\mathrm{Op_{Ber}}(a)\|_{H.S.}^2 = (a, \Lambda a)_{L^2(\Pi)}, \qquad (17.18)$$

with Λ as in (17.14). From this, and from the analogous formula (involving F_τ in place of Λ) using the active symbol rather than the contravariant one, one can see that if $\tau > -\frac{1}{2}$ any operator on $H_{\tau+1}$ with a contravariant or active symbol in $L^2(\Pi)$ is, indeed, a Hilbert-Schmidt operator.

The horocyclic calculus

We now introduce still another kind of symbol, the *horocyclic* symbol which, contrary to the four species just introduced, lives on Ξ rather than Π. It serves two purposes at once. First, it will enable us to recover the property that Hilbert-Schmidt norms of operators should just correspond to the L^2–norms of their symbols whenever this is possible. Next, it makes it possible to realize that, in some sense, the Weyl calculus, restricted to even or odd functions, is not an exception (theorem 17.6) in its role as a calculus linked to the discrete series of representations of $SL(2, \mathbb{R})$: what makes its specificity is that — provided one does not split $L^2(\mathbb{R})$ into its even and odd parts — it has an additional covariance propery, related to the Heisenberg representation.

DEFINITION 17.4. Set

$$\mathcal{R}_{\tau+1} = (2\pi)^{\frac{1}{2}} \pi^{i\pi\mathcal{E}} \frac{\Gamma(\frac{\tau}{2} + \frac{1}{4} - \frac{i\pi}{2}\mathcal{E})}{\Gamma(\frac{\tau}{2} + \frac{3}{4} + \frac{i\pi}{2}\mathcal{E})}. \tag{17.19}$$

Then, given an operator A in $H_{\tau+1}$ admitting an active symbol f in the class $L^2_0(\Pi)$ introduced above, define the horocyclic symbol of A as the function

$$h = \mathcal{R}_{\tau+1} T V f \ \in L^2(\Xi), \tag{17.20}$$

with T as defined in (4.10).

Remarks. The horocyclic symbol satisfies a covariance property analogous to the one mentioned above concerning the contravariant and covariant symbols: only, in the formula that relates the horocyclic symbol $h \circ g^{-1}$ of $\mathcal{D}_{\tau+1}(g) A \mathcal{D}_{\tau+1}(g)^{-1}$ to the horocyclic symbol h of A, it is understood that the group G should act on the space $\mathbb{R}^2 \backslash \{0\}$ (on which the present symbols live) through its linear action. In contradiction with the Berezin calculus, or the active-passive calculus, just one species of symbol is needed in the horocyclic calculus.

On the other hand, the difference (by a factor of $\pi^{i\pi\mathcal{E}}$) between our present definition of $\mathcal{R}_{\tau+1}$ and that in a previous reference [41] may be accounted for by a difference in the choice of coordinates on Ξ (so as to agree with Helgason's normalization) and is explained immediately above theorem 4.1.

THEOREM 17.5. *For every $\tau > -1$, the map $A \mapsto h$ just defined extends as an isometry from the Hilbert space of all Hilbert-Schmidt operators on $H_{\tau+1}$ onto the subspace of $L^2(\Xi) = L^2_{\text{even}}(\mathbb{R}^2)$ consisting of all functions invariant under the (unitary) symmetry*

$$\mathcal{R}_{\tau+1} T \kappa T^{-1} \mathcal{R}_{\tau+1}^{-1} = \frac{\Gamma(i\pi\mathcal{E})}{\Gamma(-i\pi\mathcal{E})} \frac{\Gamma(\tau + \frac{1}{2} - i\pi\mathcal{E})}{\Gamma(\tau + \frac{1}{2} + i\pi\mathcal{E})} \mathcal{G}, \tag{17.21}$$

in which κ was defined in (4.13).

PROOF. The theorem was stated and proven as theorem 6.1 in [41], under the unnecessary assumption that $\tau > 0$. However, let us recall the main points of the proof as it illustrates some of the difficulties inherent to the Berezin calculus. First, the verification of (17.21) involves nothing more than the duplication formula for the Gamma function. If A admits an active symbol in the class $L_0^2(\Pi)$, it also has a contravariant one a, linked to f by the equation (17.17). Then (duplication formula again)

$$h = \frac{(4\pi)^{\frac{1}{2}}}{\Gamma(\tau + 1)} (2\pi)^{i\pi\mathcal{E}} \Gamma(\tau + \frac{1}{2} - i\pi\mathcal{E}) TVa \qquad (17.22)$$

so that to prove that $\|h\|_{L^2(\Xi)} = \|A\|_{H.S.}$ we only have to use (17.14), (17.18) and theorem 4.1. The difficulty is to show that the isometry is defined on a dense subspace of the space of all Hilbert-Schmidt operators: at this point it is essential to use the active species of symbol rather than the contravariant one. Indeed, (17.9) shows that the functions $\psi_z^{\tau+1}$ ($z \in \Pi$) are permuted into one another, up to scalar factors of modulus one, under any $\mathcal{D}_{\tau+1}(g)$. As a consequence (Schur's lemma), the set of all rank-one operators $P_{w,z}\colon v \mapsto (\psi_w^{\tau+1}, v)\psi_z^{\tau+1}$, as z and w describe Π, constitutes a total set in the space of Hilbert-Schmidt operators in $H_{\tau+1}$. Now, as shown in ([39], theorem 4.3), such rank-one operators do have (explicit) active symbols: with the help of ([39], (5.10)), these can be proved to lie in $L^2(\Pi)$. But, except when $z = w$ (in which case a delta-function will do), they do not admit any contravariant symbol, not even a distribution. Note that the proof of theorem 17.8 below relies essentially on the use of the *passive* symbols of the operators $P_{w,z}$. □

We do not plan to develop the horocyclic calculus any further than necessitated by our present aim as explained in the beginning of section 17. Still, let us mention that one can develop this calculus in a smooth way. Again (this is always a good policy in pseudodifferential analysis), start from the family of coherent states $(\psi_z^{\tau+1})$ as introduced in (17.8) and the associated rank–one operators $P_{w,z}$ introduced in the proof of theorem 17.5. The horocyclic calculus can then be based on the result of the following computation.

PROPOSITION 17.6. *The horocyclic symbol $p_{w,z}$ of the rank-one operator $P_{w,z}$ just introduced is characterized, with the notations introduced in (4.21), by the equation*

$$(p_{w,z})_\lambda^\flat(s) = \gamma_\lambda (\psi_w^{\tau+1}, \psi_z^{\tau+1}) \left(\frac{i}{2}\right)^{\frac{1}{2}+\frac{i\lambda}{2}} \frac{(w - \bar{z})^{\frac{1}{2}+\frac{i\lambda}{2}}}{(w - s)^{\frac{1}{2}+\frac{i\lambda}{2}}(s - \bar{z})^{\frac{1}{2}+\frac{i\lambda}{2}}}, \qquad (17.23)$$

where the scalar product, taken in the space $H_{\tau+1}$, can be made explicit as

$$(\psi_w^{\tau+1}, \psi_z^{\tau+1}) = \frac{(\bar{w}^{-1} - w^{-1})^{\frac{\tau+1}{2}}(\bar{z}^{-1} - z^{-1})^{\frac{\tau+1}{2}}}{(\bar{z}^{-1} - w^{-1})^{\tau+1}}, \qquad (17.24)$$

and the constant γ_λ is given by

$$\gamma_\lambda = 2^{-\frac{1}{2}}(2\pi)^{-1-\frac{i\lambda}{2}}\frac{\Gamma(\frac{1}{2}+\frac{i\lambda}{2})}{\Gamma(\frac{i\lambda}{2})}\frac{\Gamma(\tau+\frac{1}{2}+\frac{i\lambda}{2})}{\Gamma(\tau+1)}. \tag{17.25}$$

PROOF (SKETCH). A "sesquiholomorphic" argument permits to reduce the computation to the case when $w = z$, in which it can be derived from (17.50) below, together with the formulas (17.14) to (17.20) connecting the various species of symbols.   □

One can then prove a composition formula generalizing the formula (5.11) from the Weyl calculus. Namely, for every $\tau > 0$, there exists a kernel $K^\tau_{\lambda_1,\lambda_2;\lambda}(s_1,s_2;s)$ with the following property. Let A_1, A_2 be two Hilbert-Schmidt operators on $H_{\tau+1}$ with horocyclic symbols h_1 and h_2 ($\in L^2(\Xi)$), and let h be the horocyclic symbol of the product $A_1 A_2$. Set

$$h_1 = \int_{-\infty}^\infty (h_1)_{\lambda_1}\,d\lambda_1, \qquad h_2 = \int_{-\infty}^\infty (h_2)_{\lambda_2}\,d\lambda_2 \tag{17.26}$$

in the sense of (4.19). Then, for (almost) all real number λ, one should have

$$h^\flat_\lambda(s) = \int_0^\infty\int_0^\infty d\lambda_1\,d\lambda_2 \int_{\mathbf{R}^2} K^\tau_{\lambda_1,\lambda_2;\lambda}(s_1,s_2;s)\,(h_1)^\flat_{\lambda_1}(s_1)\,(h_2)^\flat_{\lambda_2}(s_2)\,ds_1\,ds_2. \tag{17.27}$$

Using the result of proposition 17.6 and the covariance property, it is then easy to prove that the kernel $K^\tau_{\lambda_1,\lambda_2;\lambda}(s_1,s_2;s)$ is a linear combination, with coefficients depending on $\tau,\lambda_1,\lambda_2,\lambda$, of the two kernels $\chi^\pm_{i\lambda_1,i\lambda_2;i\lambda}(s_1,s_2;s)$. We have no expression of the coefficients, at present, as explicit in general as the one in theorem 5.3 (valid in the case of the Weyl calculus), only an integral expression which we here report for the benefit of the interested reader, if any (this result also plays a role at the very end of the present section): one has

$$K^\tau_{\lambda_1,\lambda_2;\lambda}(s_1,s_2;s) = C^\tau_{\lambda_1,\lambda_2;\lambda}(\operatorname{sign}(\tfrac{s_2-s_1}{(s_1-s)(s_2-s)}))\,\chi_{i\lambda_1,i\lambda_2;i\lambda}(s_1,s_2;s). \tag{17.28}$$

with

$$C^\tau_{\lambda_1,\lambda_2;\lambda}(\epsilon) = 2^{\frac{1}{2}}(2\pi)^{-3+\frac{i}{2}(\lambda_1+\lambda_2-\lambda)}\frac{\Gamma(\frac{1}{2}-\frac{i\lambda_1}{2})\Gamma(\frac{1}{2}-\frac{i\lambda_2}{2})\Gamma(\frac{1}{2}+\frac{i\lambda}{2})}{\Gamma(-\frac{i\lambda_1}{2})\Gamma(-\frac{i\lambda_2}{2})\Gamma(\frac{i\lambda}{2})}$$

$$\frac{\Gamma(\tau+1)\Gamma(\tau+\frac{1}{2}+\frac{i\lambda}{2})}{\Gamma(\tau+\frac{1}{2}+\frac{i\lambda_1}{2})\Gamma(\tau+\frac{1}{2}+\frac{i\lambda_2}{2})}\times I^\tau_{\lambda_1,\lambda_2;\lambda}(\epsilon) \tag{17.29}$$

and

$$I^\tau_{\lambda_1,\lambda_2;\lambda}(\epsilon) = \int_{\Pi\times\Pi}\left(\frac{|w_1|^2}{\operatorname{Im}\,w_1}\right)^{-\frac{1}{2}+\frac{i\lambda_1}{2}}\left(\frac{|w_2-\epsilon|^2}{\operatorname{Im}\,w_2}\right)^{-\frac{1}{2}+\frac{i\lambda_2}{2}}(\frac{w_2-\bar{w}_1}{2i})^{\frac{1}{2}+\frac{i\lambda}{2}}_{\text{right}}$$

$$(\cosh\frac{d(w_1,w_2)}{2})^{-2\tau-2}\,d\mu(w_1)\,d\mu(w_2), \tag{17.30}$$

where $w_{\text{right}}^{\frac{1}{2}+\frac{i\lambda}{2}}$ is associated with the principal complex determination of the logarithm of w in the plane cut from $-\infty$ to 0.

The Weyl calculus as a horocyclic calculus

Consider now the Weyl calculus on $L^2(\mathbb{R})$, which associates with each symbol $h \in L^2(\mathbb{R}^2)$ the operator $\mathrm{Op}(h)$ on $L^2(\mathbb{R})$ defined by

$$(\mathrm{Op}(h)u)(x) = \int h(\frac{x+y}{2}, \eta)\, u(y)\, e^{2i\pi(x-y)\eta}\, dy\, d\eta, \qquad u \in L^2(\mathbb{R}). \tag{17.31}$$

With $\check{u}(x) = u(-x)$ and $\check{h}(x,\eta) = h(-x,-\eta)$, the Weyl symbol of the operator $u \mapsto (\mathrm{Op}(h)\check{u})\check{\,}$ is $\check{h}$ and the Weyl symbol of the operator $u \mapsto \mathrm{Op}(h)\check{u}$ is the function $\mathcal{G}h$. Thus, even symbols in $L^2(\mathbb{R}^2)$ invariant under $\mathcal{G}$ are precisely those of all Hilbert-Schmidt operators A_1 on $L^2(\mathbb{R})$ which send the whole space to $L^2_{\text{even}}(\mathbb{R})$ and vanish on $L^2_{\text{odd}}(\mathbb{R})$: they correspond under the map $A_1 \mapsto Sq_{\text{even}}^{-1} A_1\, Sq_{\text{even}}$ (cf. proposition 17.3) to Hilbert-Schmidt operators on $H_{\frac{1}{2}}$. Considering even symbols in $L^2(\mathbb{R}^2)$ which change into their negatives under $\mathcal{G}$, the analogous result, in which $L^2_{\text{even}}(\mathbb{R})$ and $L^2_{\text{odd}}(\mathbb{R})$ have been traded for each other and the pair $(H_{\frac{3}{2}},\ Sq_{\text{odd}})$ has been substituted for the pair $(H_{\frac{1}{2}},\ Sq_{\text{even}})$, holds. Observe that the symmetry which occurs in (17.21) reduces to $\mathcal{G}$ in the case when $\tau + 1 = \frac{1}{2}$, to $-\mathcal{G}$ when $\tau + 1 = \frac{3}{2}$. The following theorem shows that the horocyclic symbol is far from being an artificial contraption.

THEOREM 17.7. *Let A be a Hilbert-Schmidt operator on the Hilbert space $H_{\frac{1}{2}}$ (resp. $H_{\frac{3}{2}}$), and let $A_1 = Sq_{\text{even}} A\, Sq_{\text{even}}^{-1}$ (resp. $Sq_{\text{odd}} A\, Sq_{\text{odd}}^{-1}$). The horocyclic symbol of A is just the Weyl symbol of A_1.*

PROOF. In the odd case, this was proved in ([41], theorem 5.2), essentially as a consequence of the formulas (13), (14) and (26) there, which make it possible to perform the computation in the case when A is a rank-one operator of the type discussed in the proof of theorem 17.5. Since the domain of validity of the latter theorem has been extended now to the case when $\tau > -1$, it also covers the even case. $\qquad\square$

The composition of passive symbols

In this subsection, we shall use the *passive* symbols, which live on Π. We denote as $f_1 \# f_2$ the passive symbol of the composition $A_1 A_2$ of two operators on $H_{\tau+1}$ with passive symbols h_1 and h_2. Also, if f lives on Π, the notation f_λ shall refer to the decomposition (4.14) of f as an integral superposition of eigenfunctions of Δ. Our aim is to construct a kernel

$\mathcal{K}^\tau_{\lambda_1, \lambda_2; \lambda}(w_1, w_2; z)$ permitting to write the formula

$$(f_1 \# f_2)_\lambda(z) = \int_0^\infty \int_0^\infty d\lambda_1 \, d\lambda_2 \int_{\Pi \times \Pi} \mathcal{K}^\tau_{\lambda_1, \lambda_2; \lambda}(w_1, w_2; z)$$
$$(f_1)_{\lambda_1}(w_1) \, (f_2)_{\lambda_2}(w_2) \, d\mu(w_1) \, d\mu(w_2). \quad (17.32)$$

Before doing this, however, let us just recall, for the sake of comparison, that one can give a *global* formula for the composition #. Indeed, for every $z \in \Pi$, denote as (r, θ) the geodesic polar coordinates on the Riemannian manifold Π centered at z and for which the half-line from z to $i\infty$ has polar angle 0. Then, if w has polar coordinates (r, θ) and if $(q, p) \in \mathbb{R}^2$ is defined through

$$q = \sinh r \cos \theta, \qquad p = \sinh r \sin \theta, \quad (17.33)$$

one can regard the map $(q, p) \mapsto w$ as defining a z-dependent chart $\Phi_z : \mathbb{R}^2 \to \Pi$. On the other hand, define on $\mathbb{R}^4$ the differential operator with constant coefficients

$$i\pi L = (4i\pi)^{-1} \left(-\frac{\partial^2}{\partial q' \partial p''} + \frac{\partial}{\partial p' \partial q''} \right). \quad (17.34)$$

Finally, consider the function E_τ, holomorphic in the plane minus the non-positive part of the real line, defined when $\zeta > 0$ by

$$E_\tau(\zeta) = 4\pi \int_0^\infty J_\tau(4\pi t) \, e^{-t^{-1}\zeta} \, dt, \quad (17.35)$$

in which J_τ is the Bessel function so denoted. Then one can define on $L^2(\mathbb{R}^4)$ the operator $E_\tau(-i\pi L)$ (choosing a spectral-theoretic definition or, what amounts to the same, using a Fourier transformation). We proved in ([42], theorem 5.1) that if τ is large enough and (f_1, f_2) is a pair of functions in $C_0^\infty(\Pi)$, then the formula

$$(f_1 \# f_2)(z) = [E_\tau(-i\pi L)((f_1 \circ \Phi_z)(q', p')(f_2 \circ \Phi_z)(q'', p''))] (q' = p' = q'' = p'' = \quad (17.$$

holds for every $z \in \Pi$. Our present problem is tantamount to extracting the "λ-component", as defined in (4.15), of $f_1 \# f_2$ in the case when f_1 and f_2 lie in $\mathcal{H}_{i\lambda_1}$ and $\mathcal{H}_{i\lambda_2}$ (the spaces defined in (4.16)) respectively.

THEOREM 17.8. *For any three points* $(w_1, w_2; z)$ *in* Π, *set*

$$\delta_{w_2, w_1}(z) = i \, \frac{(\operatorname{Im} z)^2 + (\operatorname{Re} z - w_2)(\operatorname{Re} z - \bar{w}_1)}{(\operatorname{Im} z)(w_2 - \bar{w}_1)}, \quad (17.37)$$

a complex number which reduces to $\cosh d(w_1, z)$ *in the case when* $w_2 = w_1$. *Recall that the Legendre function* $\mathfrak{P}_{-\frac{1}{2}+\frac{i\lambda}{2}}$, *defined in the usual way on* $]1, +\infty[$, *extends as a holomorphic function on the plane cut on the real line from* $-\infty$ *to* 1. *Let* f_1 *and* f_2 *lie in* $L_0^2(\Pi) \subset L^2(\Pi)$, *the set of images of compact intervals* $\subset]\frac{1}{4}, \infty[$ *under the spectral measure of* Δ. *Then, provided* τ *is large enough, formula* (17.32) *holds if one sets*

$$\mathcal{K}^\tau_{\lambda_1, \lambda_2; \lambda}(w_1, w_2; z) =$$

$$\frac{1}{8\pi} \frac{\Gamma(\frac{1}{2} + \frac{i\lambda_1}{2})\Gamma(\frac{1}{2} - \frac{i\lambda_1}{2})\Gamma(\frac{1}{2} + \frac{i\lambda_2}{2})\Gamma(\frac{1}{2} - \frac{i\lambda_2}{2})}{\Gamma(\frac{i\lambda_1}{2})\Gamma(-\frac{i\lambda_1}{2})\Gamma(\frac{i\lambda_2}{2})\Gamma(-\frac{i\lambda_2}{2})} \times$$

$$\frac{\Gamma(\frac{\tau}{2} + \frac{1}{2})\Gamma(\frac{\tau}{2} + 1)\Gamma(\frac{\tau}{2} + \frac{1}{4} + \frac{i\lambda}{4})\Gamma(\frac{\tau}{2} + \frac{1}{4} - \frac{i\lambda}{4})}{\Gamma(\frac{\tau}{2} + \frac{1}{4} + \frac{i\lambda_1}{4})\Gamma(\frac{\tau}{2} + \frac{1}{4} - \frac{i\lambda_1}{4})\Gamma(\frac{\tau}{2} + \frac{1}{4} + \frac{i\lambda_2}{4})\Gamma(\frac{\tau}{2} + \frac{1}{4} - \frac{i\lambda_2}{4})}$$

$$(\cosh \frac{d(w_1, w_2)}{2})^{-2\tau - 2} \, \mathfrak{P}_{-\frac{1}{2} + \frac{i\lambda}{2}}(\delta_{w_2, w_1}(z)), \quad (17.38)$$

whenever z *does not lie on the segment from* w_1 *to* w_2 *(in the sense of hyperbolic geometry).*

PROOF. If

$$f_1(w_1) = \int_0^\infty (f_1)_{\lambda_1}(w_1) (\frac{\pi\lambda_1}{2} \tanh \frac{\pi\lambda_1}{2}) \, d\lambda_1 \qquad (17.39)$$

as in (4.14), the contravariant symbol a_1 of A_1 (the operator with passive symbol f_1) is given as

$$a_1(w_1) = \frac{\Gamma(\tau + 1)}{2^{\tau + 1}} \pi^{-\frac{1}{2}} \int_0^\infty \frac{\frac{\pi\lambda_1}{2} \tanh \frac{\pi\lambda_1}{2}}{\Gamma(\frac{\tau}{2} + \frac{1}{4} + \frac{i\lambda_1}{4})\Gamma(\frac{\tau}{2} + \frac{1}{4} - \frac{i\lambda_1}{4})} (f_1)_{\lambda_1}(w_1) \, d\lambda_1 , \qquad (17.40)$$

as it follows from the formulas (17.15) and (17.17), which link the active symbol to the passive one for the first one, the active one to the contravariant one for the second. Since

$$\mathrm{Op}_{\mathrm{Ber}}((f_1)_{\lambda_1}) = \int_\Pi (f_1)_{\lambda_1}(w_1) \, P_{w_1, w_1} \, d\mu(w_1) \qquad (17.41)$$

and

$$P_{w_1, w_1} P_{w_2, w_2} = (\psi^{\tau+1}_{w_1}, \psi^{\tau+1}_{w_2}) \, P_{w_2, w_1}, \qquad (17.42)$$

one may write

$$A_1 A_2 = \frac{\Gamma(\tau + 1))^2}{2^{2\tau + 2} \pi}$$

$$\int_0^\infty \int_0^\infty \frac{(\frac{\pi\lambda_1}{2} \tanh \frac{\pi\lambda_1}{2})(\frac{\pi\lambda_2}{2} \tanh \frac{\pi\lambda_2}{2})}{\Gamma(\frac{\tau}{2} + \frac{1 + i\lambda_1}{4})\Gamma(\frac{\tau}{2} + \frac{1 - i\lambda_1}{4})\Gamma(\frac{\tau}{2} + \frac{1 + i\lambda_2}{4})\Gamma(\frac{\tau}{2} + \frac{1 - i\lambda_2}{4})} \, d\lambda_1 \, d\lambda_2$$

$$\int_{\Pi \times \Pi} (\psi^{\tau+1}_{w_1}, \psi^{\tau+1}_{w_2}) \, P_{w_2, w_1} \, d\mu(w_1) \, d\mu(w_2) . \qquad (17.43)$$

In order to prove theorem 17.8, all that has to be done, essentially, is computing $(f_{w_2, w_1})_\lambda$ if f_{w_2, w_1} is the passive symbol of the rank-one operator P_{w_2, w_1}.

Now the passive symbol of such a rank-one operator has been given in ([**42**], theorem 4.3): however, since slight changes have been made in our conventions, it is safer to recall from *loc.cit.* only the (change-resisting) fact that $f_{w,w}(z) = 2\,(\cosh d(w,z))^{-\tau-1}$, relying on the formula $\delta_{w,w}(z) = \cosh d(w,z)$ and on a "sesquiholomorphic argument" to get

$$f_{w_2,w_1}(z) = 2\,(\psi_{w_2}^{\tau+1}, \psi_{w_1}^{\tau+1})\,(\delta_{w_2,w_1}(z))_{\text{right}}^{-\tau-1}, \tag{17.44}$$

in which $\delta_{\text{right}}^{-\tau-1}$ is associated with the principal complex determination of the logarithm of δ in the plane cut from $-\infty$ to 0. The following fact was proven in ([42], proposition 5.3). Let $E_{w_2,w_1} \subset \mathbb{C}$ be the convex hull of the branch $\mathcal{C}$ of hyperbola with foci ± 1 such that the point $(\cosh \frac{d(w_1,w_2)}{2})^{-1}$ lies on $\mathcal{C}$; this set shrinks down to the interval $[1, \infty[$ when $w_2 = w_1$. From [39], (5.10), it follows that $\delta_{w_2,w_1}(z)$ can be real only if z lies on a (hyperbolic) line containing w_2 and w_1, and that $\text{Re}\,\delta_{w_2,w_1}(z) = (\cosh \frac{d(w_1,w_2)}{2})^{-1} \cosh d(w,z)$, where w is the geodesic middle of w_1 and w_2, so that $\delta_{w_2,w_1}(z)$ cannot be real ≤ 1 unless z lies on the segment from w_1 to w_2. Let ω be a C^2 complex-valued function in a neighborhood of E_{w_2,w_1}, holomorphic in the interior of E_{w_2,w_1} (the latter condition is void if $w_2 = w_1$). Then, as w describes Π, the function $\delta_{w_2,w_1}(z)$ describes the set E_{w_2,w_1}. Moreover, one may write

$$\Delta(\omega \circ \delta_{w_2,w_1}) = (D\omega) \circ \delta_{w_2,w_1}, \tag{17.45}$$

where

$$(D\omega)(\delta) = (1 - \delta^2)\,\omega''(\delta) - 2\delta\,\omega'(\delta). \tag{17.46}$$

We now decompose the function $\delta^{-\tau-1}$, with $\delta(z) := \cosh d(i, z)$, with respect to the spectral decomposition (4.14) of Δ. The Mehler inversion formula ([27], p.398 or [36], p.144) makes it possible to write

$$\delta^{-\tau-1} = \int_0^\infty k(\lambda)\,\mathfrak{P}_{-\frac{1}{2}+\frac{i\lambda}{2}}(\delta)\,d\lambda \tag{17.47}$$

if

$$k(\lambda) = \frac{\lambda}{4}\,(\tanh \frac{\pi\lambda}{2}) \int_1^\infty \mathfrak{P}_{-\frac{1}{2}+\frac{i\lambda}{2}}(\delta)\,\delta^{-\tau-1}\,d\delta, \tag{17.48}$$

which can be made explicit (*cf.* [36], p.146) as

$$k(\lambda) = \frac{\lambda}{4}\,(\tanh \frac{\pi\lambda}{2})\,2^{\tau-1}\,\pi^{-\frac{1}{2}}\,\frac{\Gamma(\frac{\tau}{2} + \frac{1+i\lambda}{4})\,\Gamma(\frac{\tau}{2} + \frac{1-i\lambda}{4})}{\Gamma(\tau+1)}. \tag{17.49}$$

Using G–invariance followed by a "sesquiholomorphic argument", we then get the decomposition of the function f_{w_2,w_1} as

$$(f_{w_2,w_1})_\lambda(z) =$$
$$2^{\tau-1}\pi^{-\frac{3}{2}}\frac{\Gamma(\frac{\tau}{2} + \frac{1}{4} + \frac{i\lambda}{4})\Gamma(\frac{\tau}{2} + \frac{1}{4} - \frac{i\lambda}{4})}{\Gamma(\tau+1)}\,(\psi_{w_2}^{\tau+1}, \psi_{w_1}^{\tau+1})\,\mathfrak{P}_{-\frac{1}{2}+\frac{i\lambda}{2}}(\delta_{w_2,w_1}(z)), \tag{17.50}$$

from which the conclusion of our computation easily follows after some dispensable embellishments regarding the coefficient. $\square$

We leave it to the reader to write and check directly the covariance property satisfied by the map $((f_1)_{\lambda_1}, (f_2)_{\lambda_2}) \mapsto (f_1 \# f_2)_\lambda$, starting from the observation that $\delta_{w_2, w_1}(z)$ is invariant under the diagonal action of G in $\Pi \times \Pi \times \Pi$.

Letting τ go to infinity

Looking back at (17.38), we first observe that $(\cosh \frac{d(w_1, w_2)}{2})^{-2\tau - 2}$ concentrates on the diagonal of $\Pi \times \Pi$ as $\tau \to \infty$; also, from the expression $d\mu(w) = \sinh r \, dr \, d\theta$ of the measure on Π in geodesic polar coordinates, one immediately gets

$$\int_\Pi (\cosh \frac{d(w_1, w_2)}{2})^{-2\tau - 2} \, d\mu(w_2) = \frac{4\pi}{\tau}. \qquad (17.51)$$

From this and from the Stirling equivalent

$$\frac{\Gamma(\tau + a)}{\Gamma(\tau + b)} \sim \tau^{a-b}, \qquad \tau \to \infty, \qquad (17.52)$$

the asymptotics as $\tau \to \infty$ of the integral kernel $\mathcal{K}^\tau_{\lambda_1, \lambda_2; \lambda}(w_1, w_2; z)$ that occurs in theorem 17.8 is immediate: this yields

$$\int_{\Pi \times \Pi} \mathcal{K}^\tau_{\lambda_1, \lambda_2; \lambda}(w_1, w_2; z) (f_1)_{\lambda_1}(w_1) (f_2)_{\lambda_2}(w_2) \, d\mu(w_1) \, d\mu(w_2) \sim$$

$$\frac{1}{4} \frac{\Gamma(\frac{1+i\lambda_1}{2}) \Gamma(\frac{1-i\lambda_1}{2}) \Gamma(\frac{1+i\lambda_2}{2}) \Gamma(\frac{1-i\lambda_2}{2})}{\Gamma(\frac{i\lambda_1}{2}) \Gamma(-\frac{i\lambda_1}{2}) \Gamma(\frac{i\lambda_2}{2}) \Gamma(\frac{-i\lambda_2}{2})} \times$$

$$\int_\Pi \mathfrak{P}_{-\frac{1}{2} + \frac{i\lambda}{2}}(\cosh d(w, z)) (f_1)_{\lambda_1}(w) (f_2)_{\lambda_2}(w) \, d\mu(w). \quad (17.53)$$

On the other hand, using the integral expression ((17.28)–(17.30)) of the integral kernel $K^\tau_{\lambda_1, \lambda_2; \lambda}(s_1, s_2; s)$ associated (17.27) with the composition formula in the horocyclic calculus, one can prove, with the help of (8.22),(8.23), the asymptotics (please distinguish K from $\mathcal{K}$)

$$K^\tau_{\lambda_1, \lambda_2; \lambda}(s_1, s_2; s) \sim \frac{1}{4\pi^2} \left(\frac{\tau}{2\pi}\right)^{\frac{-1+i(\lambda - \lambda_1 - \lambda_2)}{2}} \times$$

$$\frac{\Gamma(\frac{1-i(\lambda + \lambda_1 + \lambda_2)}{4}) \Gamma(\frac{1+i(\lambda - \lambda_1 + \lambda_2)}{4}) \Gamma(\frac{1+i(\lambda + \lambda_1 - \lambda_2)}{4}) \Gamma(\frac{1+i(\lambda - \lambda_1 - \lambda_2)}{4})}{\Gamma(\frac{-i\lambda_1}{2}) \Gamma(\frac{-i\lambda_2}{2}) \Gamma(\frac{i\lambda}{2})} \times$$

$$\chi_{i\lambda_1, i\lambda_2; i\lambda}(s_1, s_2; s). \quad (17.54)$$

Using the relation (a consequence of (17.20) and (17.15))

$$h^\flat_\lambda = (2\pi)^{-\frac{1}{2}} \pi^{-\frac{i\lambda}{2}} \frac{\Gamma(\frac{\tau}{2} + \frac{3}{4} + \frac{i\lambda}{4})}{\Gamma(\frac{\tau}{2} + \frac{1}{4} - \frac{i\lambda}{4})} (TVf)^\flat_\lambda$$

$$\sim 2^{-\frac{1}{2}} \left(\frac{\tau}{2\pi}\right)^{\frac{1+i\lambda}{2}} (TVf)^\flat_\lambda \qquad (17.55)$$

as well as the equation

$$(V^*T^*h_1)_{\lambda_1} = \frac{2}{\pi} \frac{\Gamma(\frac{i\lambda_1}{2})\Gamma(-\frac{i\lambda_1}{2})}{\Gamma(\frac{1+i\lambda_1}{2})\Gamma(\frac{1-i\lambda_1}{2})} V^*T^*(h_1)_{\lambda_1} \tag{17.56}$$

(a consequence of (4.15) and (4.30)), it is then a perfectly feasible, if tedious, task to verify that proposition 8.1 indeed connects the asymptotics (17.54) and (17.53) of the composition formulas relative to the horocyclic or the passive symbols.

We have not subjected ourselves to the torture the result of which would be the computation of the next term in the τ-asymptotics of the integral kernels $K^\tau_{\lambda_1,\lambda_2;\lambda}(s_1,s_2;s)$ and $\mathcal{K}^\tau_{\lambda_1,\lambda_2;\lambda}(w_1,w_2;z)$. However, since the second-main term in the product of two operators is also the main term in their commutator, it is clear that the two are connected by proposition 8.2, which links the bilinear "product" $L^-_{i\lambda_1,i\lambda_2;i\lambda}$ to Poisson brackets on the upper half-plane.

18. Moving to the forward light–cone: the Lax–Phillips theory revisited

Another model of the half–plane, viewed with its Riemannian structure, is the mass–hyperboloid of $(1+2)$ – dimensional relativistic theory. On the forward light–cone C, it is natural to consider the d'Alembertian operator $\Box$ as well as its fundamental solution Z_2 as provided by M.Riesz's theory. We show the equivalence between two boundary–value problems associated with the equation $\Box W = 0$: the first one is the classical Cauchy problem with two data on the mass–hyperboloid; the second one is a singular characteristic problem with only one trace given on the boundary ∂C of the cone. It follows from the analysis that the space of Cauchy data associated with the Lax–Phillips scattering theory [25] for the automorphic wave equation can be identified to the space of automorphic distributions on $\mathbf{R}^2$ described in section 13, the Euler operator there taking the place played by the matrix–operator A from the Lax–Phillips theory. We view this construction as advantageous in that it substitutes for the scattering theory associated with some second–order operator the one associated with a concrete first–order operator, in a natural way.

The convolution with Z_2, acting on measures supported on ∂C, is one of the tools in what precedes. The same, acting on functions (or measures) supported in C instead, gives a new construction of the integral kernel of the resolvent of Δ, in which the integral representation of that kernel is obtained at once without appealing to any previous knowledge of the hypergeometric function.

We may then go one step further, and interest ourselves in arithmetic elements of structure whose natural habitat is the forward light–cone C. Pursuing the analogy with the construction between (16.72) and (16.74), we shall be led immediately, starting from the simplest arithmetic distributions available on C, to some quadratic analogues of the Eisenstein series. It will turn out that the "forms" so obtained are nothing new, but a concept already introduced by Hejhal [18], to be complemented by Zagier [50, 51]. One of the *raisons d'être* of the present section, besides the emphasis on arithmetic (discrete) distributions, in the spirit of section 16, was our desire to get some understanding of the works just given reference to.

First, the geometry: perform a quadratic transform

$$\xi = (\xi_1, \xi_2) \mapsto \eta = (\eta_0, \eta_1, \eta_2) = (\frac{\xi_1^2 + \xi_2^2}{2}, \frac{\xi_1^2 - \xi_2^2}{2}, \xi_1 \xi_2)$$

(18.1)

from $\Xi = \mathbb{R}^2 \backslash \{0\}/(\xi \sim -\xi)$ to the set

$$\partial C = \{\eta : \eta_0 > 0, \, \eta_0^2 - \eta_1^2 - \eta_2^2 = 0\},$$

(18.2)

the boundary of the forward light-cone

$$C = \{\eta : \eta_0 > 0, \, \eta_0^2 - \eta_1^2 - \eta_2^2 > 0\}$$

(18.3)

in three-dimensional spacetime (with $\{0\}$ deleted). Accordingly, if h is a function on Ξ (an even function on $\mathbb{R}^2$), associate with it the function Qh on ∂C such that

$$h(\xi_1, \xi_2) = (Qh)(\frac{\xi_1^2 + \xi_2^2}{2}, \frac{\xi_1^2 - \xi_2^2}{2}, \xi_1 \xi_2).$$

(18.4)

Observe that

$$\|h\|_{L^2(\Xi)}^2 = \int_{\mathbb{R}^2} |h(\xi)|^2 \, d\xi = \int_{\partial C} |(Qh)(\eta)|^2 \, \frac{d\eta_1 \, d\eta_2}{\eta_0}.$$

(18.5)

Let us first transfer the action of $G = SL(2, \mathbb{R})$ on C: this is the realization of the very classical twofold covering of the Lorentz group $SO_o(1, 2) \simeq PSL(2, \mathbb{R})$ by $SL(2, \mathbb{R})$. On one hand, it may be regarded as the matrix realization of the coadjoint representation $g \mapsto {}^t Ad(g)^{-1}$ of G if the dual $\mathfrak{g}^*$ of $\mathfrak{g}$ is identified to $\mathbb{R}^3$ in the way associated with the basis (2.2) of $\mathfrak{g}$. Alternatively, taking advantage of the above–defined quadratic transform, set $\begin{pmatrix} \xi_1' \\ \xi_2' \end{pmatrix} = \begin{pmatrix} a & b \\ c & d \end{pmatrix} \begin{pmatrix} \xi_1 \\ \xi_2 \end{pmatrix}$. Since $\begin{pmatrix} \xi_1 \\ \xi_2 \end{pmatrix} (\xi_1 \, \xi_2) = \begin{pmatrix} \eta_0 + \eta_1 & \eta_2 \\ \eta_2 & \eta_0 - \eta_1 \end{pmatrix}$ when η and ξ are linked by (18.1), one has, setting $(\eta_0', \eta_1', \eta_2') = \frac{1}{2} ((\xi_1')^2 + (\xi_2')^2, (\xi_1')^2 - (\xi_2')^2, 2\xi_1' \xi_2')$ as well, the equation

$$\begin{pmatrix} \eta_0' + \eta_1' & \eta_2' \\ \eta_2' & \eta_0' - \eta_1' \end{pmatrix} = \begin{pmatrix} a & b \\ c & d \end{pmatrix} \begin{pmatrix} \eta_0 + \eta_1 & \eta_2 \\ \eta_2 & \eta_0 - \eta_1 \end{pmatrix} \begin{pmatrix} a & c \\ b & d \end{pmatrix},$$

(18.6)

which provides the transformation from ∂C to ∂C, to be extended by linearity. This last equation can of course be found in a variety of places, including ([25], p. 8).

Denote as $\Lambda(g)$ the linear transformation $\eta \mapsto \eta'$ defined in (18.6): it is Lorentz since it preserves the quadratic form $\eta_0^2 - \eta_1^2 - \eta_2^2$ which is the determinant of the matrix associated with η. Choose $\varepsilon = (1, 0, 0)$ as the base–point of C. There is a unique map $\chi : C \to \Pi$ satisfying the properties

$$\begin{cases} \chi(\varepsilon) = i \\ \chi(t\eta) = \chi(\eta) & \text{for every } \eta \in C, \, t > 0 \\ \chi(\Lambda(g)\eta) = g.\chi(\eta), & \eta \in C, \, g \in G, \end{cases}$$

(18.7)

since on one hand, $\mathbb{R}_+^\times \times SO_o(1,2)$ acts transitively on C, and on the other hand the equation $\Lambda(k)\varepsilon = \varepsilon$ implies $k \in K$, thus $k.i = i$. Since

$$\frac{ai+b}{ci+d} = \frac{ac+bd}{c^2+d^2} + \frac{i}{c^2+d^2} \tag{18.8}$$

and the equation $\eta = \Lambda(g)\varepsilon$ just means that

$$\begin{pmatrix} \eta_0 + \eta_1 & \eta_2 \\ \eta_2 & \eta_0 - \eta_1 \end{pmatrix} = \begin{pmatrix} a^2 + b^2 & ac + bd \\ ac + bd & c^2 + d^2 \end{pmatrix}, \tag{18.9}$$

one has $\chi(\eta) = \frac{\eta_2 + i}{\eta_0 - \eta_1}$ if $\eta_0^2 - \eta_1^2 - \eta_2^2 = 1$, so that

$$\chi(\eta) = \frac{\eta_2 + i\sqrt{\eta_0^2 - \eta_1^2 - \eta_2^2}}{\eta_0 - \eta_1} \tag{18.10}$$

generally. The map χ is simply the homogeneous extension to C (of degree zero) of the standard isometry between two models of G/K: the mass–hyperboloid

$$\mathcal{H} = \{\eta : \eta_0 > 0, \ \eta_0^2 - \eta_1^2 - \eta_2^2 = 1\} \tag{18.11}$$

with the metric which is the restriction of $-d\eta_0^2 + d\eta_1^2 + d\eta_2^2$, consequently the Lorentz–invariant measure $\frac{d\eta_1 \, d\eta_2}{\eta_0}$, and the Poincaré half–plane Π. We also set

$$[\eta] := (\eta_0^2 - \eta_1^2 - \eta_2^2)^{\frac{1}{2}} \tag{18.12}$$

for each $\eta \in C$, and do not distinguish between $\eta \in C$ and the matrix on the left–hand side of (18.9), so that $[\eta] = (\det \eta)^{\frac{1}{2}}$.

The inverse map χ^{-1} (more precisely, this is the inverse of the restriction of χ to $\mathcal{H}$) can be defined as $z \mapsto P_z$ with

$$P_z = \begin{pmatrix} \frac{|z|^2}{y} & \frac{x}{y} \\ \frac{x}{y} & \frac{1}{y} \end{pmatrix} = gg', \qquad (z = g.i). \tag{18.13}$$

Through the map χ^{-1}, the Laplacian Δ on Π can be transferred to the operator $\Delta_{\mathcal{H}}$ on $\mathcal{H}$, the Laplace–Beltrami operator for the Riemannian structure referred to just after (18.11). As is well–known, if ψ is a C^∞ function on C, homogeneous of degree $k \in \mathbb{C}$ and satisfying $\Box \psi = 0$, with

$$\Box = \frac{\partial^2}{\partial \eta_0^2} - \frac{\partial^2}{\partial \eta_1^2} - \frac{\partial^2}{\partial \eta_2^2}, \tag{18.14}$$

one has

$$\Delta_{\mathcal{H}}(\psi|_{\mathcal{H}}) = -k(k+1)\,\psi|_{\mathcal{H}}. \tag{18.15}$$

It is useful for the sequel to recall a very simple proof. With

$$E = \eta_0 \frac{\partial}{\partial \eta_0} + \eta_1 \frac{\partial}{\partial \eta_1} + \eta_2 \frac{\partial}{\partial \eta_2}, \tag{18.16}$$

one shows that, if $\psi \in C^\infty(C)$, $(\Box - E(E+1))\psi|_{\mathcal{H}}$ depends only on $\psi|_{\mathcal{H}}$: to see this, it suffices to observe that $(\Box - E(E+1))\psi$ is divisible by $\eta_0^2 - \eta_1^2 - \eta_2^2 - 1$ if so is ψ, which is the result of a straightforward computation. Writing $(\Box - E(E+1))\psi|_{\mathcal{H}} = A \cdot \psi|_{\mathcal{H}}$, it is clear, using Lorentz invariance, that A (a second–order differential operator) is a linear combination of $\Delta_{\mathcal{H}}$ and of the identity operator. That $A = \Delta_{\mathcal{H}}$ comes from the equation

$$(\Box - E(E+1))(\eta_0 - \eta_1)^k = -k(k+1)(\eta_0 - \eta_1)^k \tag{18.17}$$

since $\eta_0 - \eta_1$ transfers, on Π, to the function $z \mapsto (\operatorname{Im} z)^{-1}$ according to (18.13), and

$$\Delta y^{-k} = -k(k+1)y^{-k}. \tag{18.18}$$

Let Z_2 be the fundamental solution of $\Box$ supported in $\bar{C}$, as obtained by an application of M.Riesz's theory (cf. [32], p.50), i.e.,

$$Z_2 = \frac{1}{2\pi}(\eta_0^2 - \eta_1^2 - \eta_2^2)_+^{-\frac{1}{2}}, \tag{18.19}$$

where the subscript $+$ means that the whole function is to be multiplied by the characteristic function of C.

THEOREM 18.1. *Recall from* [25], *p.11, that under the map*

$$(t, z) \mapsto \begin{pmatrix} \eta_0 + \eta_1 & \eta_2 \\ \eta_2 & \eta_0 - \eta_1 \end{pmatrix} = e^t \begin{pmatrix} \frac{|z|^2}{y} & \frac{z}{y} \\ \frac{\bar{z}}{y} & \frac{1}{y} \end{pmatrix} \tag{18.20}$$

from $\mathbf{R} \times \Pi$ *to* C *and under the gauge transformation* $u \mapsto W = e^{-\frac{t}{2}}u$, *the equation* $\Box W = 0$ *inside* C *is equivalent to the wave equation*

$$\frac{\partial^2 u}{\partial t^2} + (\Delta - \frac{1}{4})u = 0, \tag{18.21}$$

in which Δ *denotes the hyperbolic Laplacian on* Π.

Consider on one hand the (classical) Cauchy problem

$$\begin{cases} \frac{\partial^2 u}{\partial t^2} + (\Delta - \frac{1}{4})u = 0 \\ u(0, z) = f_0(z) \\ \frac{\partial u}{\partial t}(0, z) = f_1(z), \end{cases} \tag{18.22}$$

on the other hand the characteristic problem

$$\begin{cases} \check{W} = 0 \text{ in } \mathbf{R}^3 \backslash C \\ \Box \check{W} = Qh. \frac{d\eta_1 \, d\eta_2}{\eta_0} \end{cases} \tag{18.23}$$

in which $\check{W}$ *denotes the function* W *extended by* 0 *in* $\mathbf{R}^3 \backslash C$, *and* Qh *is the function on* ∂C *associated by* (18.4) *to some even function* h *on* $\mathbf{R}^2$: *in particular,* (18.23) *implies that* $\Box W = 0$ *inside* C.

If f_0, f_1 and $(\Delta - \frac{1}{4})^{\frac{1}{2}} f_0$, $(\Delta - \frac{1}{4})^{-\frac{1}{2}} f_1$ lie in $L^2(\Pi)$, u is the solution of (18.22), and u and W are linked by the transformation above, then $\check{W}$ is a solution of (18.23) with h defined, with the notations of section 4, as

$$h = 2^{\frac{1}{2}} \left(TV f_1 + i\pi \, \mathcal{E} \, TV f_0 \right) : \qquad (18.24)$$

both h and $\mathcal{E}^{-1}h$ lie in $L^2_{\text{even}}(\mathbf{R}^2)$. Conversely, assume that $h \in L^2_{\text{even}}(\mathbf{R}^2)$ is such that $\mathcal{E}^{-1}h$ lies in $L^2_{\text{even}}(\mathbf{R}^2)$ as well. Then there exists a unique pair (f_0, f_1) with f_0, f_1, $(\Delta - \frac{1}{4})^{\frac{1}{2}} f_0$ and $(\Delta - \frac{1}{4})^{-\frac{1}{2}} f_1$ all lying in $L^2(\Pi)$ such that (18.24) holds.

Finally, through the correspondence $(f_0, f_1) \mapsto h$ so defined, the operator $A = \begin{pmatrix} 0 & I \\ -\Delta + \frac{1}{4} & 0 \end{pmatrix}$ the study of which plays the major part in [25] is taken to the operator $i\pi\mathcal{E}$.

PROOF. For any even function h on $\mathbf{R}^2$, denote as $\widetilde{Qh}$ the measure on $\mathbf{R}^3$, supported on ∂C, whose density with respect to the measure $\frac{d\eta_1\,d\eta_2}{\eta_0}$ is the function Qh. To solve (18.23), we just set $\check{W} = Z_2 * \widetilde{Qh}$.

Consider the case when h, to be then denoted as $h_{-i\nu}$, is homogeneous of degree $-1 - \nu$ with $\text{Re }\nu < 1$. Then $Qh_{-i\nu}$, as a function on ∂C, is homogeneous of degree $\frac{-1-\nu}{2}$, and so is $Z_2 * \widetilde{Qh_{-i\nu}}$ as a function on $\mathbf{R}^3$. Since $\square(Z_2 * \widetilde{Qh_{-i\nu}}) = 0$ in $\mathbf{R}^3 \backslash \partial C$, it follows from (18.15) that

$$\Delta_{\mathcal{H}}((Z_2 * \widetilde{Qh_{-i\nu}})|_{\mathcal{H}}) = \frac{1 - \nu^2}{4} (Z_2 * \widetilde{Qh_{-i\nu}})|_{\mathcal{H}} . \qquad (18.25)$$

Let us make the function $Z_2 * \widetilde{Qh_{-i\nu}}|_{\mathcal{H}}$ explicit in this case, in terms of the function $h^{\flat}_{-i\nu}$ defined (as in (4.21)) by $h^{\flat}_{-i\nu}(s) = h_{-i\nu}(s, 1)$. One has

$$(Z_2 * \widetilde{Qh_{-i\nu}})(\eta) = \frac{1}{2\pi} \int_{\partial C} (Qh_{-i\nu})(\eta') \, [\eta - \eta']_{+}^{-1} \, \frac{d\eta'_1 \, d\eta'_2}{\eta'_0} . \qquad (18.26)$$

Now, if on one hand $\eta = \chi^{-1}(z) = P_z$ as in (18.13), and on the other hand $\eta' \in \partial C$ is linked to $(\xi'_1, \xi'_2) \in \mathbf{R}^2$ by (18.1), it is immediate that

$$\begin{aligned}
[\eta - \eta']^2 &= (\eta_0 - \eta'_0)^2 - (\eta_1 - \eta'_1)^2 - (\eta_2 - \eta'_2)^2 \\
&= 1 - 2 \left(\eta_0 \eta'_0 - \eta_1 \eta'_1 - \eta_2 \eta'_2 \right) \\
&= 1 - \frac{|\xi'_1 - z\xi'_2|^2}{y} .
\end{aligned} \qquad (18.27)$$

We may substitute $\frac{1}{\pi} d\xi'_1 \, d\xi'_2$ for $\frac{1}{2\pi} \frac{d\eta'_1 \, d\eta'_2}{\eta'_0}$ in (18.26), integrating only on $\{(\xi'_1, \xi'_2) : \xi'_2 > 0\}$. Making the change of variable $\xi'_1 = s\xi'_2$, so that

$h_{-i\nu}(\xi') = \xi_2'^{-1-\nu} h_{-i\nu}^b(s)$, we are led to

$$(Z_2 * \widetilde{Q}h_{-i\nu})(\chi^{-1}(z)) = \frac{1}{\pi} \int_{-\infty}^{\infty} h_{-i\nu}^b(s) \, ds \times$$

$$\int_0^{\infty} \xi_2'^{-\nu} \left(1 - \frac{|z-s|^2}{y} \xi_2'^2\right)_+^{-\frac{1}{2}} d\xi_2', \qquad (18.28)$$

and the last integral is

$$\int_0^{y^{\frac{1}{2}}|z-s|^{-1}} \xi_2'^{-\nu} \left(1 - \frac{|z-s|^2}{y} \xi_2'^2\right)^{-\frac{1}{2}} d\xi_2'$$

$$= y^{\frac{1-\nu}{2}} |z-s|^{\nu-1} \int_0^1 (1-u^2)^{-\frac{1}{2}} u^{-\nu} \, du$$

$$= \frac{1}{2} y^{\frac{1-\nu}{2}} |z-s|^{\nu-1} \frac{\Gamma(\frac{1}{2})\Gamma(\frac{1-\nu}{2})}{\Gamma(\frac{2-\nu}{2})}. \qquad (18.29)$$

Thus

$$(Z_2 * \widetilde{Q}h_{-i\nu})(\chi^{-1}(z)) = \frac{1}{2\pi} \frac{\Gamma(\frac{1}{2})\Gamma(\frac{1-\nu}{2})}{\Gamma(\frac{2-\nu}{2})} \int_{-\infty}^{\infty} h_{-i\nu}^b(s) \left(\frac{|z-s|^2}{y}\right)^{\frac{\nu-1}{2}} ds$$

$$= \frac{1}{2} \pi^{-\frac{\nu}{2}} (\Gamma(\frac{2-\nu}{2}))^{-1} (\Theta_\nu h_{-i\nu}^b)(z)$$

$$= -2^{\frac{1}{2}} \nu^{-1} (V^*T^*h_{-i\nu})(z), \qquad (18.30)$$

where we have used $(2.19), (2.6)$, finally (4.44).

Next, if u and W are linked by the Lax–Phillips transformation recalled in the beginning of theorem 18.1, one has when $t = 0$ (or, equivalently, $\eta \in \mathcal{H}$)

$$\frac{\partial u}{\partial t} = \frac{\partial}{\partial t}(e^{\frac{t}{2}}W)$$

$$= \frac{\partial W}{\partial t} + \frac{1}{2}W$$

$$= (\eta_0 \frac{\partial}{\partial \eta_0} + \eta_1 \frac{\partial}{\partial \eta_1} + \eta_2 \frac{\partial}{\partial \eta_2} + \frac{1}{2})W. \qquad (18.31)$$

Since $Z_2 * \widetilde{Q}h_{-i\nu}$ is homogeneous of degree $\frac{-1-\nu}{2}$, it follows that

$$\left[(\eta_0 \frac{\partial}{\partial \eta_0} + \eta_1 \frac{\partial}{\partial \eta_1} + \eta_2 \frac{\partial}{\partial \eta_2} + \frac{1}{2})(Z_2 * \widetilde{Q}h_{-i\nu})\right](\chi^{-1}(z))$$

$$= 2^{-\frac{1}{2}} (V^*T^*h_{-i\nu})(z). \qquad (18.32)$$

Thus, if $\check{W} = Z_2 * \widetilde{Q}h$ is the solution of (18.23), and if $h = \int_{-\infty}^{\infty} h_\lambda \, d\lambda$ as in (4.19), so that

$$\frac{1}{2\pi} \mathcal{E}^{-1}h = -\int_{-\infty}^{\infty} h_\lambda \cdot \lambda^{-1} \, d\lambda, \tag{18.33}$$

the function u associated with $W = \check{W}|_C$ will be a solution of the problem (18.22) provided that

$$\begin{cases} i \int_{-\infty}^{\infty} 2^{\frac{1}{2}} (V^*T^*h_\lambda) \lambda^{-1} \, d\lambda = f_0 \\ \int_{-\infty}^{\infty} 2^{-\frac{1}{2}} (V^*T^*h_\lambda) \, d\lambda = f_1, \end{cases} \tag{18.34}$$

i.e.

$$\begin{cases} V^*T^*h = 2^{\frac{1}{2}} f_1 \\ V^*T^* \mathcal{E}^{-1}h = 2^{\frac{1}{2}} i\pi f_0. \end{cases} \tag{18.35}$$

To solve this set of equations, we rely on theorem 4.1, remembering from there that the knowledge of V^*T^*h only implies that of $\frac{1}{2}(h + T\kappa T^{-1}h) = TV(V^*T^*h)$. Now, using (4.10)–(4.13), one sees that

$$T\kappa T^{-1} = \frac{\Gamma(i\pi\mathcal{E})}{\Gamma(-i\pi\mathcal{E})} \pi^{-2i\pi\mathcal{E}} \mathcal{F}, \tag{18.36}$$

consequently that

$$T\kappa T^{-1}\mathcal{E} = -\mathcal{E} T\kappa T^{-1}. \tag{18.37}$$

From theorem 4.1, it thus follows that

$$\begin{cases} \frac{1}{2}(h + T\kappa T^{-1}h) = 2^{\frac{1}{2}} TV f_1 \\ \frac{1}{2}(h - T\kappa T^{-1}h) = 2^{\frac{1}{2}} i\pi\mathcal{E} TV f_0, \end{cases} \tag{18.38}$$

which finally leads to

$$h = 2^{\frac{1}{2}} (i\pi \mathcal{E} TV f_0 + TV f_1). \tag{18.39}$$

The computation is justified if h and $\mathcal{E}^{-1}h$ both lie in $L^2_{\text{even}}(\mathbf{R}^2)$. Since, according to theorem 4.1, the isometry TV transforms the operator $\Delta - \frac{1}{4}$ on $L^2(\Pi)$ into the operator $\pi^2\mathcal{E}^2$, this is indeed the case under the given assumptions about f_0 and f_1.

Finally, the function

$$i\pi \mathcal{E} h = 2^{\frac{1}{2}} (i\pi \mathcal{E} TV f_1 - \pi^2 \mathcal{E}^2 TV f_0)$$

$$= 2^{\frac{1}{2}} (i\pi \mathcal{E} TV f_1 + TV (\frac{1}{4} - \Delta) f_0) \tag{18.40}$$

is the one associated, through the correspondence described in the theorem, to the pair $\begin{pmatrix} f_1 \\ (\frac{1}{4} - \Delta) f_0 \end{pmatrix} = A \begin{pmatrix} f_0 \\ f_1 \end{pmatrix}$. $\qquad\square$

Recall from proposition 17.3 that the maps Sq_{even} and Sq_{odd}, from $H_{\frac{1}{2}}$ onto $L^2_{\text{even}}(\mathbb{R})$ and from $H_{\frac{3}{2}}$ onto $L^2_{\text{odd}}(\mathbb{R})$ respectively, intertwine the representation $\mathcal{D}_{\frac{1}{2}}$ (*resp.* $\mathcal{D}_{\frac{3}{2}}$) with the even ((*resp.* odd) part of the metaplectic representation.

We now relate the (even) Weyl symbol of an operator $A : L^2(\mathbb{R}) \to L^2(\mathbb{R})$ commuting with the operator $u \mapsto \check{u}$ to the pair $\{b_0, b_1\}$ of covariant (Berezin–type) symbols of the operators $Sq^{-1}_{\text{even}} A_{\text{even}} Sq_{\text{even}}$ (from $H_{\frac{1}{2}}$ to itself) and $Sq^{-1}_{\text{odd}} A_{\text{odd}} Sq_{\text{odd}}$ (from $H_{\frac{3}{2}}$ to itself) respectively, where, in conformity with the notions introduced right after (13.2), A_{even} (*resp.* A_{odd}) denotes the even–even (*resp.* odd–odd) part of the operator A. For simplicity, we shall (only temporarily) assume that b_0 and b_1 lie in the set of images under the spectral measure for Δ (the hyperbolic Laplacian on Π) of compact intervals in $]\frac{1}{4}, \infty[$.

From (17.17) and (17.14), the active symbol f of an operator on $H_{\tau+1}$ with covariant symbol b is

$$f := \frac{2^\tau \pi^{-\frac{1}{2}}}{\Gamma(\tau + 1)} \Gamma(\tfrac{1}{2}(\tau + \tfrac{3}{2} + i\sqrt{\Delta - 1/4})) \Gamma(\tfrac{1}{2}(\tau + \tfrac{3}{2} - i\sqrt{\Delta - 1/4}))$$

$$\times \frac{1}{4\pi} \frac{(\Gamma(\tau + 1))^2}{\Gamma(\tau + \tfrac{1}{2} + i\sqrt{\Delta - 1/4}) \Gamma(\tau + \tfrac{1}{2} - i\sqrt{\Delta - 1/4})} \, b \qquad (18.41)$$

or, using the duplication formula for the Gamma–function,

$$f = 2^{-\tau - 1} \pi^{-\frac{1}{2}} \frac{\Gamma(\tau + 1)}{\Gamma(\tfrac{\tau}{2} + \tfrac{1}{4} + \tfrac{1}{2}\sqrt{\Delta - 1/4}) \Gamma(\tfrac{\tau}{2} + \tfrac{1}{4} - \tfrac{1}{2}\sqrt{\Delta - 1/4})} \, b. \qquad (18.42)$$

Recall from (17.20) that the horocyclic symbol of an operator on $H_{\tau+1}$ with active symbol f is given as $\mathcal{R}_{\tau+1} TV \, f$, where the operator $\mathcal{R}_{\tau+1}$ (again a function of the Euler operator) was given in (17.19) as

$$\mathcal{R}_{\tau+1} = (2\pi)^{\frac{1}{2}} \pi^{i\pi\mathcal{E}} \frac{\Gamma(\tfrac{\tau}{2} + \tfrac{1}{4} - \tfrac{i\pi}{2}\mathcal{E})}{\Gamma(\tfrac{\tau}{2} + \tfrac{3}{4} + \tfrac{i\pi}{2}\mathcal{E})}. \qquad (18.43)$$

Using again theorem 4.1, to the effect that the isometry TV exchanges the operators $\Delta - \tfrac{1}{4}$ and $\pi^2 \mathcal{E}^2$, next the duplication formula for the Gamma function, we see that the horocyclic symbol of an operator on $H_{\tau+1}$ with covariant symbol b is the image of b under the operator

$$2^{-\tau - \frac{1}{2}} \Gamma(\tau + 1) \frac{\pi^{i\pi\mathcal{E}}}{\Gamma(\tfrac{\tau}{2} + \tfrac{3}{4} + \tfrac{i\pi}{2}\mathcal{E}) \Gamma(\tfrac{\tau}{2} + \tfrac{1}{4} + \tfrac{i\pi}{2}\mathcal{E})} TV$$

$$= \Gamma(\tau + 1) \frac{2^{-1 + i\pi\mathcal{E}} \pi^{-\frac{1}{2} + i\pi\mathcal{E}}}{\Gamma(\tau + \tfrac{1}{2} + i\pi\mathcal{E})} TV. \qquad (18.44)$$

Finally, recalling theorem 17.7, which interprets the even-even and odd-odd parts of a Weyl symbol as a horocyclic symbol when $\tau = -\tfrac{1}{2}$ or $\tau = \tfrac{1}{2}$

respectively, we get under the assumptions above that

$$\text{Weyl symbol}(A) = \frac{2^{-1+i\pi\mathcal{E}}\,\pi^{i\pi\mathcal{E}}}{\Gamma(i\pi\mathcal{E})}\,TV\,b_0 + \frac{2^{-2+i\pi\mathcal{E}}\,\pi^{i\pi\mathcal{E}}}{\Gamma(1+i\pi\mathcal{E})}\,TV\,b_1\,.$$

$$(18.45)$$

Comparing this result to (18.39), we immediately get (without having to make, working in this direction, any assumption about the spectral support of b_0, b_1) the following.

THEOREM 18.2. *Let A be a Hilbert–Schmidt operator on $L^2(\mathbf{R})$, commuting with the map $u \mapsto \check{u}$. Let b_0, b_1 be the covariant symbols of the operators $Sq_{\text{even}}^{-1}\,A_{\text{even}}\,Sq_{\text{even}}$ and $Sq_{\text{odd}}^{-1}\,A_{\text{odd}}\,Sq_{\text{odd}}$ respectively. Finally, let h be the even function on $\mathbf{R}^2$ associated by the construction of theorem 18.1 to the pair of Cauchy data $\{f_0, f_1\} := \{b_0, \frac{b_1}{2}\}$. One has*

$$h = 2^{\frac{3}{2}}\,(2\pi)^{-i\pi\mathcal{E}}\,\Gamma(1+i\pi\mathcal{E})\,.\,\text{Weyl symbol}(A)\,. \qquad (18.46)$$

Remark. The last formula may also be written as

$$h(\xi) = 2^{\frac{5}{2}}\pi \int_0^\infty e^{-2\pi t}\,\text{Weyl symbol}(A)(t^{\frac{1}{2}}\xi)\,t^{\frac{1}{2}}\,dt\,. \qquad (18.47)$$

Recall from theorem 4.1 that the role of the operator T (a function of the Euler operator) was to make TV an isometry starting from $L^2(\Pi)$. Under the assumptions of theorem 18.1, one then has

$$\|h\|_{L^2_{\text{even}}(\mathbf{R}^2)}^2 = 2\left[\pi^2\,\|\mathcal{E}\,TV\,f_0\|_{L^2_{\text{even}}(\mathbf{R}^2)}^2 + \|TV\,f_1\|_{L^2_{\text{even}}(\mathbf{R}^2)}^2\right]$$

$$= 2\pi^2\left[\|(\Delta - \tfrac{1}{4})^{\frac{1}{2}}\,f_0\|_{L^2(\Pi)}^2 + \|f_1\|_{L^2(\Pi)}^2\right]\,.$$

$$(18.48)$$

Another normalization is needed in the Γ–invariant case, for which we introduce besides T, as was done in (4.46), the operator

$$S := 2^{-\frac{1}{2}}\,\pi^{i\pi\mathcal{E}}\,(\Gamma(i\pi\mathcal{E}))^{-1}\,T\,. \qquad (18.49)$$

This amounts to introducing besides the distribution in (18.46), now denoted as $\mathfrak{T}$ rather than h, the distribution,

$$\mathfrak{S} := 2^{-i\pi\mathcal{E}}\,(2i\pi\mathcal{E})\,.\,\text{Weyl symbol}(A)\,. \qquad (18.50)$$

We display for future reference the equation

$$\mathfrak{T} = 2^{\frac{1}{2}}\,\pi^{-i\pi\mathcal{E}}\,\Gamma(i\pi\mathcal{E})\,\mathfrak{S} \qquad (18.51)$$

as well.

It would be a rather lengthy, if feasible, task, to describe in the Γ–invariant setting the composition map $\mathfrak{S} \mapsto \mathfrak{T} \mapsto \{f_0, f_1\}$, where the first map is the application $TS^{-1} = 2^{\frac{1}{2}}\,\pi^{-i\pi\mathcal{E}}\,\Gamma(i\pi\mathcal{E})$ — a still meaningful formula in the arithmetic case — and the second one would be a proper generalization of the construction in theorem 18.1. However, we may circumvent the difficulty by the use of (18.50), since covariant symbols can be defined under very general circumstances.

THEOREM 18.3. *Let* $A : \mathcal{S}(\mathbb{R}) \to \mathcal{S}'(\mathbb{R})$ *be an operator commuting with the map* $u \mapsto \breve{u}$. *Assume that the Weyl symbol of* A *is a* Γ*-invariant even tempered distribution in* $\mathbb{R}^2$, *and that the function* $(2i\pi\,\mathcal{E})$. *Weyl symbol*(A) *satisfies the hypotheses of proposition 13.2 which make the decomposition (13.41) possible: then the Weyl symbol of* A *too has some decomposition*

Weyl symbol$(A) =$

$$2^{\frac{1}{2}} c_0 + 2^{-\frac{1}{2}} c_0' \,\delta + \frac{1}{8\pi} \int_{-\infty}^{\infty} 2^{-\frac{i\lambda}{2}} \,\Psi(\lambda) \,\mathfrak{E}_{i\lambda}^{\natural} \,d\lambda + \sum_{k \neq 0} c_k \, 2^{-\frac{i\lambda_k}{2}} \,\mathfrak{M}_k^{\natural} \,.$$

Let $\mathfrak{S} \in \mathcal{S}'_{\text{even}}(\mathbb{R}^2)$ *be associated to this symbol by (18.50), i.e.,*

$$\mathfrak{S} = c_0 - c_0' \,\delta + \frac{1}{8i\pi} \int_{-\infty}^{\infty} \Psi(\lambda) \,\mathfrak{E}_{i\lambda}^{\natural} \,\lambda \,d\lambda + \frac{1}{i} \sum_{k \neq 0} c_k \,\lambda_k \,\mathfrak{M}_k^{\natural} \,. \tag{18.53}$$

Let b_0, b_1 *be the covariant symbols of the operators* $Sq_{\text{even}}^{-1} \, A_{\text{even}} \, Sq_{\text{even}}$ *and* $Sq_{\text{odd}}^{-1} \, A_{\text{odd}} \, Sq_{\text{odd}}$ *respectively, and set* $\{f_0, f_1\} := \{b_0, \frac{b_1}{2}\}$. *Then the norm as defined in definition 13.3 of* $(2i\pi\mathcal{E})$. *Weyl symbol*$(A) = 2^{i\pi\mathcal{E}} \,\mathfrak{S}$ *satisfies the equation*

$$\frac{1}{2} \, \|2^{i\pi\mathcal{E}} \,\mathfrak{S}\|_{\Gamma}^2 = \| \,|\Delta - \frac{1}{4}|^{\frac{1}{2}} \,f_0\|_{L^2(\Gamma \backslash \Pi)}^2 + \|f_1\|_{L^2(\Gamma \backslash \Pi)}^2 \,. \tag{18.54}$$

Remark. That $2^{i\pi\mathcal{E}} \,\mathfrak{S}$, rather than $\mathfrak{S}$, should occur on the left–hand side of (18.54), is due to the very definition of the Weyl calculus: indeed, it follows from this definition that the operator $\mathcal{G} = 2^{i\pi\mathcal{E}} \,\mathcal{F} \, 2^{-i\pi\mathcal{E}}$, rather than $\mathcal{F}$ (*cf.* (4.12)), plays a fundamental role as an operator on symbols, as discussed in the beginning of section 13, while it is the second of these two operators that generates the intertwining operators (*cf.* (4.28)).

PROOF. Starting from (13.3) and the definition of Sq_{even} and Sq_{odd} in proposition 17.3, it is immediate to get the formulas

$$(Sq_{\text{even}}^{-1} u_z)(s) = 2^{\frac{1}{2}} \,(\text{Im}\,(-\frac{1}{z}))^{\frac{1}{4}} \,s^{-\frac{1}{2}} \,e^{2i\pi\frac{s}{z}}$$

$$(Sq_{\text{odd}}^{-1} u_z^1)(s) = 4\pi^{\frac{1}{2}} \,(\text{Im}\,(-\frac{1}{z}))^{\frac{3}{4}} \,s^{\frac{1}{2}} \,e^{2i\pi\frac{s}{z}} \,. \tag{18.55}$$

On the other hand, (17.8) defines the functions $\psi_z^{\tau+1}$ associated with the Hilbert space $H_{\tau+1}$. This leads to the formulas

$$u_z = Sq_{\text{even}} \,\psi_z^{\frac{1}{2}} \qquad \text{and} \qquad u_z^1 = Sq_{\text{odd}} \,\psi_z^{\frac{3}{2}} \tag{18.56}$$

which relate the functions u_z and u_z^1 used in connection with the even–even and odd–odd parts of the Weyl calculus to those used in connection with the spaces $H_{\frac{1}{2}}$ and $H_{\frac{3}{2}}$.

In the definition (17.13) of covariant symbols in the $H_{\tau+1}$-calculus, one must of course use, on the right-hand side, the scalar product on this Hilbert space. Since Sq_{even} (resp. Sq_{odd}) is an isometry in each case, one can also write

$$(u_z|Au_z) = (\psi^{\frac{1}{2}}|Sq_{\text{even}}^{-1} A_{\text{even}} Sq_{\text{even}} \psi^{\frac{1}{2}})_{H_{\frac{1}{2}}}$$
$$= (\text{Symb}_{\text{Ber}}(Sq_{\text{even}}^{-1} A_{\text{even}} Sq_{\text{even}}))(z)$$
$$= b_0(z)$$
$$= f_0(z) \tag{18.57}$$

and

$$(u_z^1|Au_z^1) = (\psi^{\frac{3}{2}}|Sq_{\text{odd}}^{-1} A_{\text{odd}} Sq_{\text{odd}} \psi^{\frac{3}{2}})_{H_{\frac{3}{2}}}$$
$$= (\text{Symb}_{\text{Ber}}(Sq_{\text{odd}}^{-1} A_{\text{odd}} Sq_{\text{odd}}))(z)$$
$$= b_1(z)$$
$$= 2 f_1(z). \tag{18.58}$$

Using propositions 13.2 and 13.4, we are done, noting first that applying $2|\Delta - \frac{1}{4}|^{\frac{1}{2}}$ to f_0 and f_1 simultaneously amounts to substituting $|\lambda| \Psi(\lambda)$ and $|\lambda_k| c_{\pm k}$ for $\Psi(\lambda)$ and $c_{\pm k}$ respectively, on the right-hand sides of (13.42) and (13.43). $\qquad\square$

Remark. Formula (18.54) should of course be compared to (18.48).

Theorem 18.3 actually expresses that the Lax-Phillips Hilbert space of data $\{f_0, f_1\}$ as defined in [25], p.103, has a dense subspace which is isometric to the space of automorphic distributions on $\mathbf{R}^2$ as introduced in theorem 13.2 and definition 13.3; under this isometry, the basic matrix-operator of the Lax-Phillips theory transfers to the Euler operator $i\pi\mathcal{E}$.

Statements in [25] are often more transparent when rephrased through theorem 18.3.

For instance, corollary 7.23 there, to the effect that the eigenvectors of the operator $A = \begin{pmatrix} 0 & I \\ -\Delta + \frac{1}{4} & 0 \end{pmatrix}$ which correspond to the eigenvalues $\pm\frac{1}{2}$ are the pairs $\{f_0, f_1\} = \{1, \frac{1}{2}\}$ or $\{1, -\frac{1}{2}\}$, becomes the following: in the Hilbert space of automorphic distributions on $\mathbf{R}^2$, the eigendistributions of the operator $i\pi\mathcal{E}$ which correspond to the eigenvalues $\pm\frac{1}{2}$ are the distributions 1 and $\frac{1}{2}\delta$. To check that the two assertions are indeed equivalent, it suffices, using theorem 18.3, to remember the well-known fact that $\frac{1}{2}\delta$, as a distribution on $\mathbf{R}^2$, is the symbol of the operator $u \mapsto \check{u}$.

We now characterize, in terms of the distribution $\mathfrak{S}$ that occurs in (18.53), or $\mathfrak{T}$ related to $\mathfrak{S}$ as in (18.51), *i.e.*

$$\mathfrak{T} = 2^{\frac{1}{2}} c_0 + 2^{\frac{3}{2}} \pi c_0' \delta + \frac{2^{\frac{1}{2}}}{8i\pi} \int_{-\infty}^{\infty} \pi^{\frac{i\lambda}{2}} \Gamma(-\frac{i\lambda}{2}) \Psi(\lambda) \mathfrak{E}_{i\lambda}^{\sharp} \lambda \, d\lambda$$

$$+ \frac{2^{\frac{1}{2}}}{i} \sum_{k \neq 0} c_k \pi^{\frac{i\lambda_k}{2}} \Gamma(-\frac{i\lambda_k}{2}) \lambda_k \mathfrak{M}_k^{\sharp}, \tag{18.59}$$

the spaces orthogonal to the Lax–Phillips incoming and outgoing spaces D_- and D_+ respectively: the definition of these two spaces depends on some arbitrary constant $a > 1$. According to [25], p.123, the space of data $\{f_0, f_1\}$ orthogonal to D_- is characterized by the validity, whenever $y > a$, of the relation

$$\mathrm{Coeff}_0(f_1)(y) = -y^{\frac{3}{2}} \frac{d}{dy}(y^{-\frac{1}{2}} \mathrm{Coeff}_0(f_0)(y)) \tag{18.60}$$

and the orthogonality to D_+ by

$$\mathrm{Coeff}_0(f_1)(y) = y^{\frac{3}{2}} \frac{d}{dy}(y^{-\frac{1}{2}} \mathrm{Coeff}_0(f_0)(y)) : \tag{18.61}$$

in these two relations $\mathrm{Coeff}_0(f_i)(y)$ denotes the so–called constant term in the Fourier expansion with respect to x of the automorphic function f_i. Now, given any even tempered distribution $\mathfrak{T}$ on $\mathbf{R}^2$, invariant under the linear transformation associated with the matrix $\left(\begin{smallmatrix} 1 & 1 \\ 0 & 1 \end{smallmatrix}\right)$, one may define its "0–th Fourier coefficient" too: as a substitute for what would be, if $\mathfrak{T}$ were a function, the function (in the definition of which ξ_1 is arbitrary)

$$C_0(\xi_2) = \int_{-\frac{1}{2}}^{\frac{1}{2}} \mathfrak{T}(\xi_1 + t\xi_2, \xi_2) \, dt, \tag{18.62}$$

we take the distribution $C_0(\mathfrak{T})$ on $\mathbf{R} \backslash \{0\}$ defined by

$$< C_0(\mathfrak{T}), w > = < \mathfrak{T}, \xi \mapsto |\xi_2|^{-1} \phi(\frac{\xi_1}{\xi_2}) w(\xi_2) >, \tag{18.63}$$

where ϕ is any function in the space $C_0^{\infty}(\mathbf{R})$ such that $\sum_{n \in \mathbf{Z}} \phi(x - n) = 1$ for every x: that this definition does not depend on the choice of ϕ may be checked by inserting the extra factor $1 = \sum_{n \in \mathbf{Z}} \psi(\frac{\xi_1 - n\xi_2}{\xi_2})$ under the right-hand side of (18.63), ψ satisfying the same condition as ϕ; obviously, too, the distribution $C_0(\mathfrak{T})$ is even, and the definition agrees with (18.62) in the case when $\mathfrak{T}$ is a function.

PROPOSITION 18.4. *Assume that A satisfies the hypotheses of theorem 18.3 and let $\mathfrak{T}$ be the Γ–invariant distribution on $\mathbf{R}^2$ linked to $\mathfrak{S}$ by (18.51). Then the pair of Cauchy data $\{f_0, f_1\}$ is orthogonal to D_+, i.e. (18.61) is satisfied, if and only if the distribution $C_0(\mathfrak{T})$ defined in (18.63) has its support disjoint from $]-a^{-\frac{1}{2}}, a^{-\frac{1}{2}}[$. Orthogonality to D_- is given by the same condition after $\mathfrak{S}$ has been replaced by $\mathcal{F}\mathfrak{S}$.*

PROOF. Under the assumptions of theorem 18.3, one has, if $w \in C_0^\infty(\mathbb{R}\backslash\{0\})$ is an even function,

$$
< C_0(\mathfrak{T}), w > = 2^{\frac{1}{2}} c_0 \int_{\mathbb{R}^2} h(\xi) \, d\xi + \frac{2^{\frac{1}{2}}}{8i\pi} \int_{-\infty}^{\infty} \pi^{\frac{i\lambda}{2}} \Gamma(-\frac{i\lambda}{2}) \Psi(\lambda) < \mathfrak{E}_{i\lambda}^{\#}, h > \lambda \, d\lambda
$$

$$
+ \frac{2^{\frac{1}{2}}}{i} \sum_{k \neq 0} \pi^{\frac{i\lambda_k}{2}} \Gamma(-\frac{i\lambda_k}{2}) c_k \lambda_k < \mathfrak{M}_k^{\#}, h >, \tag{18.64}
$$

where

$$
h(\xi_1, \xi_2) := |\xi_2|^{-1} \phi(\frac{\xi_1}{\xi_2}) w(\xi_2). \tag{18.65}
$$

In the sense of (4.20), one has

$$
h_{-\lambda}^b(s) = \frac{1}{2\pi} \int_0^\infty t^{-i\lambda} h(ts, t) \, dt
$$

$$
= \frac{1}{2\pi} \phi(s) \int_0^\infty t^{-1-i\lambda} w(t) \, dt. \tag{18.66}
$$

As shown by (13.18),

$$
< \mathfrak{E}_{i\lambda}^{\#}, h >= 2 \Big(\int_0^\infty t^{-1-i\lambda} w(t) \, dt \Big) < \mathfrak{E}_{i\lambda}, \phi > \tag{18.67}
$$

and a similar equation holds with $\mathfrak{M}_k^{\#}$ in place of $\mathfrak{E}_{i\lambda}^{\#}$.

Then, if $\mathrm{Re}\ \nu < -1$, (3.4) yields

$$
< \mathfrak{E}_\nu, \phi >= \sum_{m \geq 1, n \in \mathbb{Z}} m^{\nu-1} \phi(\frac{n}{m}) = \zeta(-\nu) \tag{18.68}
$$

under the assumptions relative to ϕ, and the result can be analytically continued to $\nu = i\lambda$. For $n \in \mathbb{Z}^\times$, one has

$$
\hat{\phi}(n) = \sum_{m \in \mathbb{Z}} \int_0^1 e^{2i\pi n x} \phi(x + m) \, dx = 0, \tag{18.69}
$$

and $\hat{\phi}(0) = 1$. From (13.23), we get $< \mathfrak{M}_k, \phi >= 0$, then $< \mathfrak{M}_k^{\#}, h >= 0$ too. Substituting this result together with (18.68) and the observation that

$$
\int_{\mathbb{R}^2} h(\xi) \, d\xi = \hat{\phi}(0) \, \hat{w}(0) \tag{18.70}
$$

in (18.64), we get

$$
< C_0(\mathfrak{T}), w > = 2^{\frac{1}{2}} c_0 \, \hat{w}(0)
$$

$$
+ \frac{2^{\frac{3}{2}}}{8i\pi} \Big(\int_0^\infty t^{-1-i\lambda} w(t) \, dt \Big) \int_{-\infty}^{\infty} \pi^{\frac{i\lambda}{2}} \Gamma(-\frac{i\lambda}{2}) \zeta(-i\lambda) \lambda \, \Psi(\lambda) \, d\lambda. \tag{18.71}
$$

This shows that $C_0(\mathfrak{T})$ is actually an (even) function in $\mathbb{R} \backslash \{0\}$,

$$(C_0(\mathfrak{T}))(\xi_2) = 2^{\frac{1}{2}} c_0 + \frac{2^{\frac{1}{2}}}{8i\pi} \int_{-\infty}^{\infty} \pi^{\frac{i\lambda}{2}} \Gamma(-\frac{i\lambda}{2}) \zeta(-i\lambda) \lambda \Psi(\lambda) |\xi_2|^{-1-i\lambda} d\lambda, \tag{18.72}$$

or

$$(C_0(\mathfrak{T}))(y^{-\frac{1}{2}}) = 2^{\frac{1}{2}} c_0 + \frac{2^{\frac{1}{2}}}{8i\pi} \int_{-\infty}^{\infty} \lambda \Psi(\lambda) \zeta^*(-i\lambda) y^{\frac{1+i\lambda}{2}} d\lambda. \tag{18.73}$$

On the other hand, (18.57) and (13.44) (in which the Weyl symbol of A is substituted for the distribution $\mathfrak{S}$ that occurs there) imply

$$f_0(z) = 2^{\frac{1}{2}} (c_0 + c_0') + \frac{2^{\frac{1}{2}}}{8\pi} \int_{-\infty}^{\infty} \Psi(\lambda) E^*_{\frac{1-i\lambda}{2}}(z) d\lambda$$
$$+ 2^{\frac{1}{2}} \sum_{k \neq 0} c_k \mathcal{M}_{|k|}(z) \tag{18.74}$$

and in a similar way, substituting (18.58) for (18.57) and (13.45) for (13.44),

$$2 f_1(z) = 2^{\frac{1}{2}} (c_0 - c_0') - \frac{2^{\frac{1}{2}} i}{8\pi} \int_{-\infty}^{\infty} \lambda \Psi(\lambda) E^*_{\frac{1-i\lambda}{2}}(z) d\lambda$$
$$- 2^{\frac{1}{2}} i \sum_{k \neq 0} c_k \lambda_k \mathcal{M}_{|k|}(z). \tag{18.75}$$

Thus the constant terms $\mathrm{Coeff}_0(f_0)$ and $\mathrm{Coeff}_0(f_1)$ which occur in (18.60) and (18.61) are given by

$\mathrm{Coeff}_0(f_0) =$

$$2^{\frac{1}{2}} (c_0 + c_0') + \frac{2^{\frac{1}{2}}}{8\pi} \int_{-\infty}^{\infty} [\zeta^*(1 - i\lambda) y^{\frac{1-i\lambda}{2}} + \zeta^*(1 + i\lambda) y^{\frac{1+i\lambda}{2}}] \Psi(\lambda) d\lambda \tag{18.76}$$

and

$\mathrm{Coeff}_0(f_1) =$

$$2^{-\frac{1}{2}} (c_0 - c_0') + \frac{2^{-\frac{1}{2}}}{8\pi i} \int_{-\infty}^{\infty} [\zeta^*(1 - i\lambda) y^{\frac{1-i\lambda}{2}} + \zeta^*(1 + i\lambda) y^{\frac{1+i\lambda}{2}}] \lambda \Psi(\lambda) d\lambda. \tag{18.77}$$

It follows that

$$\mathrm{Coeff}_0(f_1) - y^{\frac{3}{2}} \frac{d}{dy}(y^{-\frac{1}{2}} \mathrm{Coeff}_0(f_0))$$

$$= 2^{\frac{1}{2}} c_0 + \frac{2^{\frac{1}{2}}}{8\pi i} \int_{-\infty}^{\infty} \zeta^*(1 + i\lambda) y^{\frac{1+i\lambda}{2}} \lambda \Psi(\lambda) d\lambda, \tag{18.78}$$

thus (18.61) is equivalent, in view of (18.73), to the condition that $C_0(\mathfrak{T})(y^{-\frac{1}{2}}) = 0$ for $y > a$.

Since the pair of covariant symbols $\{b_0, -b_1\}$ is associated to the operator $Ch\,A$ where Ch denotes the operator $u \mapsto \check{u}$, and the symbol of this latter operator is ($cf.$ (13.2) and what follows) the image of that of A under $\mathcal{G}$, the condition (18.60) is the same as (18.61) after $\mathcal{G}$ Weyl symbol(A) has been substituted for Weyl symbol(A), $i.e.$ after

$$2^{1-i\pi\mathcal{E}}\,(i\pi\mathcal{E})\,\mathcal{G}\,(i\pi\mathcal{E})^{-1}\,2^{-1+i\pi\mathcal{E}}\,\mathfrak{S} = -2^{-2i\pi\mathcal{E}}\,\mathcal{G}\mathfrak{S} = -\mathcal{F}\mathfrak{S} \tag{18.79}$$

has been substituted for $\mathfrak{S}$.   □

We now switch to a different subject, the starting point of which, however, has some similarity with the construction in theorem 18.1. Namely, we interest ourselves to convolutions $Z_2 * \Psi$ where, this time, Ψ is no longer a measure carried by ∂C, but a function (or distribution) supported in $\bar{C}$.

If Ψ is homogeneous of degree $\frac{-5-\nu}{2}$, $Z_2 * \Psi$, as a function on C, is homogeneous of degree $\frac{-1-\nu}{2}$ so that, as a consequence of the equation

$$(\Box - E(E+1))\psi|_{\mathcal{H}} = \Delta_{\mathcal{H}}\cdot\psi|_{\mathcal{H}} \tag{18.80}$$

reported right after (18.15), one has

$$(\Delta_{\mathcal{H}} - \frac{1-\nu^2}{4})\left((Z_2 * \Psi)|_{\mathcal{H}}\right) = \Psi|_{\mathcal{H}}. \tag{18.81}$$

A first consequence is that if $f \in L^2(\Pi)$ and $\mathrm{Re}\,\nu < 0$, the following recipe gives a new way of computing $(\Delta - \frac{1-\nu^2}{4})^{-1}f$: extend $f \circ \chi$, a function on $\mathcal{H}$, to a function Ψ on C homogeneous of degree $\frac{-5-\nu}{2}$, then restrict $Z_2 * \Psi$ to $\mathcal{H}$, finally compose this restriction with χ^{-1}.

Making the net result of this recipe explicit will yield a possibly new derivation of the integral kernel of the resolvent $(\Delta_{\mathcal{H}} - \frac{1-\nu^2}{4})^{-1}$.

Indeed, as $d(t\zeta) = t^2\,dt\,\frac{d\zeta_1\,d\zeta_2}{\zeta_0}$ $(t > 0, \zeta \in \mathcal{H})$,

$$((\Delta_{\mathcal{H}} - \frac{1-\nu^2}{4})^{-1}f)(\chi(\eta)) = \frac{1}{2\pi}\int_0^\infty t^{\frac{-1-\nu}{2}}\,dt$$

$$\int_{\mathcal{H}}[(\eta_0 - t\zeta_0)^2 - (\eta_1 - t\zeta_1)^2 - (\eta_2 - t\zeta_2)^2]_+^{-\frac{1}{2}}\,f(\chi(\zeta))\,\frac{d\zeta_1\,d\zeta_2}{\zeta_0}. \tag{18.82}$$

With

$$< \eta, J\zeta >: = \eta_0\,\zeta_0 - \eta_1\,\zeta_1 - \eta_2\,\zeta_2, \tag{18.83}$$

the condition $\eta - t\zeta \in C$ is equivalent to the pair of conditions

$$\begin{cases} t^2 - 2t < \eta, J\zeta > +1 > 0 \\ t < \frac{\eta_0}{\zeta_0} \end{cases} \tag{18.84}$$

and, since

$$\frac{\eta_0}{\zeta_0} \le \ <\eta, J\zeta> + (<\eta, J\zeta>^2 -1)^{\frac{1}{2}} \tag{18.85}$$

when both η and ζ lie in $\mathcal{H}$ (an inequality useful in relativistic pseudodifferential analysis: cf. [43], p.70), it follows from (18.83) that, on the support of $Z_2(\eta - t\,\zeta)$, one has

$$t \le \ <\eta, J\zeta> - (<\eta, J\zeta>^2 -1)^{\frac{1}{2}}. \tag{18.86}$$

We then set $\chi(\eta) = z$, $\chi(\zeta) = z'$ so that $\frac{d\zeta_1 \, d\zeta_2}{\zeta_0} = d\mu(z')$ and

$$<\eta, J\zeta> = \frac{y^2 + y'^2 + (x - x')^2}{2yy'}$$

$$= \cosh d(z, z'). \tag{18.87}$$

Thus, the integral kernel of the resolvant $(\Delta_\mathcal{H} - \frac{1-\nu^2}{4})^{-1}$ is given, with Re $\nu < 0$, as

$$k_{\frac{1-\nu}{2}}(z, z') = \frac{1}{2\pi} \int_0^{\exp -d(z,z')} t^{\frac{-\nu-1}{2}} (1 - 2t \cosh d(z, z') + t^2)^{-\frac{1}{2}} \, dt. \tag{18.88}$$

Abbreviating $d(z, z')$ as d, we may write this (using an elementary change of variable and [27], p.54) as

$$\frac{1}{2\pi} \int_0^{e^{-d}} t^{\frac{-\nu-1}{2}} [(1 - te^d)(1 - te^{-d})]^{-\frac{1}{2}} \, dt$$

$$= \frac{1}{2\pi} \frac{\Gamma(\frac{1-\nu}{2})\,\Gamma(\frac{1}{2})}{\Gamma(\frac{2-\nu}{2})} e^{\frac{(\nu-1)d}{2}} \,_2F_1(\frac{1}{2}, \frac{1-\nu}{2}; \frac{2-\nu}{2}; e^{-2d}). \tag{18.89}$$

Using the quadratic transformation formula in ([27], p.50) involving the map (in this reference's notation) $z \mapsto \frac{4z}{(1+z)^2}$, this is the same as

$$\frac{1}{2\pi} \frac{\Gamma(\frac{1-\nu}{2})\,\Gamma(\frac{1}{2})}{\Gamma(\frac{2-\nu}{2})} (2 \cosh d)^{\frac{\nu-1}{2}} \,_2F_1(\frac{1-\nu}{4}, \frac{3-\nu}{4}; \frac{2-\nu}{2}; \frac{1}{\cosh^2 d}), \tag{18.90}$$

which can be transformed (for the sake of comparison with other references), using a second quadratic transformation formula on the same page, this one involving the map $z \mapsto \frac{2\sqrt{z}}{1+\sqrt{z}}$, to

$$\frac{1}{4\pi} \frac{(\Gamma(\frac{1-\nu}{2}))^2}{\Gamma(1 - \nu)} (\cosh^2 \frac{d}{2})^{\frac{\nu-1}{2}} \,_2F_1(\frac{1-\nu}{2}, \frac{1-\nu}{2}; 1 - \nu; \frac{1}{\cosh^2 \frac{d}{2}}). \tag{18.91}$$

According to [27], p. 153, the right–hand side of (18.89) is just $\frac{1}{2\pi} \mathfrak{Q}_{-\frac{1}{2}-\frac{\nu}{2}}(\cosh d)$ in terms of Legendre functions.

A most interesting situation occurs in the case when we substitute for Ψ, as introduced just before (18.79), a $PSL(2, \mathbf{Z})$ – invariant homogeneous distribution (where $PSL(2, \mathbf{Z}) \subset PSL(2, \mathbf{R})$ is identified with a subgroup of $SO_o(1, 2)$) with support contained in C, generated as a cone by a discrete subset Σ of $\mathcal{H}$: for, then, the equation

$$\Delta_{\mathcal{H}}((Z_2 * \Psi)|_{\mathcal{H}}) = \frac{1 - \nu^2}{4} (Z_2 * \Psi)|_{\mathcal{H}}, \qquad (18.92)$$

holds on $\mathcal{H}$, in the distribution sense, in the complementary of Σ. In other words, $(Z_2 * \Psi)|_{\mathcal{H}} \circ \chi^{-1}$ is *almost* an eigenfunction of Δ in Π, *i.e.*, it satisfies the eigenfunction equation for the eigenvalue $\frac{1-\nu^2}{4}$ outside the set $\chi(\Sigma) \subset \Pi$.

Before we specialize to this situation, we need to describe, at least in a formal sense, how one can decompose distributions on C as "sums" (generally, both a continuous and a discrete part may appear, but only the first one is apparent in the formal setting) of homogeneous distributions. To do this, we proceed in analogy with what has been done in section 16 (from (16.72) to (16.74)). First, functions f on C, square–integrable with respect to the Lebesgue measure $d\eta$, should be decomposed this time as

$$f = \int_{-\infty}^{\infty} f_\lambda \, d\lambda, \qquad (18.93)$$

with

$$f_\lambda(\eta) = \frac{1}{2\pi} \int_0^{\infty} t^{\frac{1}{2} + i\lambda} f(t\eta) \, dt, \qquad (18.94)$$

a function homogeneous of degree $-\frac{3}{2} - i\lambda$. This is the three–dimensional analogue of $(4.19), (4.20)$. Defining then (compare (4.21))

$$f_\lambda^\flat := f_\lambda \Big|_{\mathcal{H}}, \qquad (18.95)$$

one has, using $d\eta = t^2 dt \frac{d\zeta_1 d\zeta_2}{\zeta_0}$ if $\eta = t\zeta$, $t > 0$, $\zeta \in \mathcal{H}$,

$$\int_C |f(\eta)|^2 \, d\eta = 2\pi \int_{-\infty}^{\infty} d\lambda \int_{\mathcal{H}} |f_\lambda^\flat(\zeta)|^2 \frac{d\zeta_1 d\zeta_2}{\zeta_0}, \qquad (18.96)$$

i.e.,

$$\|f\|_{L^2(C)}^2 = 2\pi \int_{-\infty}^{\infty} \|f_\lambda^\flat\|_{L^2(\mathcal{H})}^2 \, d\lambda, \qquad (18.97)$$

the analogue of (4.22).

In just the same way as the one that led from (16.72) to (16.74), we then define, if $\mathfrak{S}$ is a distribution on C and $u \in C_0^{\infty}(\mathcal{H})$,

$$< \mathfrak{S}_\lambda^\flat, u > = \frac{1}{2\pi} < \mathfrak{S}, \eta \mapsto [\eta]^{-\frac{3}{2} + i\lambda} u(\frac{\eta}{[\eta]}) >, \qquad (18.98)$$

i.e.,

$$< \mathfrak{S}_\lambda^\flat, u > = \frac{1}{2\pi} < \mathfrak{S}, u^\sharp >, \qquad (18.99)$$

with

$$u^{\natural}(\eta) = [\eta]^{-\frac{3}{2}+i\lambda} u(\frac{\eta}{[\eta]}), \tag{18.100}$$

the homogeneous extension to C of degree $-\frac{3}{2} + i\lambda$ of u. All this must be understood in a rather formal way, since we do not plan to develop a general frame which would make it meaningful, only examine an interesting particular case.

With the same understanding, the distribution $\mathfrak{S}$ can be written as

$$\mathfrak{S} = \int_{-\infty}^{\infty} \mathfrak{S}_\lambda \, d\lambda \tag{18.101}$$

with $\mathfrak{S}_\lambda$ homogeneous of degree $-\frac{3}{2} - i\lambda$ in the distribution sense, given as

$$\begin{aligned}
< \mathfrak{S}_\lambda, f > &= < \mathfrak{S}, f_{-\lambda} > \\
&= < \mathfrak{S}, \eta \mapsto [\eta]^{-\frac{3}{2}+i\lambda} f_{-\lambda}^{\flat}(\frac{\eta}{[\eta]}) \\
&= 2\pi < \mathfrak{S}_\lambda^{\flat}, f_{-\lambda}^{\flat} > .
\end{aligned} \tag{18.102}$$

Obviously, we increase the chances of the preceding constructions being meaningful if we substitute for λ, a real number, a complex number μ in some suitable half–plane.

We now come to examples of arithmetic interest. Two $PSL(2,\mathbf{Z})$–invariant discrete subsets of C of special interest are the following ones:

$$\mathbf{Z}_C = \mathbf{Z}^3 \cap C = \{\eta = \begin{pmatrix} p & q \\ q & r \end{pmatrix} : p,q,r \in \mathbf{Z}, \quad pr - q^2 > 0, \, p + r > 0\} \tag{18.103}$$

and

$$\mathbf{Z}_C^{\text{even}} = \{\eta = \begin{pmatrix} 2p & q \\ q & 2r \end{pmatrix} : p,q,r \in \mathbf{Z}, \quad 4pr - q^2 > 0, \, p + r > 0\} : \tag{18.104}$$

to say that η lies in the second of these sets means that the quadratic form $\xi \mapsto p\,\xi_1^2 + q\,\xi_1\xi_2 + r\,\xi_2^2$ is positive definite and even integral (*i.e.*, takes even values on $\mathbf{Z}^2$). The set $\mathbf{Z}_C^{\text{even}}$ decomposes as

$$\mathbf{Z}_C^{\text{even}} = \bigcup_{\substack{D<0 \\ D \equiv 0 \text{ or } 1 \bmod 4}} A_D, \tag{18.105}$$

where

$$A_D = \{\eta^j = \begin{pmatrix} 2p & q \\ q & 2r \end{pmatrix} : p,q,r \in \mathbf{Z}, \quad 4pr - q^2 = -D, \, p + r > 0\} : \tag{18.106}$$

D is the discriminant of the above–mentioned quadratic form, and the $PSL(2, \mathbf{Z})$–invariant set A_D plays a major role in Zagier's paper [50]. Set

$$\zeta^j = \frac{\eta^j}{[\eta^j]} = (-\frac{1}{D})^{\frac{1}{2}} \eta^j . \tag{18.107}$$

We now wish to consider, in analogy with the Dirac comb (16.1) on $\mathbf{R}^2$, the Dirac–like distribution $\mathcal{D}_{\mathbf{Z}_C^{\mathrm{even}}}$ defined as (2π) times the sum of the Dirac masses on the points of $\mathbf{Z}_C^{\mathrm{even}}$.

PROPOSITION 18.5. With $\mathcal{D}_{\mathbf{Z}_C^{\mathrm{even}}}$ as just defined, the distribution $(\mathcal{D}_{\mathbf{Z}_C^{\mathrm{even}}})_\mu$ as defined when Im μ is large by (18.101), can be written as

$$(\mathcal{D}_{\mathbf{Z}_C^{\mathrm{even}}})_\mu = \sum_{\substack{D<0 \\ D \equiv 0 \,\mathrm{or}\, 1 \,\mathrm{mod}\, 4}} |D|^{-\frac{3}{2}+i\mu} \; \Psi^D_{-\frac{3}{2}+i\mu} , \tag{18.108}$$

in which the distribution $\Psi^D_{\frac{\nu-1}{2}}$ on C, homogeneous of degree $\frac{-5-\nu}{2}$, can be characterized by

$$(Z_2 * \Psi^D_{\frac{\nu-1}{2}})(\eta) = \sum_{\eta^j \in A_D} k_{\frac{1-\nu}{2}}(z, z_j) \tag{18.109}$$

where $z_j := \chi(\zeta^j) = \chi(\frac{\eta^j}{[\eta^j]})$ and $k_{\frac{1-\nu}{2}}$ is the integral kernel of the operator $(\Delta - \frac{1-\nu^2}{4})^{-1}$.

PROOF. As a consequence of (18.101), (18.93)–(18.94), one has

$$< (\mathcal{D}_{\mathbf{Z}_C^{\mathrm{even}}})_\mu, f >= \sum_{\substack{D<0 \\ D \equiv 0 \,\mathrm{or}\, 1 \,\mathrm{mod}\, 4}} \sum_{\eta^j \in A_D} [\eta^j]^{-\frac{3}{2}+i\mu} \int_0^\infty t^{\frac{1}{2}-i\mu} f(t\zeta^j) \, dt \tag{18.110}$$

or

$$(\mathcal{D}_{\mathbf{Z}_C^{\mathrm{even}}})_\mu = \sum_{\substack{D<0 \\ D \equiv 0 \,\mathrm{or}\, 1 \,\mathrm{mod}\, 4}} |D|^{-\frac{3}{2}+i\mu} \; \Psi^D_{-\frac{3}{2}+i\mu} , \tag{18.111}$$

if we define the distribution $\Psi^D_{\frac{\nu-1}{2}}$ on C, homogeneous of degree $\frac{-5-\nu}{2}$, by the equation

$$< \Psi^D_{\frac{\nu-1}{2}}, f >: = \int_0^\infty t^{\frac{-\nu-1}{2}} \, dt \sum_{\eta^j \in A_D} f(t \zeta^j). \tag{18.112}$$

In accordance with the program initiated right before (18.79), we now compute (for complex ν with Re $\nu < -1$, as it will turn out)

$$(Z_2 * \Psi^D_{\frac{\nu-1}{2}})(\eta) =< \Psi^D_{\frac{\nu-1}{2}}, \eta \mapsto Z_2(\eta - \eta') >$$

$$= \int_0^\infty t^{\frac{-\nu-1}{2}} \, dt \sum_{\eta^j \in A_D} Z_2(\eta - t\zeta^j). \tag{18.113}$$

As a consequence of the computations that precede,

$$(Z_2 * \Psi^D_{\frac{\nu-1}{2}})(\eta) = \sum_{\eta^j \in A_D} k_{\frac{1-\nu}{2}}(z, z_j), \tag{18.114}$$

where we recall that $z_j = \chi(\zeta^j) = \chi(\frac{\eta^j}{[\eta^j]})$.

$\square$

According to [50], p. 281, we partition the set $\chi(A_D) \subset \Pi$ as

$$\chi(A_D) = \Gamma.z_1 \cup \cdots \cup \Gamma.z_{h(D)} \tag{18.115}$$

for some choice of an appropriate finite subset $\{\eta^1, \ldots, \eta^{h(D)}\}$ of A_D. Then

$$(Z_2 * \Psi^D_{\frac{\nu-1}{2}})(\eta) = \sum_{\ell=1}^{h(D)} \frac{1}{\#(\Gamma_\ell)} \sum_{\gamma \in \Gamma} k_{\frac{1-\nu}{2}}(z, \gamma.z_\ell), \tag{18.116}$$

where Γ_ℓ denotes the stabilizer of z_ℓ in Γ. From the theory of the automorphic Green function (*cf.* [20], chapter 5 or [18], p. 104, with the references there), it follows that, given z and $z' \in \Pi$ with $z \notin \Gamma.z'$, the series

$$G_{\frac{1-\nu}{2}}(z; z') := \frac{1}{2} \sum_{\gamma \in \Gamma} k_{\frac{1-\nu}{2}}(z, \gamma.z') \tag{18.117}$$

converges absolutely for Re $\nu < -1$, and has a meromorphic extension to the plane. The singularities in the half–plane Re $\nu \leq 0$ can be only simple poles at $\nu = -1$ or $\nu = i\lambda_k$ with $\frac{1+\lambda_k^2}{4}$ in the discrete spectrum of the modular Laplacian. This is just a quotation from Hejhal's paper [18], and a proof can be derived from the Roelcke–Selberg expansion (for Re $\nu < -1$) of the function $z \mapsto G_{\frac{1-\nu}{2}}(z; z')$ (see *e.g.* [20], chapter 7, or [36], p. 271, or [47], p. 25). The factor $\frac{1}{2}$ in the definition of the automorphic Green kernel is of course due to the fact that our Γ is a subgroup of $SL(2, \mathbb{R})$, not of $PSL(2, \mathbb{R})$.

It also follows from the theory of the Fourier expansion of the automorphic Green function ([20], p. 81) that for fixed z', and $\nu \neq 0$ in some neighbourhood of the half–plane Re $\nu \leq 0$ (this implies in particular that $\zeta^*(1 - \nu) \neq 0$ if this neighbourhood is well–chosen), one has when the left–hand side is well–defined in the above sense

$$G_{\frac{1-\nu}{2}}(z; z') = -\nu^{-1} y^{\frac{1+\nu}{2}} E_{\frac{1-\nu}{2}}(z') + O(e^{-2\pi y}) \tag{18.118}$$

as $y = \text{Im } z \to \infty$. Thus

$$(Z_2 * \Psi^D_{\frac{\nu-1}{2}})(\chi^{-1}(z)) = 2 \sum_{\ell=1}^{h(D)} \frac{1}{\#(\Gamma_\ell)} G_{\frac{1-\nu}{2}}(z; z_\ell), \tag{18.119}$$

a Γ–invariant function of z, is rapidly decreasing at infinity in the fundamental domain if the condition

$$\sum_{\ell=1}^{h(D)} \frac{1}{\#(\Gamma_\ell)} E_{\frac{1-\nu}{2}}(z_\ell) = 0 \qquad (18.120)$$

is satisfied. The function $(Z_2 * \Psi_{\frac{D}{\nu-1}}) \circ \chi^{-1}$ is not a cusp–form, however, since it is singular on the discrete set $\chi(A_D) \subset \Pi$: still, it does satisfy the eigenfunction equation for the eigenvalue $\frac{1-\nu^2}{4}$ outside this set. This is (at least in the case when the class number $h(D)$ is 1: *cf.* Hejhal's slightly cryptic remark on the bottom of page 105) the spirit of the argument in Hejhal's paper [18], where — for reasons having to do with the explanation of puzzling facts concerning the numerical analysis of the modular Laplacian — the main emphasis is put on the case when $D = -3$, in which $h(D) = 1$ and z_1 may be taken as $z_1 = e^{\frac{i\pi}{3}}$.

Since, for Re $\nu < -1$ and any ℓ,

$$\zeta(1-\nu)\, E_{\frac{1-\nu}{2}}(z_\ell) = \frac{1}{2} \sum_{|m|+|n| \neq 0} \left(\frac{|mz_\ell + n|^2}{\text{Im } z_\ell} \right)^{\frac{\nu-1}{2}}$$

$$= \frac{1}{2} \sum_{|m|+|n| \neq 0} (\eta^\ell(m,n))^{\frac{\nu-1}{2}}, \qquad (18.121)$$

where η^ℓ is the quadratic form associated with the matrix P_{z_ℓ} (*cf.* (18.13)), one may write the left–hand side of (18.120), using also Zagier's definition ([50], p. 281)

$$\zeta(s,D): = \sum_{\ell=1}^{h(D)} \frac{1}{\#(\Gamma_\ell)} \sum_{|m|+|n| \neq 0} (\eta^\ell(m,n))^{\frac{\nu-1}{2}}, \qquad (18.122)$$

as

$$\sum_{\ell=1}^{h(D)} \frac{1}{\#(\Gamma_\ell)} E_{\frac{1-\nu}{2}}(z_\ell) = \frac{1}{2} \frac{\zeta(\frac{1-\nu}{2}, D)}{\zeta(1-\nu)}, \qquad (18.123)$$

an equation valid in some neighbourhood of the half–plane Re $\nu \leq 0$. A detailed study of the function $\zeta(s, D)$ can be found in ([51], p. 26). Since, according to *loc.cit.* or ([50], p.281), the ratio $\frac{\zeta(\frac{1-\nu}{2},D)}{\zeta(\frac{1-\nu}{2})}$ is an entire function of ν, it follows from Hejhal's argument and (18.123) that whenever $\frac{1-\nu}{2}$ is a non–trivial zero of the zeta function, the function $(Z_2 * \Psi_{\frac{D}{\nu-1}}) \circ \chi^{-1}$ is a pseudo–cusp–form in Hejhal's sense (a concept explained right after (18.120)), substituting of course the set $\chi(A_D)$ for the set $\Gamma.e^{\frac{i\pi}{3}}$ specially emphasized in [18].

Let us mention that Zagier's paper contains interesting quite different applications of his result concerning the vanishing of (18.120) whenever $\frac{1-\nu}{2}$ is a non-trivial zero of the zeta function. Our sole point in discussing the distribution $\mathcal{D}_{Z_C^{even}}$ was to show that its decomposition into homogeneous terms yields interesting "almost modular" forms in just the same way that, as shown by proposition 16.1, Eisenstein distributions arise from the decomposition of the Dirac comb.

19. Automorphic functions associated with quadratic $PSL(2,\mathbb{Z})$–orbits in $P_1(\mathbb{R})$

Warning: In this section and the next one, the map $\rho \mapsto \bar{\rho}$ shall denote the conjugation within some fixed real quadratic extension of $\mathbb{Q}$: accordingly, when necessary in section 20, we shall then denote the complex conjugation by a star.

The construction in the present section is not unrelated, in its motivation, to that of "pseudo–cusp forms" briefly reported to at the end of last section. Only, real quadratic fields will take the place of imaginary quadratic fields. As a consequence, the automorphic functions that will appear presently will have their singularities (of a mild type: they will be continuous and piecewise C^∞) not on some discrete subset of Π, but spread out on the union of a locally finite set of lines. Another background for the present study was of course the Maass construction of cusp–forms (not for Γ, for congruence subgroups [3, p. 112]) associated with suitable quadratic fields: some elements of this latter construction will show up here too, in a natural way.

The genuine starting point, however, of the present investigations was the observation that, in the modular distribution setting (*cf.* proposition 3.3), the Eisenstein series which are the basic material of the classical theory of non–holomorphic modular forms are associated with $P_1(\mathbb{Q})$, the $PSL(2,\mathbb{Z})$–orbit of ∞ in $P_1(\mathbb{R})$. Trying to substitute another orbit for the simplest one just referred to led to the consideration of quadratic orbits as the next simplest case.

Consider the function

$$t \mapsto \frac{1}{2}\left[\mathfrak{P}_{\frac{\nu-1}{2}}(it) + \mathfrak{P}_{\frac{\nu-1}{2}}(-it)\right] \tag{19.1}$$

on the real line, where, if $t \geq 0$, resp. $t \leq 0$), $\mathfrak{P}_{\frac{\nu-1}{2}}(it)$ is defined by complex continuation from the Legendre function (initially defined on $[1, +\infty[$ above (resp. below) the half–line $]-\infty, 1]$: this is an analytic function, as it follows from the fact ([27, p. 166]) that $\mathfrak{P}_{\frac{\nu-1}{2}}(x+i0) = \mathfrak{P}_{\frac{\nu-1}{2}}(x-i0)$ for $-1 < x < 1$. Alternatively ([27, p. 153]), the function in (19.1) can be written as

$$t \mapsto \frac{1}{2}\left[{}_2F_1\left(\frac{1-\nu}{2}, \frac{1+\nu}{2}; 1; \frac{1-it}{2}\right) + {}_2F_1\left(\frac{1-\nu}{2}, \frac{1+\nu}{2}; 1; \frac{1+it}{2}\right)\right], \tag{19.2}$$

and the hypergeometric function extends as a single–valued analytic function in $\mathbb{C}\backslash[1,\infty[$.

We then define, in Π, the function

$$h_\nu(x+iy) := \frac{1}{2}[\mathfrak{P}_{\frac{\nu-1}{2}}(i\frac{x}{y}) + \mathfrak{P}_{\frac{\nu-1}{2}}(-i\frac{x}{y})] \qquad (19.3)$$

and observe, after a trivial calculation using the Legendre equation

$$(1-s^2)\,\mathfrak{P}''_{\frac{\nu-1}{2}}(s) - 2s\,\mathfrak{P}'_{\frac{\nu-1}{2}}(s) + \frac{\nu^2-1}{4}\,\mathfrak{P}_{\frac{\nu-1}{2}}(s) = 0, \qquad (19.4)$$

that it satisfies the equation

$$\Delta h_\nu = \frac{1-\nu^2}{4}\,h_\nu. \qquad (19.5)$$

Also, observe that $h_\nu(az) = h_\nu(z)$ for every $a > 0$: it may be handy to extend this to $a \neq 0$ by the appropriate extension of h_ν to the lower half-plane.

Let now ρ be an irrational quadratic number in some real field $K = \mathbb{Q}(\sqrt{D})$, and consider the function $z \mapsto h_\nu(\frac{-z+\bar\rho}{z-\rho})$, where $\bar\rho$ stands for the conjugate of ρ in K: we assume that $\rho > \bar\rho$. This function is invariant under the subgroup Γ_ρ of Γ consisting of all matrices $\begin{pmatrix} a & b \\ c & d \end{pmatrix} \in \Gamma$ such that $\frac{a\rho+b}{c\rho+d} = \rho$.

The problem we want to tackle with in this section is trying to extract something from the divergent series

$$f_\nu(z) = \sum_{\left(\begin{smallmatrix} n_1 & n_2 \\ m_1 & m_2 \end{smallmatrix}\right)\in\Gamma/\Gamma_\rho} h_\nu\left(\frac{-z+\frac{n_1\bar\rho+n_2}{m_1\bar\rho+m_2}}{z+\frac{n_1\rho+n_2}{m_1\rho+m_2}}\right). \qquad (19.6)$$

If this can be achieved, noting that, for every $\left(\begin{smallmatrix} a & b \\ c & d \end{smallmatrix}\right) \in \Gamma$, the ratio $\frac{-(m_1\bar\rho+m_2)z+(n_1\bar\rho+n_2)}{(m_1\rho+m_2)z-(n_1\rho+n_2)}$ transforms, under the associated fractional–linear transform, to $\frac{-(m_1'\bar\rho+m_2')z+(n_1'\bar\rho+n_2')}{(m_1'\rho+m_2')z-(n_1'\rho+n_2')}$ with

$$\begin{pmatrix} n_1' & n_2' \\ m_1' & m_2' \end{pmatrix} = \begin{pmatrix} d & -b \\ -c & a \end{pmatrix}\begin{pmatrix} n_1 & n_2 \\ m_1 & m_2 \end{pmatrix}, \qquad (19.7)$$

one would expect f_ν to be a non–holomorphic modular form for the full group Γ and the eigenvalue $\frac{1-\nu^2}{4}$.

Observe that this "construction" is analogous to that of Eisenstein series, with Γ_ρ substituted for $\Gamma_\infty = \{\pm g \in \Gamma : g \in \Gamma_\infty^\circ = \Gamma \cap N\}$ (the stabilizer of ∞ in Γ) and a Γ_ρ–invariant function (namely $z \mapsto h_\nu(\frac{-z+\bar\rho}{z-\rho})$) being substituted for the Γ_∞–invariant function $z \mapsto (\text{Im } z)^{\frac{1-\nu}{2}}$. However, much more work is required towards studying the series (19.6) supposed to define f_ν. We first need to recall a few facts relative to continued fractions: the use of such in view of the construction of fundamental units (in the sense of Dirichlet's theorem) in the rings of integers of quadratic fields is very classical.

We use the same (classical) notations as in [16, 19]. Given integers $a_0, \ldots, a_n$, all positive except possibly a_0, we set

$$[a_0, \ldots, a_n] = a_0 + \cfrac{1}{a_1+}\cfrac{1}{a_2+} \cdots \cfrac{1}{a_n} : \tag{19.8}$$

one associates with this system the set of partial quotients $\frac{p_m}{q_m}$ $(0 \le m \le n)$ in the usual way: defining inductively

$$\begin{matrix} p_{-1} = 1 & p_0 = a_0 & p_m = a_m p_{m-1} + p_{m-2} \\ q_{-1} = 0 & q_0 = 1 & q_m = a_m q_{m-1} + q_{m-2} \end{matrix}, (1 \le m \le n), \tag{19.9}$$

one gets the "lowest terms" expressions of the partial quotients. One has

$$p_m q_{m-1} - p_{m-1} q_m = (-1)^{m-1} \tag{19.10}$$

for all $m \ge 0$ ([16, theorem 150]). One may also consider the fractional–linear transform (from $\dot{\mathbb{R}}$ to itself)

$$s \mapsto [a_0, \ldots, a_n, s] = \frac{p_n s + p_{n-1}}{q_n s + q_{n-1}}. \tag{19.11}$$

Set $\Gamma_\pm = \Gamma \cup \left(\begin{smallmatrix} -1 & 0 \\ 0 & 1 \end{smallmatrix}\right) \Gamma$, the group of matrices with integral entries and determinant ± 1. We shall refer to the matrix $\left(\begin{smallmatrix} p_n & p_{n-1} \\ q_n & q_{n-1} \end{smallmatrix}\right) \in \Gamma_\pm$ as to *the matrix associated with* $\{a_0, \ldots, a_n\}$ *by the process in* (19.9): it lies in Γ if and only if $n+1$ (the length of the system $\{a_0, \ldots, a_n\}$) is even. The composition of any two such fractional–linear transformations may be expressed [16, p.130] by the formula

$$[a_0, \ldots, a_n, s] = [a_0, \ldots, a_{m-1}, [a_m, \ldots, a_n, s]]. \tag{19.12}$$

Irrational numbers are associated with (uniquely defined) arbitrary infinite sequences $\{a_0, \ldots, a_n, \ldots\}$ of integers, all of which, except possibly the first one, are positive. It is a fundamental fact ([16, theorem 175]) that the continued fraction representations of two such numbers have the same tail (what remains after you have deleted a few initial digits, not necessarily the same number thereof in both) if and only if one is the image of the other under the fractional–linear transformation associated with a matrix in $\Gamma_\pm$.

Quadratic (irrational) numbers are exactly the numbers with periodic continued fraction expansions, *i.e.*, with the notations in [16, p.144],

$$[a_0, \ldots, a_{L-1}, \dot{a}_L, a_{L+1}, \ldots, \dot{a}_{L+k-1}] =$$
$$[a_0, \ldots, a_{L-1}, a_L, a_{L+1}, \ldots, a_{L+k-1}, a_L, a_{L+1}, \ldots, a_{L+k-1}, \ldots] : \tag{19.13}$$

we shall isolate the "period" (chosen with a minimal length $k \geq 1$)

$$\varpi : = \{a_L, \ldots, a_{L+k-1}\}, \tag{19.14}$$

denoted in what follows as

$$\varpi = \{b_0, \ldots, b_{k-1}\}, \tag{19.15}$$

and write the preceding number as $[a_0, \ldots, a_{L-1}, \varpi, \varpi, \ldots,]$ or even $[a_0, \ldots, a_{L-1}, \dot{\varpi}]$. Set

$$\rho = [\dot{\varpi}] = [\varpi, \varpi, \ldots]. \tag{19.16}$$

It is a quadratic irrational number, since it satisfies the equation

$$\rho = [b_0, \ldots, b_{k-1}, \rho]$$
$$= \frac{\alpha \rho + \beta}{\gamma \rho + \delta}, \tag{19.17}$$

where the matrix $A = \begin{pmatrix} \alpha & \beta \\ \gamma & \delta \end{pmatrix}$ is that associated with ϖ by the process in (19.9). The numbers the continued fraction expansions of which admit the period ϖ are exactly the real numbers in the orbit of ρ under the group of fractional–linear transformations with matrices in $\Gamma_\pm$. The number ρ itself needs not lie in the ring $\mathfrak{o}_K$ of integers of $K = \mathbb{Q}(\rho)$: for instance, when $\varpi = \{1, 3\}$, one has $\rho = \frac{3+\sqrt{21}}{6}$, a number not in the (euclidean) ring $\mathfrak{o}_K$, $K = \mathbb{Q}(\sqrt{21})$.

Observe that if $k = 1$, $A = \begin{pmatrix} b_0 & 1 \\ 1 & 0 \end{pmatrix}$ for some $b_0 > 0$. In all cases, $\det A = (-1)^k$; in view of (19.9), one always has $\alpha \geq \gamma \geq \delta$ and $\alpha > \beta$. Finally, all entries of A are positive if $k \neq 1$.

LEMMA 19.1. *Let* $k \geq 1$, *and* $\varpi = \{b_0, \ldots, b_{k-1}\}$, *with* $b_j > 0$ *for every* j. *Let* $A = \begin{pmatrix} \alpha & \beta \\ \gamma & \delta \end{pmatrix} \in \Gamma_\pm$ *be the matrix associated with the period* ϖ *by the process in* (19.9), *and let* $\rho = [\dot{\varpi}]$: *in other words* ρ *is the positive root of the (fixed-point) equation* $\gamma \rho^2 + (\delta - \alpha)\rho - \beta = 0$. *Set*

$$\eta = \gamma \rho + \delta, \qquad \bar{\eta} = \gamma \bar{\rho} + \delta, \tag{19.18}$$

and note for easy reference that

$$\alpha \rho + \beta = \rho \eta$$
$$\gamma \rho - \alpha = \frac{(-1)^{k-1}}{\eta}$$
$$\delta \rho - \beta = (-1)^k \frac{\rho}{\eta}. \tag{19.19}$$

If $\theta = \pm 1$, *and a relation*

$$\begin{pmatrix} n_1 & n_2 \\ m_1 & m_2 \end{pmatrix} = \theta \begin{pmatrix} n_1' & n_2' \\ m_1' & m_2' \end{pmatrix} \begin{pmatrix} \alpha & \beta \\ \gamma & \delta \end{pmatrix}^{-M}, \qquad M = 0, 1, 2, \ldots \tag{19.20}$$

holds in $\Gamma_\pm$, one has

$$m_1\rho + m_2 = \theta(m_1'\rho + m_2')\eta^{-M}$$
$$m_1\bar{\rho} + m_2 = \theta(m_1'\bar{\rho} + m_2')\bar{\eta}^{-M}$$
$$n_1\rho + n_2 = \theta(n_1'\rho + n_2')\eta^{-M}$$
$$n_1\bar{\rho} + n_2 = \theta(n_1'\bar{\rho} + n_2')\bar{\eta}^{-M}. \tag{19.21}$$

Given any matrix $\begin{pmatrix} n_1 & n_2 \\ m_1 & m_2 \end{pmatrix} \in \Gamma_\pm$, one can choose θ and M so that the matrix $\begin{pmatrix} n_1' & n_2' \\ m_1' & m_2' \end{pmatrix}$ characterized by (19.20) should be the one associated with some system $\{a_0, \ldots, a_n\}$ ($a_n \in \mathbb{Z}$ for all n, $a_n \geq 1$ for $n \neq 0$) by the process in (19.9).

The group of matrices $\begin{pmatrix} n_1 & n_2 \\ m_1 & m_2 \end{pmatrix} \in \Gamma_\pm$ such that $\frac{n_1\rho + n_2}{m_1\rho + m_2} = \rho$ consists of all matrices $\pm \begin{pmatrix} \alpha & \beta \\ \gamma & \delta \end{pmatrix}^{-M}$, $M \in \mathbb{Z}$.

PROOF. One has $\rho = \frac{\alpha - \delta + \sqrt{D}}{2\gamma}$ with $D = (\alpha + \delta)^2 + 4(-1)^{k+1}$: we also set $\bar{\rho} = \frac{\alpha - \delta - \sqrt{D}}{2\gamma}$. Clearly, $\rho > 1$ (since $\rho = [\bar{\omega}]$) and $\bar{\rho} < 0$; also, $\bar{\rho} > -1$ since $\gamma - \delta + \alpha - \beta > 0$, as it follows from the observation that precedes the statement of the present lemma. The eigenvalues of the matrix A are the numbers $\frac{1}{2}(\alpha + \delta \pm \sqrt{D})$, i.e., $\eta = \gamma\rho + \delta$, $\bar{\eta} = \gamma\bar{\rho} + \delta$. Since $\eta\bar{\eta} = (-1)^k$, they are units of the ring o_K, $K = \mathbb{Q}(\sqrt{D})$: they may not be fundamental units since if $\bar{\omega} = \{4\}$, one has $\eta = (\frac{1+\sqrt{5}}{2})^3$; in this case, fundamental units are associated with $\bar{\omega} = \{1\}$.

One may diagonalize A as

$$A = (\rho - \bar{\rho})^{-1} \begin{pmatrix} \rho & \bar{\rho} \\ 1 & 1 \end{pmatrix} \begin{pmatrix} \eta & 0 \\ 0 & \bar{\eta} \end{pmatrix} \begin{pmatrix} 1 & -\bar{\rho} \\ -1 & \rho \end{pmatrix}, \tag{19.22}$$

which yields (19.21) after a trivial calculation. One should also note that

$$m_1' = \theta(\rho - \bar{\rho})^{-1}[(m_1\rho + m_2)\eta^M - (m_1\bar{\rho} + m_2)\bar{\eta}^M],$$
$$m_2' = \theta(\rho - \bar{\rho})^{-1}[-\bar{\rho}(m_1\rho + m_2)\eta^M + \rho(m_1\bar{\rho} + m_2)\bar{\eta}^M]. \tag{19.23}$$

Since $\eta > 1 > (-1)^k\bar{\eta} > 0$, one gets as M goes to infinity, choosing $\theta = \text{sign}(m_1\rho + m_2)$, the equivalents

$$m_1' \sim (\rho - \bar{\rho})^{-1}|m_1\rho + m_2|\eta^M$$
$$m_2' \sim -\bar{\rho}\, m_1', \tag{19.24}$$

so that for large M one has $m_1' > m_2' \geq 1$ (recall that $-1 < \bar{\rho} < 0$). It then follows from ([**19**], theorem 10.5.2) that the fractional–linear transformation given by the matrix $\begin{pmatrix} n_1' & n_2' \\ m_1' & m_2' \end{pmatrix}$ is the one associated with some system $\{a_0, \ldots, a_n\}$ by the process in (19.9). If, moreover, the fractional–linear transformation given by the matrix $\begin{pmatrix} n_1 & n_2 \\ m_1 & m_2 \end{pmatrix}$ fixes ρ, then so does the one associated with $\begin{pmatrix} n_1' & n_2' \\ m_1' & m_2' \end{pmatrix}$, which shows that $\{a_0, \ldots, a_n\} = \{\varpi \varpi \ldots \varpi\}$ (the period being repeated any number of times) since $\frac{n_1' \rho + n_2'}{m_1' \rho + m_2'} = [a_0, \ldots, a_n, \dot{\varpi}]$ and $\rho = [\dot{\varpi}]$; but, then, the matrix associated with the system $\{a_0, \ldots, a_n\}$ by the process in (19.9) has to be plus or minus a power of A, as a consequence of (19.12). $\qquad\square$

We now come back to the problem of trying to give (19.6) a meaning. More generally, we wish to consider, for any fixed $q \in \mathbf{Z}$, the series

$$f_{\nu,q}(z) = \sum_{\left(\begin{smallmatrix} n_1 & n_2 \\ m_1 & m_2 \end{smallmatrix}\right) \in \Gamma/\Gamma_\rho} h_\nu \left(\frac{-z + \frac{n_1 \bar{\rho} + n_2}{m_1 \bar{\rho} + m_2}}{z + \frac{n_1 \rho + n_2}{m_1 \rho + m_2}} \right) \left| \frac{m_1 \rho + m_2}{m_1 \bar{\rho} + m_2} \right|^{\frac{i\pi q}{\kappa \log \eta}} . \tag{19.25}$$

One can also define the function $f_{\nu,q,\text{sign}}(z)$ which has the same definition as $f_{\nu,q}(z)$, only with an extra factor $\text{sign}(\frac{m_1 \rho + m_2}{m_1 \bar{\rho} + m_2})$ on the right–hand side. To spare notations, we shall refer to this function only when new ideas are required for its study (*cf.* for instance the proof of theorem 20.3).

Contrary to (19.6) (the case when $q = 0$, no sign), these series will not permit to build new automorphic functions: however, their introduction is quite natural; also, they have interesting Fourier coefficients (*cf.* theorem 19.9), the analytic continuation of which, by the method in section 20, can be obtained without extra effort.

That one can "mod out" Γ_ρ in (19.25) comes from the fact (a consequence of lemma 19.1) that, if

$$\begin{pmatrix} n_1' & n_2' \\ m_1' & m_2' \end{pmatrix} = \pm \begin{pmatrix} n_1 & n_2 \\ m_1 & m_2 \end{pmatrix} \begin{pmatrix} \alpha & \beta \\ \gamma & \delta \end{pmatrix}^{\kappa j}, \tag{19.26}$$

one has

$$\frac{m_1' \rho + m_2'}{m_1' \bar{\rho} + m_2'} = \eta^{2\kappa j} \frac{m_1 \rho + m_2}{m_1 \bar{\rho} + m_2}, \tag{19.27}$$

together with the fact that $\eta^{2\kappa j \cdot \frac{i\pi q}{\kappa \log \eta}} = 1$. The new extra factor plays a role related to that of Hecke's character [**3**, p.78] in the theory of Maass cusp–forms for congruence subgroups: also, the exponent $\frac{i\pi q}{\kappa \log \eta}$, $q \in \mathbf{Z}$, is the same (*loc. cit.*, p. 112) as the corresponding spectral parameter there. Also, in some "dual" context (that of section 20), these extra factors will arise on their own (*cf.* theorem 20.1), without our having done anything to introduce them.

In the case when k is even, *i.e.*, $\det \left(\begin{smallmatrix} \alpha & \beta \\ \gamma & \delta \end{smallmatrix} \right) = 1$, the stabilizer of ρ in $\Gamma_\pm$, as described in lemma 19.1, is contained in Γ, thus coincides with Γ_ρ; if k is odd, Γ_ρ is the set of matrices $\pm \left(\begin{smallmatrix} \alpha & \beta \\ \gamma & \delta \end{smallmatrix} \right)^{-M}$, $M \in 2\mathbb{Z}$. We set $\kappa = 1$ in the first case, $\kappa = 2$ in the second one: then $\bar\eta^\kappa = (\gamma\bar\rho + \delta)^\kappa$ is just the inverse of $\eta^\kappa = (\gamma\rho + \delta)^\kappa$. We also set $\Gamma_\rho^o = \{ \left(\begin{smallmatrix} \alpha & \beta \\ \gamma & \delta \end{smallmatrix} \right)^{\kappa j}$, $j \in \mathbb{Z} \}$, so that $\Gamma_\rho = \{ \pm g, \ g \in \Gamma_\rho^o \}$: the coset space Γ/Γ_ρ^o, even though somewhat redundant, is easier to describe (*cf. infra*) than the coset space Γ/Γ_ρ.

With

$$\tau = \frac{n_1 \rho + n_2}{m_1 \rho + m_2}, \qquad g = \begin{pmatrix} n_1 & n_2 \\ m_1 & m_2 \end{pmatrix} \in \Gamma, \qquad (19.28)$$

one has

$$\begin{vmatrix} \mathrm{Re}\ \frac{-z+\bar\tau}{z-\tau} \\ \mathrm{Im}\ \frac{-z+\bar\tau}{z-\tau} \end{vmatrix} = \left| \frac{-y^2 - (x - \tau)(x - \bar\tau)}{y(\tau - \bar\tau)} \right|$$

$$= |\phi(z; g)|, \qquad (19.29)$$

where

$$\phi(z; g) = \frac{y^2 + [x - \frac{1}{2}(\frac{n_1\rho+n_2}{m_1\rho+m_2} + \frac{n_1\bar\rho+n_2}{m_1\bar\rho+m_2})]^2 - [\frac{\rho-\bar\rho}{2\,(m_1\rho+m_2)(m_1\bar\rho+m_2)}]^2}{y\,\frac{\rho-\bar\rho}{(m_1\rho+m_2)(m_1\bar\rho+m_2)}}, \qquad (19.30)$$

a function we want to get some control of as $\left(\begin{smallmatrix} n_1 & n_2 \\ m_1 & m_2 \end{smallmatrix} \right)$ describes Γ/Γ_ρ.

Observe that a set of representatives of Γ/Γ_ρ^o is given, as a consequence of (19.21), as the set S of all matrices $\left(\begin{smallmatrix} n_1 & n_2 \\ m_1 & m_2 \end{smallmatrix} \right)$ with

$$\eta^{-2\kappa} < \left| \frac{m_1\rho + m_2}{m_1\bar\rho + m_2} \right| \leq 1 \qquad (19.31)$$

(recall that $\kappa = 1$ or 2 has just been defined). We also denote as S^o the set of pairs m_1, m_2 satisfying (19.31) together with $(m_1, m_2) = 1$. It can be identified with the set of classes of matrices $\left(\begin{smallmatrix} n_1 & n_2 \\ m_1 & m_2 \end{smallmatrix} \right) \in S$ in which one does not distinguish between $\left(\begin{smallmatrix} n_1 & n_2 \\ m_1 & m_2 \end{smallmatrix} \right) \in S$ and $\left(\begin{smallmatrix} n_1+jm_1 & n_2+jm_2 \\ m_1 & m_2 \end{smallmatrix} \right) \in S$, $j \in \mathbb{Z}$, in other words with the double coset space $\Gamma_\infty^o \backslash \Gamma / \Gamma_\rho^o$.

Also observe that one gets whenever g and g' lie in Γ, or even $\Gamma_\pm$ with the natural extension of (19.30), the identity

$$\phi(g'.z; g'g) = \phi(z; g) \qquad (19.32)$$

(the dot . meaning that g' is to act on z by means of the associated fractional-linear transformation), a consequence of

$$\frac{-g'.z + (g'g).\bar{\rho}}{g'.z - (g'g).\rho} = \frac{-z + g.\bar{\rho}}{z - g.\rho} \times \frac{c'(g.\rho) + d'}{c'(g.\bar{\rho}) + d'}. \tag{19.33}$$

LEMMA 19.2. *Fix an integer $\ell \geq 0$ such that $\frac{\rho}{|\bar{\rho}|} \leq \eta^{2\kappa\ell}$. For any matrix $\begin{pmatrix} n_1 & n_2 \\ m_1 & m_2 \end{pmatrix} \in S$, one has*

$$|m_1\bar{\rho} + m_2| \geq |\bar{\rho}| (|m_1| + |m_2|), \tag{19.34}$$

and

$$|m_1\rho + m_2| \geq C^{-1} (|m_1| + |m_2|), \tag{19.35}$$

where $C = 2C_1C_2$, $C_1 = \eta^{\kappa(\ell+1)}\max(\rho^{-1}, 1)$, and C_2 is the largest of the entries of the matrix $\begin{pmatrix} \alpha & \beta \\ \gamma & \delta \end{pmatrix}^{-\kappa(\ell+1)}$.

PROOF. Recall that $\rho > 1$, $-1 < \bar{\rho} < 0$, in particular $\rho + \bar{\rho} > 0$. To prove the first inequality, it suffices to prove that $m_1 m_2 \leq 0$, and we may assume that $m_1\bar{\rho} + m_2 < 0$. Since $|m_1\rho + m_2| \leq |m_1\bar{\rho} + m_2|$, one has

$$m_1 + \frac{m_2}{\bar{\rho}} \geq \frac{\rho}{|\bar{\rho}|} |m_1 + \frac{m_2}{\rho}| > |m_1 + \frac{m_2}{\rho}|. \tag{19.36}$$

Thus $m_2 (\frac{1}{\bar{\rho}} - \frac{1}{\rho}) > 0$, hence $m_2 < 0$, and $m_1 (1 + \frac{\rho}{|\bar{\rho}|}) \geq m_2 (-\frac{1}{\bar{\rho}} - \frac{1}{|\bar{\rho}|}) = 0$, thus $m_1 \geq 0$.

To prove the second inequality, introduce the matrix

$$\begin{pmatrix} n'_1 & n'_2 \\ m'_1 & m'_2 \end{pmatrix} = \begin{pmatrix} n_1 & n_2 \\ m_1 & m_2 \end{pmatrix} \begin{pmatrix} \alpha & \beta \\ \gamma & \delta \end{pmatrix}^{\kappa(\ell+1)}. \tag{19.37}$$

Then

$$\begin{aligned} \left|\frac{m'_1\rho + m'_2}{m'_1\bar{\rho} + m'_2}\right| &= \eta^{2\kappa(\ell+1)} \left|\frac{m_1\rho + m_2}{m_1\bar{\rho} + m_2}\right| \\ &> \eta^{2\kappa\ell} \\ &\geq \frac{\rho}{|\bar{\rho}|}, \end{aligned} \tag{19.38}$$

a consequence of $(19.21), (19.31)$ and the definition of ℓ. We then show that $m'_1 m'_2 > 0$, for which we may assume that $m'_1\rho + m'_2 > 0$: we then write, for the last estimate,

$$m'_1 + \frac{m'_2}{\rho} > |m'_1 + \frac{m'_2}{\bar{\rho}}|, \tag{19.39}$$

so that $m'_2 (\frac{1}{\rho} - \frac{1}{\bar{\rho}}) > 0$, hence $m'_2 > 0$, and $2 m'_1 > m'_2 \frac{\rho + \bar{\rho}}{-\rho\bar{\rho}} > 0$. Thus $|m'_1\rho + m'_2| = |m'_1|\rho + |m'_2|$ and ((19.21) again)

$$\begin{aligned} |m_1\rho + m_2| &= \eta^{-\kappa(\ell+1)} [|m'_1|\rho + |m'_2|] \\ &\geq C_1^{-1} (|m'_1| + |m'_2|) \end{aligned} \tag{19.40}$$

with C_1 as indicated; next,

$$|m_1| + |m_2| \leq 2 C_2 (|m_1'| + |m_2'|), \qquad (19.41)$$

with C_2 as indicated as well. $\square$

LEMMA 19.3. *There exists a constant $C_3 > 0$ such that, under the assumptions of lemma 19.2, the inequality*

$$|(m_1\rho + m_2)(n_1\rho + n_2)| \geq C_3^{-1} (|m_1| + |m_2|)(|n_1| + |n_2|) \qquad (19.42)$$

always holds.

PROOF. In the case when

$$\eta^{-4\kappa} < \left| \frac{n_1\rho + n_2}{n_1\bar\rho + n_2} \right| \leq 1, \qquad (19.43)$$

exactly the same proof as that of lemma 19.2, replacing everywhere m_1, m_2 by n_1, n_2, shows that

$$|n_1\rho + n_2| \geq C'^{-1}(|n_1| + |n_2|), \qquad (19.44)$$

where C' is just the same as C, with $\ell + 1$ substituted for ℓ (this is needed because we have replaced $\eta^{-2\kappa}$ by $\eta^{-4\kappa}$ when going from (19.31) to (19.43)): using (19.35), we are done in this case.

In the case when $\left| \frac{n_1\rho + n_2}{n_1\bar\rho + n_2} \right| > 1$, writing

$$n_1 = \frac{(n_1\rho + n_2) - (n_1\bar\rho + n_2)}{\rho - \bar\rho}, \qquad n_2 = \frac{-\bar\rho(n_1\rho + n_2) + \rho(n_1\bar\rho + n_2)}{\rho - \bar\rho}, \qquad (19.45)$$

one gets

$$|n_1| < \frac{2 |n_1\rho + n_2|}{\rho - \bar\rho} \qquad \text{and} \qquad |n_2| < |n_1\rho + n_2|, \qquad (19.46)$$

which is just what is needed.

Finally, assume $\left| \frac{n_1\rho + n_2}{n_1\bar\rho + n_2} \right| \leq \eta^{-4\kappa}$. Since

$$\frac{m_1\rho + m_2}{m_1\bar\rho + m_2} - \frac{n_1\rho + n_2}{n_1\bar\rho + n_2} = \frac{\bar\rho - \rho}{(m_1\bar\rho + m_2)(n_1\bar\rho + n_2)}, \qquad (19.47)$$

one has (using (19.31) as well)

$$\frac{\rho - \bar\rho}{|(m_1\bar\rho + m_2)(n_1\bar\rho + n_2)|} > \eta^{-2\kappa} - \eta^{-4\kappa}, \qquad (19.48)$$

thus

$$|n_1\bar\rho + n_2| \leq \frac{\rho - \bar\rho}{(\eta^{-2\kappa} - \eta^{-4\kappa}) |m_1\bar\rho + m_2|} \qquad (19.49)$$

and

$$|(m_1\rho + m_2)(n_1\bar{\rho} + n_2)| \leq \frac{\rho - \bar{\rho}}{\eta^{-2\kappa} - \eta^{-4\kappa}} \left| \frac{m_1\rho + m_2}{m_1\bar{\rho} + m_2} \right|$$

$$\leq \frac{\rho - \bar{\rho}}{\eta^{-2\kappa} - \eta^{-4\kappa}}. \tag{19.50}$$

Then, using again (19.35), (19.45) and (19.50),

$$C^{-1}(|m_1| + |m_2|)(|n_1| + |n_2|) \leq |(m_1\rho + m_2)(|n_1| + |n_2|)|$$

$$\leq \frac{1 + |\bar{\rho}|}{\rho - \bar{\rho}} |(m_1\rho + m_2)(n_1\rho + n_2)| + \frac{\rho + 1}{\eta^{-2\kappa} - \eta^{-4\kappa}}, \tag{19.51}$$

and the second term on the right–hand side can be absorbed in the left–hand side except possibly for a finite number of matrices $g = \left(\begin{smallmatrix} n_1 & n_2 \\ m_1 & m_2 \end{smallmatrix} \right)$. $\qquad \square$

LEMMA 19.4. *Only a finite number of Γ/Γ_ρ-transforms of the straight line (in the sense of hyperbolic geometry) through $\{\rho, \bar{\rho}\}$ intersects any given compact subset K of Π. Given $C > 0$ and K, the inequality $|\phi(z; g)| \leq C$, with $\phi(z; g)$ given in (19.30), is possible with some $z \in K$ only for a finite number of classes g mod Γ_ρ.*

PROOF. If $g = \left(\begin{smallmatrix} n_1 & n_2 \\ m_1 & m_2 \end{smallmatrix} \right) \in \Gamma$, the equation of the line through $\{g.\rho, g.\bar{\rho}\}$ is

$$\text{Re} \, \frac{-z + \frac{n_1\bar{\rho} + n_2}{m_1\bar{\rho} + m_2}}{z - \frac{n_1\rho + n_2}{m_1\rho + m_2}} = 0. \tag{19.52}$$

Assuming (19.31)

$$\eta^{-2\kappa} < \left| \frac{m_1\rho + m_2}{m_1\bar{\rho} + m_2} \right| \leq 1, \tag{19.53}$$

we have to show that, given a compact subset K of Π, the equation (19.52) is possible with $z \in K$ only for a finite number of matrices g. In view of (19.29), it is then clear that the first part of lemma 19.4 is a consequence of the second one, which we now proceed to prove.

Using when $m_2 \neq 0$ the equation $n_1 = \frac{m_1}{m_2}n_2 + \frac{1}{m_2}$, one can transform a certain expression that occurs within (19.30) as

$$\frac{1}{2} \left(\frac{n_1\rho + n_2}{m_1\rho + m_2} + \frac{n_1\bar{\rho} + n_2}{m_1\bar{\rho} + m_2} \right) = \frac{n_2}{m_2} + \frac{1}{2m_2} \left(\frac{\rho}{m_1\rho + m_2} + \frac{\bar{\rho}}{m_1\bar{\rho} + m_2} \right) \tag{19.54}$$

and observe for future reference that, for a given pair m_1, m_2, $\frac{n_2}{m_2}$ only varies through a sequence $\frac{n_2^o}{m_2} + j$, $j \in \mathbb{Z}$; when $m_2 = 0$, one can write the same expression as $\frac{n_1}{m_1} - \frac{\rho + \bar{\rho}}{2\rho\bar{\rho}}$, with $m_1 = \pm 1$.

Using S, defined in (19.31), as a set of representatives of Γ/Γ_ρ^o, one may apply the result of lemma 19.2: it is then immediate that, when z varies through K, the function $\phi(z;g)$ can only remain bounded when so do the entries of the matrix g. Indeed, as a consequence of (19.34), (19.35), one has

$$[\frac{\rho - \bar\rho}{2\,(m_1\rho + m_2)(m_1\bar\rho + m_2)}]^2 \le \frac{1}{2}\,y^2 \qquad (19.55)$$

when $|m_1| + |m_2|$ is large, in which case

$$|\phi(z;g)| \ge \frac{y}{2\,(\rho - \bar\rho)}\,|(m_1\rho + m_2)(m_1\bar\rho + m_2)|. \qquad (19.56)$$

On the other hand, it follows from (19.54) (and the particular case $m_2 = 0$ just below (19.54)) that $|\phi(z;g)|$ is large also when $|\frac{n_2}{m_2}|$ (or $|\frac{n_1}{m_1}|$ in the case when $m_2 = 0$) is large. $\qquad\square$

Coming back to (19.2), (19.3) and assuming that $\nu \notin \mathbf{Z}$, we note that one has for $t \ne 0$ the identity ([27, p. 48])

$$_2F_1(\frac{1-\nu}{2}, \frac{1+\nu}{2}; 1; \frac{1-it}{2}) =$$

$$\frac{\Gamma(\nu)}{(\Gamma(\frac{1+\nu}{2}))^2}\,(-\frac{1}{2} + \frac{it}{2})^{\frac{\nu-1}{2}}\,{}_2F_1(\frac{1-\nu}{2}, \frac{1-\nu}{2}; 1-\nu; \frac{2}{1-it})$$

$$+ \frac{\Gamma(-\nu)}{(\Gamma(\frac{1-\nu}{2}))^2}\,(-\frac{1}{2} + \frac{it}{2})^{\frac{-\nu-1}{2}}\,{}_2F_1(\frac{1+\nu}{2}, \frac{1+\nu}{2}; 1+\nu; \frac{2}{1-it})$$

$$(19.57)$$

in which we recall that the definition of the hypergeometric function necessitates a cut along $[1, \infty[$: in particular, each of the two terms on the right-hand side of (19.57), though analytic on $\mathbb{R}\backslash\{0\}$, fails to be extendable as a continuous function on $\mathbb{R}$.

Split h_ν as

$$h_\nu(z) = h_\nu^+(z) + h_\nu^-(z), \qquad (19.58)$$

in which

$$h_\nu^+(x + iy) = Q_\nu^+(\frac{x}{y}) \qquad (19.59)$$

with

$$Q_\nu^+(t) = \frac{1}{2}\frac{\Gamma(\nu)}{(\Gamma(\frac{1+\nu}{2}))^2}[(-\frac{1}{2} + \frac{it}{2})^{\frac{\nu-1}{2}}\,{}_2F_1(\frac{1-\nu}{2}, \frac{1-\nu}{2}; 1-\nu; \frac{2}{1-it})$$

$$+ (-\frac{1}{2} - \frac{it}{2})^{\frac{\nu-1}{2}}\,{}_2F_1(\frac{1-\nu}{2}, \frac{1-\nu}{2}; 1-\nu; \frac{2}{1+it})], \quad (19.60)$$

and h_ν^- is the same with $-\nu$ substituted for ν.

We shall then define

$$f_{\nu,q}^{\pm}(z) = \sum_{\left(\begin{smallmatrix} n_1 & n_2 \\ m_1 & m_2 \end{smallmatrix}\right)\in\Gamma/\Gamma_\rho} h_\nu^{\pm}\left(\frac{-z+\frac{n_1\bar\rho+n_2}{m_1\bar\rho+m_2}}{z-\frac{n_1\rho+n_2}{m_1\rho+m_2}}\right) \left|\frac{m_1\rho+m_2}{m_1\bar\rho+m_2}\right|^{\frac{i\pi q}{\kappa\log\eta}}, \tag{19.61}$$

concentrating, for the sake of comparison to the Eisenstein series, on $f_{\nu,q}^{+}(z)$. We abreviate $f_{\nu,0}^{\pm}$ as $f_\nu^{\pm}$.

What we shall do is showing first that $f_{\nu,q}^{+}(z)$ has a meaning for Re $\nu <$ -1, then — after some not inconsiderable work — that it can be extended to the half–plane Re $\nu < 0$. The same could be done with $f_{\nu,q}^{-}(z)$, extending it from the half–plane Re $\nu > 1$ to the half–plane Re $\nu > 0$, and the answer to the question raised after (19.25) (giving $f_{\nu,q}$ a meaning) is that this meaning is that of a hyperfunction. However, just as in theorem 7.4 (but not theorem 7.3), it will be just as well to consider only one of the two functions $f_{\nu,q}^{\pm}$, say $f_{\nu,q}^{+}$.

THEOREM 19.5. *Let* $\Sigma \subset \Pi$ *be the union of the locally finite (lemma 19.4) collection of* Γ/Γ_ρ*–transforms of the straight line through* $\{\rho,\bar\rho\}$. *For any* $z \in \Pi\backslash\Sigma$, $f_{\nu,q}^{+}(z)$, *as a function of* ν, *is holomorphic in the open set* $\{\nu \in \mathbb{C}:$ Re $\nu < -1, \nu \notin \mathbb{Z}\}$. *For any such* ν, $f_\nu^{+} = f_{\nu,0}^{+}$ *is* Γ*–invariant. Outside* Γ, $f_{\nu,q}^{+}$ *satisfies the equation* $(\Delta - \frac{1-\nu^2}{4})f_{\nu,q}^{+} = 0$.

PROOF. Let $\chi \in C^\infty(\mathbb{R})$ be an even non–negative function chosen so that 0 should not belong to the support of $\chi(t)$, and so that $1 - \chi$ should have compact support. Set

$$f_{\nu,q}^{+} = f_{\nu,q;\chi}^{+} + f_{\nu,q;1-\chi}^{+} \tag{19.62}$$

where $f_{\nu,q;\chi}^{+}(z)$ has the same definition as $f_{\nu,q}^{+}(z)$, only with the extra factor $\chi(\frac{\text{Re } w}{\text{Im } w})$, $w = \frac{-z+\frac{n_1\bar\rho+n_2}{m_1\bar\rho+m_2}}{z-\frac{n_1\rho+n_2}{m_1\rho+m_2}}$, under the summation sign.

Recalling from (19.29) that $\frac{\text{Re } w}{\text{Im } w}$ is just $\phi(z;g)$, we see from lemma 19.4 that, for z in a fixed compact subset of Π, the series defining $f_{\nu,q;1-\chi}^{+}(z)$ has only finitely many non–zero terms. The function $f_{\nu,q;1-\chi}^{+}$ is C^∞ outside Σ since, from (19.60), $Q_\nu^{+}(\frac{\text{Re } w}{\text{Im } w})$ is C^∞ when Re $w \neq 0$, and the equation Re $\frac{-z+\bar\rho}{z-\rho} = 0$ is that of the straight line through $\{\rho,\bar\rho\}$.

On the other hand, each of the terms in the series which defines $f_{\nu,q;\chi}^{+}$ is a C^∞ function throughout Π. We now show that the series defining $f_{\nu,q;\chi}^{+}(z)$ for $z \notin \Sigma$ is convergent. Starting from

$$|Q_\nu^{+}(t)| \le C\,\chi(t)\,|t|^{\frac{\text{Re } \nu-1}{2}}, \tag{19.63}$$

it has to be proven that

$$\sum_{g\in\Gamma/\Gamma_\rho} \chi(\phi(z;g))\,|\phi(z;g)|^{\frac{\mathrm{Re}\;\nu-1}{2}} < \infty \qquad (19.64)$$

for $\mathrm{Re}\;\nu < -1$. Using (19.54), rewrite (19.30), when $m_2 \neq 0$, as

$$\phi(z;g) = \frac{y^2 + [x - \frac{n_2}{m_2} - \frac{1}{2m_2}\left(\frac{\rho}{m_1\rho+m_2} + \frac{\bar\rho}{m_1\bar\rho+m_2}\right)]^2 - [\frac{\rho-\bar\rho}{2\,(m_1\rho+m_2)(m_1\bar\rho+m_2)}]^2}{y\,\frac{\rho-\bar\rho}{(m_1\rho+m_2)(m_1\bar\rho+m_2)}}. \qquad (19.65)$$

We may substitute S for Γ/Γ_ρ in the sum, provided we do not forget the extra factor $\frac{1}{2}$, since S is really $\Gamma/\Gamma_\rho^\circ \sim \Gamma/\Gamma_\rho \times \{\pm1\}$: then, for any pair $\{m_1, m_2\} \in S^\circ$, it has been observed right after (19.54) that $\frac{n_2}{m_2}$ (or $\frac{n_1}{m_1}$ in the case when $m_2 = 0$) varies though a translate of $\mathbb{Z}$. Comparing series to integrals, we then only have to remark, when $|m_1| + |m_2|$ is bounded, that

$$\int_{-\infty}^{\infty} [y^2 + (x - t)^2]^{\frac{\mathrm{Re}\;\nu-1}{2}}\, dt < \infty; \qquad (19.66)$$

when $|m_1| + |m_2|$ is large enough so that $\left|\frac{\rho-\bar\rho}{2\,(m_1\rho+m_2)(m_1\bar\rho+m_2)}\right| \leq \frac{y}{2}$, we note that

$$\sum_{\{m_1,m_2\}\in S^\circ} \int_{-\infty}^{\infty} \left[\frac{y^2 + (x - t)^2}{y\,\frac{\rho-\bar\rho}{|(m_1\rho+m_2)(m_1\bar\rho+m_2)|}}\right]^{\frac{\mathrm{Re}\;\nu-1}{2}} dt < \infty \qquad (19.67)$$

for $\mathrm{Re}\;\nu < -1$, a consequence of lemma 19.2.

The Γ-invariance of f_ν^+ is a consequence of (19.32).

Finally, to prove the equation $(\Delta - \frac{1-\nu^2}{4})\,f_{\nu,q}^+ = 0$, it suffices to remark that each of the two hypergeometric terms on the right-hand side of (19.57) satisfies the same second-order differential equation as the Legendre function itself. $\qquad\square$

We now analyze the singularities of $f_{\nu,q}^+$ on Σ: recall that, on each compact subset of Π, they originate from finitely many terms in the sum (19.61).

THEOREM 19.6. *For $\mathrm{Re}\;\nu < -1$, $\nu \notin \mathbb{Z}$, $f_{\nu,q}^+$ extends as a continuous function throughout Π: its singularities are supported in Σ, and consist in jumps of the normal derivative on each Γ/Γ_ρ-transform of the straight line (in the sense of hyperbolic geometry) through $\rho, \bar\rho$. Denote as $\delta(\mathrm{Re}\;\frac{-z+\bar\rho}{z-\rho})$ the measure $\frac{dt}{\sin t}$ carried by this line when parametrized as $z = \frac{\rho+\bar\rho}{2} + \frac{\rho-\bar\rho}{2}\,e^{it}$ and, for $g = \left(\begin{smallmatrix} n_1 & n_2 \\ m_1 & m_2 \end{smallmatrix}\right) \in \Gamma$, let $\delta\left(\mathrm{Re}\;\frac{-z+\frac{n_1\bar\rho+n_2}{m_1\bar\rho+m_2}}{z-\frac{n_1\rho+n_2}{m_1\rho+m_2}}\right)$ be the direct image of the preceding measure by the map $z \mapsto g.z = \frac{n_1 z+n_2}{m_1 z+m_2}$. Then, in the distribution* *sense,*

$$(\Delta - \frac{1-\nu^2}{4}) f^+_{\nu,q}(z) = C(\nu) \sum_{g \in \Gamma/\Gamma_\rho} \left| \frac{m_1\rho + m_2}{m_1\bar{\rho} + m_2} \right|^{\frac{i\pi q}{\kappa \log \eta}} \times$$

$$\delta \left(\mathrm{Re} \, \frac{-z + \frac{n_1\bar{\rho}+n_2}{m_1\bar{\rho}+m_2}}{z - \frac{n_1\rho+n_2}{m_1\rho+m_2}} \right), \quad (19.68)$$

where $C(\nu)$ is the constant

$$C(\nu) = \frac{\pi(1-\nu^2)}{4} \frac{\Gamma(\nu)\Gamma(1-\nu)}{(\Gamma(\frac{1-\nu}{2})\Gamma(\frac{1+\nu}{2}))^2} \, {}_2F_1(\frac{3-\nu}{2}, \frac{3+\nu}{2}; 2; \frac{1}{2}). \quad (19.69)$$

PROOF. We must examine $Q^+_\nu(t)$, as defined in (19.60), as $t \to 0^+$ or 0^-. This requires a transformation of the logarithmic case ${}_2F_1(a, a; c; z)$ of the hypergeometric equation and yields [27, p. 48]

$$Q^+_\nu(t) = \sum_{\varepsilon = \pm 1} \frac{1}{2} \frac{\Gamma(\nu)\Gamma(1-\nu)}{(\Gamma(\frac{1-\nu}{2})\Gamma(\frac{1+\nu}{2}))^2} \sum_{n \geq 0} \frac{\Gamma(\frac{1-\nu}{2}+n)}{\Gamma(\frac{1-\nu}{2})} \frac{\Gamma(\frac{1+\nu}{2}+n)}{\Gamma(\frac{1+\nu}{2})} \frac{1}{(n!)^2} \times$$

$$(\frac{1 - i\varepsilon t}{2})^n \left[\log(\frac{2}{i\varepsilon t - 1}) + \frac{2\Gamma'(1+n)}{\Gamma(1+n)} - \frac{\Gamma'(\frac{1-\nu}{2}+n)}{\Gamma(\frac{1-\nu}{2}+n)} - \frac{\Gamma'(\frac{1-\nu}{2}-n)}{\Gamma(\frac{1-\nu}{2}-n)} \right],$$

$$(19.70)$$

where $|\arg(\frac{2}{i\varepsilon t-1})| < \pi$. When $\varepsilon t \to 0^+$ (*resp.* 0^-), the logarithm goes to $\log 2 - i\pi$ (*resp.* $\log 2 + i\pi$), and $Q^+_\nu(0)$ has a well–defined value, which makes Q^+_ν continuous throughout $\mathbb{R}$. Its first derivative, however, has a jump through 0, given as

$$\left(\frac{d}{dt} Q^+_\nu \right)(0+) - \left(\frac{d}{dt} Q^+_\nu \right)(0-) =$$

$$- 2\pi \frac{\Gamma(\nu)\Gamma(1-\nu)}{(\Gamma(\frac{1-\nu}{2})\Gamma(\frac{1+\nu}{2}))^2} \sum_{n \geq 0} \frac{\Gamma(\frac{1-\nu}{2}+n)}{\Gamma(\frac{1-\nu}{2})} \frac{\Gamma(\frac{1+\nu}{2}+n)}{\Gamma(\frac{1+\nu}{2})} \frac{n \cdot 2^{-n}}{(n!)^2}. \quad (19.71)$$

From [27, p. 37–38],

$$(\Gamma(\frac{1-\nu}{2})\Gamma(\frac{1+\nu}{2}))^{-1} \sum_{n \geq 1} \frac{\Gamma(\frac{1-\nu}{2}+n)\Gamma(\frac{1+\nu}{2}+n)}{\Gamma(n)} \frac{2^{-n}}{n!}$$

$$= \frac{1-\nu^2}{8} \, {}_2F_1(\frac{3-\nu}{2}, \frac{3+\nu}{2}; 2; \frac{1}{2}) \quad (19.72)$$

so that the above jump can be written as

$$- \frac{\pi(1-\nu^2)}{4} \frac{\Gamma(\nu)\Gamma(1-\nu)}{(\Gamma(\frac{1-\nu}{2})\Gamma(\frac{1+\nu}{2}))^2} \, {}_2F_1(\frac{3-\nu}{2}, \frac{3+\nu}{2}; 2; \frac{1}{2}).$$

$$(19.73)$$

Since the function $\zeta \mapsto h^+_\nu(\frac{\mathrm{Re}\,\zeta}{\mathrm{Im}\,\zeta})$ satisfies the equation $(\Delta - \frac{1-\nu^2}{4}) h^+_\nu(\frac{\mathrm{Re}\,\zeta}{\mathrm{Im}\,\zeta}) = 0$ in the complementary of the line $\mathrm{Re}\,\zeta = 0$, it is immediate to get, in the distribution sense,

$$(\Delta - \frac{1-\nu^2}{4})(\zeta \mapsto h_\nu^+(\frac{\text{Re } \zeta}{\text{Im } \zeta})) = C(\nu)\,\delta(\text{Re } \zeta), \qquad (19.74)$$

where $\delta(\text{Re } \zeta)$ denotes the measure $\frac{d(\text{Im } \zeta)}{\text{Im } \zeta}$ carried by the (half)–line $\text{Re } \zeta = 0$, and $C(\nu)$ is the negative of the constant (19.73).

Finally, the direct image of the measure $\delta(\text{Re } \zeta)$ under the map $\zeta \mapsto z = \frac{\rho\zeta+\bar\rho}{\zeta+1}$ (this is equivalent to $\zeta = \frac{-z+\bar\rho}{z-\rho}$) is easily seen, if $z = \frac{\rho+\bar\rho}{2} + \frac{\rho-\bar\rho}{2}e^{it}$, so that $e^{it} = \frac{\zeta-1}{\zeta+1}$, to be just $\frac{dt}{\sin t}$. This, together with (19.61), proves theorem 19.6.

Remark. The measure $\delta(\text{Re } \frac{-z+\bar\rho}{z-\rho})$ also occurs in [51, p. 282], where it is traced back to a certain formula of Hecke regarding the zeta function of ideal classes in quadratic fields: the same measure will occur in (20.40), for quite similar reasons. $\qquad\square$

COROLLARY 19.7. *If the function $f_\nu^+(z)$, as a function of ν, extends as a meromorphic function to some neighbourhood of the closed left half–plane, with a pole at some point $\nu^\circ \notin \mathbb{Z}$, the coefficients of negative orders in the Laurent expansion of $f_\nu^+(z)$ at $\nu = \nu^\circ$ furnish non–holomorphic modular forms.*

PROOF. First observe that this statement bears some similarity to the remark which follows the proof of theorem 15.4. It suffices, in order to prove the corollary, to show that the distribution $(\Delta - \frac{1-\nu^2}{4})\,f_\nu^+$ extends as a holomorphic function of ν for $\nu \notin \mathbb{Z}$. Now, given any compact subset K of Π, it is clear, as a consequence of lemma 19.4, that the image under $\Delta - \frac{1-\nu^2}{4}$ of each term in the decomposition of f_ν^+, except for finitely many ones, vanishes in some neighbourhood of K.

$\qquad\square$

In order to analytically continue the function $f_{\nu,q}^+$, which is Γ–invariant in the case when $q = 0$ but not an eigenfunction of Δ, we first approximate it by some function $\tilde{f}_{\nu,q}^+$, an eigenfunction of Δ for the eigenvalue $\frac{1-\nu^2}{4}$ but not a Γ–invariant function, even for $q = 0$. What we demand is that the difference $f_{\nu,q}^+ - \tilde{f}_{\nu,q}^+$ should admit a holomorphic continuation to the half–plane $\text{Re } \nu < 1$, $\nu \notin \mathbb{Z}$: with the notation in the proof of theorem 19.5, it is just as well that $f_{\nu,q;x}^+ - \tilde{f}_{\nu,q}^+$ should admit such a continuation. Set

$$\psi(z;g) = \frac{y^2 + [x - \frac{1}{2}\left(\frac{n_1\rho+n_2}{m_1\rho+m_2} + \frac{n_1\bar\rho+n_2}{m_1\bar\rho+m_2}\right)]^2}{y\,\frac{\rho-\bar\rho}{(m_1\rho+m_2)(m_1\bar\rho+m_2)}} \qquad (19.75)$$

so that

$$\phi(z;g) = \psi(z;g) - y^{-1}\frac{\rho-\bar\rho}{4\,(m_1\rho+m_2)(m_1\bar\rho+m_2)}. \qquad (19.76)$$

The function $\psi(z;g)$ still depends only on z and on the class $g\,\Gamma_\rho$; then, the error term (the second one on the right–hand side of the preceding equation) is less than $C\,(|m_1| + |m_2|)^{-2}$ for z in any compact subset of Π, in view of lemma 19.2. Next, we retain on the right-hand side of (19.60) only the constant term 1 from the Taylor expansions at 0 of the hypergeometric functions involved: with

$$\mathrm{Err}_\nu^+(\frac{|x|}{y}) : = \chi(\frac{x}{y})\,h_\nu^+(x+iy) - 2^{\frac{1-\nu}{2}}\,\frac{\Gamma(\nu)}{(\Gamma(\frac{\nu+1}{2}))^2}\,\chi(\frac{|x|}{y})\,(\frac{|x|}{y})^{\frac{\nu-1}{2}}\,\cos\frac{\pi(1-\nu)}{4},$$
(19.77)

one then has

$$|\mathrm{Err}_\nu^+(\frac{|x|}{y})| \le C\,\chi(\frac{|x|}{y})\,(\frac{|x|}{y})^{\frac{\mathrm{Re}\,\nu-3}{2}}$$
(19.78)

and

$$\sum_{g=(\begin{smallmatrix} n_1 & n_2 \\ m_1 & m_2 \end{smallmatrix})\in\Gamma/\Gamma_\rho} |\mathrm{Err}_\nu^+(|\phi(z;g)|)| \le C \sum_{g\in\Gamma/\Gamma_\rho} \chi(|\phi(z;g)|)\,|\phi(z;g)|^{\mathrm{Re}\,\nu-3},$$
(19.79)

a series which converges (uniformly for z in any compact subset of Π) if $\mathrm{Re}\,\nu < 1$. As a consequence of this, together with (19.76) and lemma 19.2, we get:

PROPOSITION 19.8. *Set*

$$\tilde{f}_{\nu,q}^+(z) = \frac{2^{\frac{\nu-1}{2}}\,\pi^{\frac{1}{2}}\,\Gamma(\frac{\nu}{2})}{\Gamma(\frac{\nu+1}{2})\,\Gamma(\frac{3-\nu}{4})\,\Gamma(\frac{1+\nu}{4})} \times$$

$$\sum_{g=(\begin{smallmatrix} n_1 & n_2 \\ m_1 & m_2 \end{smallmatrix})\in\Gamma/\Gamma_\rho} |\psi(z;g)|^{\frac{\nu-1}{2}} \left|\frac{m_1\rho + m_2}{m_1\bar\rho + m_2}\right|^{\frac{i\pi q}{\kappa\log\eta}}.$$
(19.80)

Then the difference $f_{\nu,q}^+ - \tilde{f}_{\nu,q}^+$, *as a distribution on* Π *(or as a function on* $\Pi\backslash\Sigma$*), extends as a holomorphic function of* ν *for* $\mathrm{Re}\,\nu < 1$, $\nu \notin \mathbb{Z}$.

Observe that, for $\mathrm{Re}\,\nu < -1$, $\tilde{f}_{\nu,q}^+(z)$ is still $\mathbb{Z}$–periodic.

THEOREM 19.9. *The Fourier coefficients* $c_m^+(\nu, q)$ *of* $\tilde{f}_{\nu,q}^+(z)$, *which make the expansion*

$$\tilde{f}_{\nu,q}^+(x + iy) = c_0^+(\nu, q)\,y^{\frac{1+\nu}{2}} + \sum_{m\in\mathbb{Z}^\times} c_m^+(\nu, q)\,y^{\frac{1}{2}}\,K_{\frac{\nu}{2}}(2\pi|m|y)\,e^{-2i\pi m x}$$
(19.81)

valid for $\mathrm{Re}\,\nu < -1$, *are given by the equation*

$$c_m^+(\nu, q) = \frac{2^{\frac{\nu+1}{2}}\,\pi^{\frac{2-\nu}{2}}\,\Gamma(\frac{\nu}{2})}{\Gamma(\frac{1-\nu}{2})\,\Gamma(\frac{1+\nu}{2})\,\Gamma(\frac{3-\nu}{4})\,\Gamma(\frac{1+\nu}{4})}\,(\rho - \bar\rho)^{\frac{1-\nu}{2}}\,|m|^{-\frac{\nu}{2}}$$

$$\sum_{(\begin{smallmatrix} n_1 & n_2 \\ m_1 & m_2 \end{smallmatrix})\in\Gamma_\infty\backslash\Gamma/\Gamma_\rho} |N(m_1\rho + m_2)|^{\frac{\nu-1}{2}}\,e^{i\pi m\,\mathrm{Tr}(\frac{n_1\rho+n_2}{m_1\rho+m_2})} \left|\frac{m_1\rho + m_2}{m_1\bar\rho + m_2}\right|^{\frac{i\pi q}{\kappa\log\eta}}$$
(19.82)

for $m \neq 0$, *and*

$$c_0^+(\nu, q) = \frac{2^{\frac{\nu-1}{2}} \pi \, \Gamma(\frac{\nu}{2}) \Gamma(\frac{-\nu}{2})}{\Gamma(\frac{1-\nu}{2}) \Gamma(\frac{1+\nu}{2}) \Gamma(\frac{3-\nu}{4}) \Gamma(\frac{1+\nu}{4})} (\rho - \bar{\rho})^{\frac{1-\nu}{2}}$$

$$\sum_{\left(\begin{smallmatrix} n_1 & n_2 \\ m_1 & m_2 \end{smallmatrix}\right) \in \Gamma_\infty \backslash \Gamma / \Gamma_\rho} |N(m_1\rho + m_2)|^{\frac{\nu-1}{2}} \left| \frac{m_1\rho + m_2}{m_1\bar{\rho} + m_2} \right|^{\frac{i\pi q}{\kappa \log \eta}}, \quad (19.83)$$

in which the norm and trace from the field $\mathbb{Q}$ *to the field* $\mathbb{Q}(\rho)$ *occur.*

PROOF. Recall that

$$N(m_1\rho + m_2) = (m_1\rho + m_2)(m_1\bar{\rho} + m_2) \tag{19.84}$$

and that

$$\text{Tr}\left(\frac{n_1\rho + n_2}{m_1\rho + m_2}\right) = \frac{n_1\rho + n_2}{m_1\rho + m_2} + \frac{n_1\bar{\rho} + n_2}{m_1\bar{\rho} + m_2}. \tag{19.85}$$

That one can "mod out" Γ_∞ in the sum on the right–hand side of (19.82) or (19.83) is immediate from the last equation; that the same applies to Γ_ρ follows from lemma 19.1. In the whole proof, the factor $\left| \frac{m_1\rho+m_2}{m_1\bar{\rho}+m_2} \right|^{\frac{i\pi q}{\kappa \log \eta}}$ is carried through without modification: to simplify notations, we shall therefore consider only the case when $q = 0$. We start from the expansion

$$\frac{2^{\frac{1-\nu}{2}} \Gamma(\frac{\nu+1}{2}) \Gamma(\frac{3-\nu}{4}) \Gamma(\frac{1+\nu}{4})}{\pi^{\frac{1}{2}} \Gamma(\frac{\nu}{2})} \, \tilde{f}_{\nu,0}(x + iy) = \sum_{m \in \mathbb{Z}} a_m(y) \, e^{-2i\pi mx}, \tag{19.86}$$

with

$$a_m(y) = \frac{1}{2} \sum_{\substack{\{m_1, m_2\} \in S^\circ \\ j \in \mathbb{Z}}} \int_0^1 \left[\frac{y^2 + [x - j - \frac{1}{2}\left(\frac{n_1^0\rho + n_2^0}{m_1\rho + m_2} + \frac{n_1^0\bar{\rho} + n_2^0}{m_1\bar{\rho} + m_2}\right)]^2}{y \frac{\rho - \bar{\rho}}{|(m_1\rho + m_2)(m_1\bar{\rho} + m_2)|}} \right]^{\frac{\nu-1}{2}} e^{2i\pi mx} \, dx \tag{19.87}$$

in which $\left(\begin{smallmatrix} n_1^0 & n_2^0 \\ m_1 & m_2 \end{smallmatrix}\right)$ is any matrix in Γ with a specified second row $\{m_1, m_2\}$: for, then, the general such matrix is obtained by substituting $\{n_1^0 + jm_1, n_2^0 + jm_2\}$, $j \in \mathbb{Z}$, for $\{n_1^0, n_2^0\}$ (as has been indicated right after (19.31), S° is a set of representatives of $\Gamma_\infty^\circ \backslash \Gamma / \Gamma_\rho^\circ$, which covers $\Gamma_\infty \backslash \Gamma / \Gamma_\rho = \Gamma_\infty^\circ \backslash \Gamma / \Gamma_\rho$ twice, whence the extra factor $\frac{1}{2}$). Performing first the $\sum_j$ summation so as to replace the integral $\int_0^1$ by an integral over $\mathbb{R}$ on the right–hand side, one gets the expression

$$a_m(y) = \frac{1}{2} \sum_{\{m_1, m_2\} \in S^\circ} \left[\frac{y(\rho - \bar{\rho})}{|(m_1\rho + m_2)(m_1\bar{\rho} + m_2)|} \right]^{\frac{1-\nu}{2}}$$

$$e^{i\pi m\left(\frac{n_1^0\rho + n_2^0}{m_1\rho + m_2} + \frac{n_1^0\bar{\rho} + n_2^0}{m_1\bar{\rho} + m_2}\right)} \int_{-\infty}^{\infty} (x^2 + y^2)^{\frac{\nu-1}{2}} e^{2i\pi mx} \, dx \tag{19.88}$$

or, using (4.39),

$$
a_m(y) = \frac{\pi^{\frac{1-\nu}{2}}}{\Gamma(\frac{1-\nu}{2})} \, y^{\frac{1}{2}} \, |m|^{-\frac{\nu}{2}} \, K_{\frac{\nu}{2}}(2\pi|m|y)
$$

$$
\sum_{\substack{(m_1,m_2)=1 \\ \eta^{-2\kappa} < |\frac{m_1\rho+m_2}{m_1\bar{\rho}+m_2}| \leq 1}} \left(\frac{|N(m_1\rho+m_2)|}{\rho-\bar{\rho}} \right)^{\frac{\nu-1}{2}} e^{i\pi m \, \mathrm{Tr}(\frac{n_1^0\rho+n_2^0}{m_1\rho+m_2})} \quad (19.89)
$$

for $m \neq 0$, and

$$
a_0(y) = \frac{1}{2} \pi^{\frac{1}{2}} \frac{\Gamma(\frac{-\nu}{2})}{\Gamma(\frac{1-\nu}{2})} \, y^{\frac{1+\nu}{2}} \sum_{\substack{(m_1,m_2)=1 \\ \eta^{-2\kappa} < |\frac{m_1\rho+m_2}{m_1\bar{\rho}+m_2}| \leq 1}} \left(\frac{|N(m_1\rho+m_2)|}{\rho-\bar{\rho}} \right)^{\frac{\nu-1}{2}} . \quad (19.90)
$$

$\square$

We now have to face the problem of analytically continuing, with respect to ν, the series on the right–hand sides of (19.82) and (19.83).

20. Quadratic orbits: a dual problem

The series the study of which we have been left with after theorem 19.9 occur as Fourier coefficients of the usual type of the unusual functions $f^{\pm}_{\nu,q}$. We shall presently show that related series occur as Fourier coefficients of an unusual type — ρ-adapted, as we shall refer to them — of functions already understood, to wit Eisenstein series and Selberg's non–holomorphic Poincaré series: as has been pointed out to us, a calculation relative to Eisenstein series, quite similar to theorem 20.1 that follows, was already made by Siegel ([34], p.107–128). This will make it possible to analytically continue the function $f^{+}_{\nu,q}$ of last section to the half–plane Re $\nu < 1$. Recall that it is only when $q = 0$ that $f^{+}_{\nu,q}$ is Γ–invariant: this does not prevent the residues at appropriate poles of the function $f^{+}_{\nu,q}$ in general from being modular forms (theorem 20.5).

With the same notations as in last section, we represent a point $z \in \Pi$ by the pair $\{s,\theta\} \in \mathbf{R}\times]0,\pi[$ such that

$$\zeta = \frac{-z + \bar{\rho}}{z - \rho} = \eta^{2\kappa s}\, e^{i\theta} . \tag{20.1}$$

The map $z \mapsto \begin{pmatrix} \alpha & \beta \\ \gamma & \delta \end{pmatrix} . z = \frac{\alpha z + \beta}{\gamma z + \delta}$ transforms ζ into

$$\frac{\frac{-z+\bar{\rho}}{(\gamma z+\delta)(\gamma\bar{\rho}+\delta)}}{\frac{z-\rho}{(\gamma z+\delta)(\gamma\rho+\delta)}} = \frac{-z+\bar{\rho}}{z-\rho} \times \frac{\eta}{\bar{\eta}} ; \tag{20.2}$$

consequently the map $z \mapsto \begin{pmatrix} \alpha & \beta \\ \gamma & \delta \end{pmatrix}^{\kappa} . z$ transforms ζ into $\eta^{2\kappa}\zeta$. Thus, if f is any Γ–invariant function on Π, and if we set

$$\tilde{f}(s,\theta) = f(\frac{\rho\zeta + \bar{\rho}}{\zeta + 1}), \qquad \zeta = \eta^{2\kappa s}\, e^{i\theta} , \tag{20.3}$$

the function $s \mapsto \tilde{f}(s,\theta)$ is $\mathbf{Z}$–periodic, and can be expanded as

$$\tilde{f}(s,\theta) = \sum_{q\in\mathbf{Z}} b_q(\theta)\, e^{2i\pi q s} , \tag{20.4}$$

with

$$b_q(\theta) = \int_0^1 \tilde{f}(s',\theta)\, e^{-2i\pi q s'}\, ds' . \tag{20.5}$$

Note our deliberate choice of the variable q (the same as the one introduced for a quite distinct purpose in (19.25)) to parametrize the set of characters of the unit circle.

THEOREM 20.1. *Asume* $\operatorname{Re} \nu < -1$. *Then, if*

$$\zeta = \frac{-z + \bar{\rho}}{z - \rho} = r e^{i\theta} \qquad (r > 0, \ 0 < \theta < \pi), \tag{20.6}$$

one has the expansion

$$E_{\frac{1-\nu}{2}}(z) = \sum_{q \in \mathbb{Z}} b_q(\nu, \theta) \, r^{\frac{i\pi q}{\kappa \log \eta}} \tag{20.7}$$

with

$$b_q(\nu, \theta) = \frac{1}{2\kappa \log \eta} (\rho - \bar{\rho})^{\frac{1-\nu}{2}} (\sin \theta)^{\frac{1-\nu}{2}} \sum_{\left(\begin{smallmatrix} n_1 & n_2 \\ m_1 & m_2 \end{smallmatrix}\right) \in \Gamma_\rho \backslash \Gamma / \Gamma_\infty} |m_1 \rho - n_1|^{\nu - 1}$$

$$\int_0^\infty \left[\frac{t^2 + 2\sigma_1 t \cos \theta + \sigma_1^2}{t} \right]^{\frac{\nu-1}{2}} t^{-\frac{i\pi q}{\kappa \log \eta}} \frac{dt}{t}, \tag{20.8}$$

where $\sigma_1 = \frac{m_1 \bar{\rho} - n_1}{m_1 \rho - n_1}$.

In particular, the function $\zeta(1 - \nu) b_q(\nu, \frac{\pi}{2})$ extends as a meromorphic function to the whole complex ν-plane with poles only at $\nu = 0$ and -1, and satisfies the functional equation which expresses the invariance under $\nu \mapsto -\nu$ of the function $\pi^{\frac{\nu-1}{2}} \Gamma(\frac{1-\nu}{2}) \zeta(1-\nu) b_q(\nu, \frac{\pi}{2})$. It can be made explicit as

$$\zeta(1 - \nu) b_q(\nu, \frac{\pi}{2}) = \frac{\zeta(1 - \nu)}{4\kappa \log \eta} \frac{\Gamma(\frac{1-\nu}{4} - \frac{i\pi q}{2\kappa \log \eta}) \Gamma(\frac{1-\nu}{4} + \frac{i\pi q}{2\kappa \log \eta})}{\Gamma(\frac{1-\nu}{2})} (\rho - \bar{\rho})^{\frac{1-\nu}{2}}$$

$$\sum_{\left(\begin{smallmatrix} n_1 & n_2 \\ m_1 & m_2 \end{smallmatrix}\right) \in \Gamma_\rho \backslash \Gamma / \Gamma_\infty} |N(m_1 \rho - n_1)|^{\frac{\nu-1}{2}} \left| \frac{m_1 \rho - n_1}{m_1 \bar{\rho} - n_1} \right|^{\frac{i\pi q}{\kappa \log \eta}}. \tag{20.9}$$

Observe from (20.6) that, when $\theta = \frac{\pi}{2}$, z describes the straight line, in the sense of hyperbolic geometry, from ρ to $\bar{\rho}$.

PROOF. From

$$z = \frac{\rho \zeta + \bar{\rho}}{\zeta + 1} = \frac{(\rho r e^{i\theta} + \bar{\rho})(r e^{-i\theta} + 1)}{(r \cos \theta + 1)^2 + r^2 \sin^2 \theta}, \tag{20.10}$$

we get

$$\operatorname{Im} z = \frac{(\rho - \bar{\rho}) r \sin \theta}{1 + 2r \cos \theta + r^2} \tag{20.11}$$

and

$$|m_1 z - n_1|^2 = \frac{[(m_1\rho - n_1)r + (m_1\bar\rho - n_1)e^{i\theta}][(m_1\rho - n_1)r + (m_1\bar\rho - n_1)e^{-i\theta}]}{1 + 2r\cos\theta + r^2}$$

$$= (m_1\rho - n_1)^2 \frac{(r + \sigma_1 e^{i\theta})(r + \sigma_1 e^{-i\theta})}{1 + 2r\cos\theta + r^2},$$

(20.12)

so that, from (3.13) and (3.14),

$$E_{\frac{1-\nu}{2}}(z) = (\rho-\bar\rho)^{\frac{1-\nu}{2}} (\sin\theta)^{\frac{1-\nu}{2}} \sum_{g\in\Gamma/\Gamma_\infty} |m_1\rho - n_1|^{\nu-1} \left[\frac{(r + \sigma_1 e^{i\theta})(r + \sigma_1 e^{-i\theta})}{r}\right]^{\frac{\nu-1}{2}}.$$

(20.13)

Also note that, if $r = \eta^{2\kappa s}$, one has $e^{2i\pi q s} = r^{\frac{i\pi q}{\kappa \log\eta}}$. From (20.5), setting $\sigma_1' = \frac{m_1' \bar\rho - n_1'}{m_1'\rho - n_1'}$ in association with $g' = \begin{pmatrix} n_1' & n_2' \\ m_1' & m_2' \end{pmatrix} \in \Gamma$, we get

$$b_q(\nu,\theta) = (\rho - \bar\rho)^{\frac{1-\nu}{2}} (\sin\theta)^{\frac{1-\nu}{2}} \sum_{g'=\begin{pmatrix} n_1' & n_2' \\ m_1' & m_2' \end{pmatrix}\in\Gamma/\Gamma_\infty} |m_1'\rho - n_1'|^{\nu-1} \times$$

$$\frac{1}{2\kappa \log\eta} \int_1^{\eta^{2\kappa}} \left[\frac{(r' + \sigma_1' e^{i\theta})(r' + \sigma_1' e^{-i\theta})}{r'}\right]^{\frac{\nu-1}{2}} r'^{-\frac{i\pi q}{\kappa \log\eta}} \frac{dr'}{r'}.$$

(20.14)

Substitute for the sum over $g' \in \Gamma/\Gamma_\infty$ the sum $\sum_{j\in\mathbb{Z}} \sum_{g\in\Gamma_\rho\backslash\Gamma/\Gamma_\infty}$ with $g' = \begin{pmatrix} \alpha & \beta \\ \gamma & \delta \end{pmatrix}^{\kappa j} g$, recalling that $\Gamma_\rho\backslash\Gamma/\Gamma_\infty = \Gamma_\rho^o\backslash\Gamma/\Gamma_\infty$. Using (19.19), it is easy to get the relations $m_1'\rho - n_1' = \eta^{-\kappa j}(m_1\rho - n_1)$, $m_2'\rho - n_2' = \eta^{-\kappa j}(m_2\rho - n_2)$, $\sigma_1' = \eta^{2\kappa j}\sigma_1$. Alternatively, one may note that the relation between g and g' can be written as

$$\begin{pmatrix} m_2' & -n_2' \\ -m_1' & n_1' \end{pmatrix} = \begin{pmatrix} m_2 & -n_2 \\ -m_1 & n_1 \end{pmatrix} \begin{pmatrix} \alpha & \beta \\ \gamma & \delta \end{pmatrix}^{-\kappa j}$$

(20.15)

and use directly (19.21). Thus

$$b_q(\nu,\theta) = \frac{1}{2\kappa \log\eta} (\rho-\bar\rho)^{\frac{1-\nu}{2}} (\sin\theta)^{\frac{1-\nu}{2}} \sum_{j\in\mathbb{Z}} \sum_{g\in\Gamma_\rho\backslash\Gamma/\Gamma_\infty} \eta^{\kappa j(1-\nu)} |m_1\rho - n_1|^{\nu-1} \times$$

$$\int_1^{\eta^{2\kappa}} \left[\frac{(r' + \eta^{2\kappa j}\sigma_1 e^{i\theta})(r' + \eta^{2\kappa j}\sigma_1 e^{-i\theta})}{r'}\right]^{\frac{\nu-1}{2}} r'^{-\frac{i\pi q}{\kappa \log\eta}} \frac{dr'}{r'}.$$

(20.16)

Setting $r' = \eta^{2\kappa j}t$, the measure $\frac{dr'}{r'}$ on $[1, \eta^{2\kappa}]$ is replaced by $\frac{dt}{t}$ over $[\eta^{-2\kappa j}, \eta^{2\kappa(1-j)}]$ and there is an extra factor of $(\eta^{2\kappa j})^{\frac{\nu-1}{2}} \times (\eta^{2\kappa j})^{-\frac{i\pi q}{\kappa \log\eta}}$ which just cancels out the apparent factor $\eta^{\kappa j(1-\nu)}$ since $\eta^{-\frac{2i\pi q}{\log\eta}} = 1$. This leads to (20.8).

There is no need for the application we have in mind (corollary 20.2) to analyze for general θ the hypergeometric function given as the integral on the right–hand side of the last equation. However, its value for $\theta = \frac{\pi}{2}$ is immediately seen to be

$$|\sigma_1|^{\frac{\nu-1}{2} - \frac{i\pi q}{\kappa \log \eta}} \frac{\Gamma(\frac{1-\nu}{4} - \frac{i\pi q}{2\kappa \log \eta}) \, \Gamma(\frac{1-\nu}{4} + \frac{i\pi q}{2\kappa \log \eta})}{\Gamma(\frac{1-\nu}{2})}, \tag{20.17}$$

from which the expression for $\zeta(1-\nu) \, b_q(\nu, \frac{\pi}{2})$ follows.

□

Remark. In the case when $q = 0$ and D, as defined in the beginning of the proof of lemma 19.1, is the discriminant of the field $\mathbb{Q}(\sqrt{D})$, it follows from the relation ([19], theorems 16.13.1 and 16.13.2) between classes of quadratic forms and ideal classes in quadratic fields that the functional equation indicated in theorem 20.1 reduces to Hecke's functional equation [24, p. 254]. Series like the one on the right–hand side of (20.9) also occur (in the case when $q = 0$) in [50, p. 28]. Their analogues, in the case when an imaginary quadratic field takes the place of our present $\mathbb{Q}(\sqrt{D})$, were quoted in (18.122).

COROLLARY 20.2. *For any $q \in \mathbb{Z}$, the constant Fourier coefficient $c_0^+(\nu, q)$ in the expansion of the function $f_{\nu,q}^+(z)$ introduced in (19.61) extends as a holomorphic function of ν for* $\mathrm{Re}\,\nu < 1$, $\nu \notin \mathbb{Z}$, *outside the set of zeros of the zeta function.*

PROOF. The series on the right–hand sides of (20.9) and of (19.83) are the same, as seen by changing g to g^{-1}. On the other hand, the left–hand side of (20.9) arises (*cf.* (20.5)) from integrating $\zeta(1-\nu) \, E_{\frac{1-\nu}{2}}$ against some measure carried by some compact path in Π. □

THEOREM 20.3. *Corollary 20.2 extends if we substitute for $f_{\nu,q}^+(z)$ the function $f_{\nu,q,\mathrm{sign}}^+(z)$ defined, as explained after (19.25), by putting on the right–hand side the extra factor* $\mathrm{sign}(\frac{m_1 \rho + m_2}{m_1 \bar{\rho} + m_2})$.

PROOF. For $\mathrm{Re}\,\nu < -1$, and $\varepsilon = \pm 1$, consider the (hypergeometric) integral

$$F(\nu; q; \varepsilon) = \int_0^\infty (t^2 + \varepsilon t + 1)^{\frac{\nu-1}{2}} \, t^{\frac{1-\nu}{2} - \frac{i\pi q}{\kappa \log \eta}} \, \frac{dt}{t}. \tag{20.18}$$

Since

$$\varepsilon = \frac{F(\nu; q; \varepsilon) - F(\nu; q; -\varepsilon)}{F(\nu; q; 1) - F(\nu; q; -1)}, \tag{20.19}$$

one gets from (20.8) (using (20.19) with $\varepsilon = \mathrm{sign}\,\sigma_1$) that

$$\frac{\zeta(1-\nu)\left[b_q(\nu,\frac{\pi}{3}) - b_q(\nu,\frac{2\pi}{3})\right]}{F(\nu;q;1) - F(\nu;q;-1)} = \frac{1}{2\kappa \log \eta} \zeta(1-\nu)(\rho - \bar{\rho})^{\frac{1-\nu}{2}} \left(\frac{\sqrt{3}}{2}\right)^{\frac{1-\nu}{2}}$$

$$\sum_{g \in \Gamma_\rho \backslash \Gamma / \Gamma_\infty} |N(m_1\rho - n_1)|^{\frac{\nu-1}{2}} \left|\frac{m_1\rho - n_1}{m_1\bar{\rho} - n_1}\right|^{\frac{i\pi q}{\kappa \log \eta}} \mathrm{sign}\left(\frac{m_1\rho - n_1}{m_1\bar{\rho} - n_1}\right),$$

$$(20.20)$$

initially defined for $\mathrm{Re}\,\nu < -1$, extends as a holomorphic function for $\nu \in \mathbb{C}\backslash\{-1,0,1\}$, from which the proposition is easily derived.

Note that a functional equation as in theorem 20.1 can still be written since $\pi^{\frac{\nu-1}{2}} \Gamma(\frac{1-\nu}{2}) \zeta(1-\nu) b_q(\nu,\theta)$ is invariant under $\nu \mapsto -\nu$. □

In order to get at the complex continuation of the coefficients $c_m^+(\nu,q)$, $m \neq 0$, from theorem 19.8, we must appeal to the non–holomorphic Poincaré series, introduced in [33], also used by [12, 7] in investigations relative to the Kloosterman sums.

For any integer $m > 0$, and $\nu \in \mathbb{C}$ with $\mathrm{Re}\,\nu < -1$, set

$$U_m(z,\frac{1-\nu}{2}) = \sum_{g = \left(\begin{smallmatrix} n_1 & n_2 \\ m_1 & m_2 \end{smallmatrix}\right) \in \Gamma/\Gamma_\infty} \left(\frac{|m_1 z - n_1|^2}{y}\right)^{\frac{\nu-1}{2}} \exp(2i\pi m \frac{m_2 z - n_2}{-m_1 z + n_1}).$$

$$(20.21)$$

The function $U_0(z, \frac{1-\nu}{2})$, though meaningful, reduces to $E_{\frac{1-\nu}{2}}(z)$. It is better to discard this case, would it be only in order to take advantage of the fact that $U_m(.,\frac{1-\nu}{2})$ lies in $L^2(\Gamma\backslash\Pi)$. It is not an eigenfunction of Δ but, as observed by Selberg, one has

$$\left(\Delta + \frac{\nu^2 - 1}{4}\right) U_m(.,\frac{1-\nu}{2}) = 2\pi m(1-\nu) U_m(.,\frac{3-\nu}{2}).$$

$$(20.22)$$

A consequence (also due to Selberg) of this equation is that $U_m(z,\frac{1-\nu}{2})$ extends as a meromorphic function of ν in the whole complex plane.

Since $U_m(.,\frac{1-\nu}{2})$ is rapidly decreasing at infinity in the fundamental domain, it is easy to find the coefficients of its Roelcke–Selberg expansion, using the Rankin–Selberg trick and the formula [27, p. 92]

$$\int_0^\infty e^{-2\pi my} K_{\frac{i\lambda}{2}}(2\pi my)\, y^{-\frac{\nu}{2} - 1}\, dy$$

$$= \pi^{\frac{1}{2}}(4\pi m)^{\frac{\nu}{2}} \frac{\Gamma(-\frac{\nu}{2} - \frac{i\lambda}{2})\Gamma(-\frac{\nu}{2} + \frac{i\lambda}{2})}{\Gamma(\frac{1-\nu}{2})}. \quad (20.23)$$

As indicated in [12, p. 247], one has (the star denoting the complex conjugation)

$$\int_{\Gamma\backslash\Pi} U_m(z, \frac{1-\nu}{2}) \, \mathcal{M}^*(z) \, d\mu(z)$$

$$= b_m^* \times \pi^{\frac{1}{2}} (4\pi m)^{\frac{\nu}{2}} \frac{\Gamma(-\frac{\nu}{2} - \frac{i\lambda_k}{2})\Gamma(-\frac{\nu}{2} + \frac{i\lambda_k}{2})}{\Gamma(\frac{1-\nu}{2})} \quad (20.24)$$

if $\mathcal{M}$ is a cusp–form corresponding to the eigenvalue $\frac{1+\lambda_k^2}{4}$ with the Fourier expansion

$$\mathcal{M}(z) = y^{\frac{1}{2}} \sum_{n\neq 0} b_n \, K_{\frac{i\lambda_k}{2}}(2\pi|n|y) \, e^{2i\pi nx} . \quad (20.25)$$

On the other hand (loc.cit. or [7, p. 246]),

$$\int_{\Gamma\backslash\Pi} U_m(z, \frac{1-\nu}{2}) \, E^*_{\frac{1-i\lambda}{2}}(z) \, d\mu(z)$$

$$= \frac{2^{\nu+1} \pi^{\frac{2+\nu+i\lambda}{2}}}{\zeta(1+i\lambda)} \, m^{\frac{\nu-i\lambda}{2}} \, \sigma_{i\lambda}(m) \frac{\Gamma(-\frac{\nu}{2} - \frac{i\lambda}{2})\Gamma(-\frac{\nu}{2} + \frac{i\lambda}{2})}{\Gamma(\frac{1-\nu}{2})\Gamma(\frac{1+i\lambda}{2})} . \quad (20.26)$$

We now turn to the study of the ρ–adapted Fourier expansion of the function $U_m(., \frac{1-\nu}{2})$, assuming that $\mathrm{Re}\,\nu < -1$ to start with.

PROPOSITION 20.4. *In terms of the variables (r, θ) related to z by the equation (20.6), one has*

$$U_m(z, \frac{1-\nu}{2}) = \sum_{q\in\mathbf{Z}} c_q(m, \nu, \theta) \, r^{\frac{i\pi q}{\kappa \log \eta}} \quad (20.27)$$

with

$$c_q(m, \nu, \theta) = \frac{1}{2\kappa \log \eta} \sum_{g\in\Gamma_\rho\backslash\Gamma/\Gamma_\infty} |m_1\rho - n_1|^{\nu-1} ((\rho - \bar\rho) \sin \theta)^{\frac{1-\nu}{2}}$$

$$\exp(2i\pi m \frac{m_2\rho - n_2}{-m_1\rho + n_1}) \int_0^\infty (t^2 + 2\sigma_1 t \cos \theta + \sigma_1^2)^{\frac{\nu-1}{2}}$$

$$\exp(2i\pi m \frac{m_2\rho - n_2}{-m_1\rho + n_1} \frac{\sigma_2 - \sigma_1}{\sigma_1 + te^{i\theta}}) \, t^{\frac{1-\nu}{2} - \frac{i\pi q}{\kappa \log \eta}} \frac{dt}{t} , \quad (20.28)$$

where

$$\sigma_1 = \frac{m_1\bar\rho - n_1}{m_1\rho - n_1} \qquad \sigma_2 = \frac{m_2\bar\rho - n_2}{m_2\rho - n_2} . \quad (20.29)$$

The function $c_q(m, \nu, \frac{\pi}{2}) - \tilde{c}_q(m, \nu, \frac{\pi}{2})$ *extends as a holomorphic function for* Re $\nu < 1$ *if*

$$\tilde{c}_q(m, \nu, \frac{\pi}{2}) = \frac{1}{4\kappa \log \eta} \frac{\Gamma(\frac{1-\nu}{4} - \frac{i\pi q}{2\kappa \log \eta}) \Gamma(\frac{1-\nu}{4} + \frac{i\pi q}{2\kappa \log \eta})}{\Gamma(\frac{1-\nu}{2})} (\rho - \bar{\rho})^{\frac{1-\nu}{2}}$$

$$\sum_{\left(\begin{smallmatrix} n_1 & n_2 \\ m_1 & m_2 \end{smallmatrix}\right) \in \Gamma_\rho \backslash \Gamma / \Gamma_\infty} |N(m_1 \rho - n_1)|^{\frac{\nu-1}{2}} \left|\frac{m_1 \rho - n_1}{m_1 \bar{\rho} - n_1}\right|^{\frac{i\pi q}{\kappa \log \eta}} e^{i\pi m} \operatorname{Tr}(\frac{m_2 \rho - n_2}{-m_1 \rho + n_1}).$$

$$(20.30)$$

PROOF. Recall from $(20.11), (20.12)$ that

$$\frac{|m_1 z - n_1|^2}{y} = (m_1 \rho - n_1)^2 \frac{r^2 + 2\sigma_1 r \cos\theta + \sigma_1^2}{(\rho - \bar{\rho}) r \sin\theta}.$$

$$(20.31)$$

Also, from (20.10),

$$\frac{m_2 z - n_2}{-m_1 z + n_1} = \frac{m_2(\rho r e^{i\theta} + \bar{\rho}) - n_2(r e^{i\theta} + 1)}{-m_1(\rho r e^{i\theta} + \bar{\rho}) + n_1(r e^{i\theta} + 1)}$$

$$= \frac{m_2 \rho - n_2}{-m_1 \rho + n_1} \times \frac{r e^{i\theta} + \sigma_2}{r e^{i\theta} + \sigma_1}$$

$$= \frac{m_2 \rho - n_2}{-m_1 \rho + n_1} \times [1 + \frac{\sigma_2 - \sigma_1}{\sigma_1 + r e^{i\theta}}], \qquad (20.32)$$

so that

$$U_m(z, \frac{1-\nu}{2}) = \sum_{g \in \Gamma/\Gamma_\infty} |m_1 \rho - n_1|^{\nu-1} [\frac{r^2 + 2\sigma_1 r \cos\theta + \sigma_1^2)}{(\rho - \bar{\rho}) r \sin\theta}]^{\frac{\nu-1}{2}}$$

$$\exp(2i\pi m \frac{m_2 \rho - n_2}{-m_1 \rho + n_1}) \exp(2i\pi m \frac{m_2 \rho - n_2}{-m_1 \rho + n_1} \frac{\sigma_2 - \sigma_1}{\sigma_1 + r e^{i\theta}}). \quad (20.33)$$

The argument from (20.14) to (20.17) then extends and yields (20.28).

As a set of representatives of the coset space $\Gamma_\rho^\circ \backslash \Gamma$, we take $S^{-1} = \{g = \left(\begin{smallmatrix} n_1 & n_2 \\ m_1 & m_2 \end{smallmatrix}\right) \in \Gamma: g^{-1} \in S\}$, where S was defined in (19.31): note when applying lemma 19.3 that (19.42), for instance, reads $|(m_1 \rho - n_1)(m_2 \rho - n_2)| \geq C_3^{-1}(|m_1| + |n_1|)(|m_2| + |n_2|)$ when used with $g \in S^{-1}$. It is an immediate consequence of (20.29) that

$$\sigma_2 - \sigma_1 = \frac{\rho - \bar{\rho}}{(m_1 \rho - n_1)(m_2 \rho - n_2)}: \qquad (20.34)$$

in particular, it stays bounded when $g \in S^{-1}$, and (from (20.29) and (19.31), not forgetting to change g to g^{-1} in the last quoted inequality), so does $\sigma_1^{\pm 1}$.

With $\varepsilon = \mathrm{sign}(\sigma_1)$, one may also write, using (20.28) and making the change of variables $t \mapsto |\sigma_1| t$,

$$
c_q(m, \nu, \frac{\pi}{2}) = \frac{1}{2\kappa \log \eta} \sum_{g \in \Gamma_\rho \backslash \Gamma / \Gamma_\infty} |N(m_1\rho - n_1)|^{\frac{\nu-1}{2}} \left| \frac{m_1\rho - n_1}{m_1\bar\rho - n_1} \right|^{\frac{i\pi q}{\kappa \log \eta}}
$$

$$
(\rho - \bar\rho)^{\frac{1-\nu}{2}} \exp(2i\pi m \frac{m_2\rho - n_2}{-m_1\rho + n_1})
$$

$$
\int_0^\infty (t^2 + 1)^{\frac{\nu-1}{2}} \exp(2i\pi \frac{m(\bar\rho - \rho)}{N(m_1\rho - n_1)(1 + i\varepsilon t)}) t^{\frac{1-\nu}{2} - \frac{i\pi q}{\kappa \log \eta}} \frac{dt}{t}.
$$

$$(20.35)$$

We now remark that there are two modifications one can make in the last expression of $c_q(m, \nu, \frac{\pi}{2})$ without affecting it, up to an error term admitting a complex continuation to the half–plane $\mathrm{Re}\ \nu < 1$. The first one consists in forgetting the exponential in the integral (a convergent one if $\mathrm{Re}\ \nu < 1$), since

$$
\sum_{g \in \Gamma_\rho \backslash \Gamma / \Gamma_\infty} |N(m_1\rho - n_1)|^{\frac{\mathrm{Re}\ \nu-3}{2}} < \infty \tag{20.36}
$$

if $\mathrm{Re}\ \nu < 1$, a consequence of lemma 19.2 again. The second one consists in substituting $\exp(i\pi m\, \mathrm{Tr}(\frac{m_2\rho - n_2}{-m_1\rho + n_1}))$ for $\exp(2i\pi m \frac{m_2\rho - n_2}{-m_1\rho + n_1})$, since

$$
\frac{m_2\rho - n_2}{-m_1\rho + n_1} - \frac{m_2\bar\rho - n_2}{-m_1\bar\rho + n_1} = \frac{\rho - \bar\rho}{N(m_1\rho - n_1)}. \tag{20.37}
$$

Making the new integral explicit, one finds (20.30). □

Warning: Contrary to what has been found necessary in sections 14 and 15, there is no need in the sequel to consider separately the even and odd eigenvalues. As a consequence, we denote as $(\frac{1+\lambda_k^2}{4})_{k \geq 1}$ the set of all eigenvalues of Δ, and denote as $\{\mathcal{M}_{k,\ell}\}_{1 \leq \ell \leq n_k}$ an orthonormal basis of Mass–Hecke forms for such an eigenvalue, be it even or odd, or both if any such should exist. Just like in the beginning of section 14, we fully normalize $\mathcal{M}_{k,\ell}$ (only determined up to some coefficient of absolute value 1) by the fact that it should the product of $\mathcal{N}_{k,\ell}$ by the real positive number $\|\mathcal{N}_{k,\ell}\|^{-1}$. Then, the coefficients $b_{n;k,\ell}$ below (cf. (20.45)) are real.

Given a Mass–Hecke form $\mathcal{M}_{k,\ell}$, let, for $q \in \mathbf{Z}$, $a_q(\rho; k, \ell; \theta)$ be the q–th coefficient of the ρ–adapted (cf. (20.4)) Fourier expansion of $\mathcal{M}_{k,\ell}$, in other words

$$
a_q(\rho; k, \ell; \theta) = \frac{1}{2\kappa \log \eta} \int_1^{\eta^{2\kappa}} \mathcal{M}_{k,\ell}(z)\, r^{-\frac{i\pi q}{\kappa \log \eta}} \frac{dr}{r} \tag{20.38}
$$

with z linked to r, θ by the equation $\frac{-z+\bar\rho}{z-\rho} = r\, e^{i\theta}$. We are particularly interested in the coefficient $a_q(\rho; k, \ell; \frac{\pi}{2})$. Recall that, when $\theta = \frac{\pi}{2}$, z describes the straight line (in the sense of hyperbolic geometry) from ρ to $\bar\rho$

as $r \in (0, \infty)$. In order not to forget it, we may set

$$z = \frac{\rho + \bar{\rho}}{2} + \frac{\rho - \bar{\rho}}{2} e^{it}, \qquad z = \frac{\rho ir + \bar{\rho}}{ir + 1}, \qquad (20.39)$$

so that $e^{it} = \frac{ir-1}{ir+1}$, getting after an elementary calculation

$$a_q(\rho; k, \ell; \frac{\pi}{2}) = \frac{1}{2\kappa \log \eta} \int_{2 \arctan(\eta^{-2\kappa})}^{\frac{\pi}{2}} M_{k,\ell}\left(\frac{\rho + \bar{\rho}}{2} + \frac{\rho - \bar{\rho}}{2} e^{it}\right)$$

$$(\tan \frac{t}{2})^{\frac{i\pi q}{\kappa \log \eta}} \frac{dt}{\sin t}. \quad (20.40)$$

We may then consider, for any eigenvalue λ_k, the Mass cusp–form

$$\mathcal{M}_{\rho,q;k} = \sum_{\ell} a_q(\rho; k, \ell; \frac{\pi}{2}) M_{k,\ell}. \qquad (20.41)$$

This definition is somewhat similar to the definition (15.37) of $\mathcal{M}_{\nu;k}$ in that it builds some specific linear combination of the $M_{k,\ell}$'s depending only on the Δ–eigenvalue $\frac{1+\lambda_k^2}{4}$, with the following two differences: here, the even and odd forms are not separated; next and more important, the coefficients $a_q(\rho; k, \ell; \frac{\pi}{2})$ of the linear combination under consideration cannot be read directly from the values of some L–function.

THEOREM 20.5. *The function $f_{\nu,q}^+$ introduced in* (19.61) *for* Re $\nu < -1$, $\nu \notin \mathbf{Z}$, *extends as a meromorphic function of ν for* Re $\nu < 1$, $\nu \notin \mathbf{Z}$. *Its only possible poles there are the zeros of the zeta function, and the numbers $i\lambda_k$ with $\frac{1+\lambda_k^2}{4}$ in the discrete spectrum of the modular Laplacian. The latter ones are at most simple poles, and the residues of $f_{\nu,q}^+$ are given as*

$$\mathrm{Res}_{\nu=i\lambda_k} f_{\nu,q} = -2\,\pi^{\frac{3}{2}} \frac{\Gamma(-\frac{i\lambda_k}{2}) \Gamma(\frac{i\lambda_k}{2})}{\Gamma(\frac{3-i\lambda_k}{4}) \Gamma(\frac{3+i\lambda_k}{4}) (\Gamma(\frac{1+i\lambda_k}{4}))^2} \times$$

$$\frac{4\kappa \log \eta}{\Gamma(\frac{1-i\lambda_k}{4} - \frac{i\pi q}{2\kappa \log \eta}) \Gamma(\frac{1-i\lambda_k}{4} + \frac{i\pi q}{2\kappa \log \eta})} \mathcal{M}_{\rho,q;k}, \quad (20.42)$$

where $\mathcal{M}_{\rho,q;k}$ was defined in (20.41).

PROOF. In view of proposition 19.8, we may substitute $\tilde{f}_{\nu,q}^+$ for $f_{\nu,q}^+$ without changing the significance of the theorem: then, $\tilde{f}_{\nu,q}^+$ has its Fourier coefficients $c_m^+(\nu, q)$ (of the usual type) given by (19.82), or (19.83) in the case when $m = 0$. We already observed in corollary 20.2 that, for Re $\nu < -1$, $\nu \notin \mathbf{Z}$, the coefficient $c_0^+(\nu, q)$ could be continued as a meromorphic function, without any pole outside the set of zeros of the zeta function. In particular, the residues of $f_{\nu,q}^+$ at poles on the pure imaginary line can only be cusp–forms.

Comparing (19.82) and (20.30), we get when $\mathrm{Re}\,\nu < -1$, $\nu \notin \mathbf{Z}$, and $m \geq 1$, the formula

$$c_m^+(\nu, q) = \frac{2^{\frac{\nu+1}{2}}\,\pi^{\frac{2-\nu}{2}}\,\Gamma(\frac{\nu}{2})}{\Gamma(\frac{1+\nu}{2})\,\Gamma(\frac{3-\nu}{4})\,\Gamma(\frac{1+\nu}{4})} \times \frac{4\kappa\,\log\eta}{\Gamma(\frac{1-\nu}{4} - \frac{i\pi q}{2\kappa\,\log\eta})\,\Gamma(\frac{1-\nu}{4} + \frac{i\pi q}{2\kappa\,\log\eta})}$$

$$\times\,|m|^{-\frac{\nu}{2}}\,\tilde{c}_q(m, \nu, \tfrac{\pi}{2}) \tag{20.43}$$

in which $\tilde{c}_q(m, \nu, \frac{\pi}{2})$ differs by a function holomorphic in the half–plane $\mathrm{Re}\,\nu < 1$ from the q-th ρ–related Fourier coefficient $c_q(m, \nu, \frac{\pi}{2})$ of the Selberg function $U_m(z, \frac{1-\nu}{2})$.

On the other hand, (20.24) and (20.26) give the coefficients of the Roelcke–Selberg decomposition of the function $U_m(\,.\,, \frac{1-\nu}{2})$. Keeping in mind that, if $\nu \in i\mathbf{R}$, a singularity such as the one arising from the factor $\Gamma(-\frac{\nu}{2} + \frac{i\lambda}{2})$ from the integral term can be avoided by a contour deformation, we see that $U_m(\,.\,, \frac{1-\nu}{2})$ extends as a meromorphic function of ν for $\mathrm{Re}\,\nu < 1$. The poles can only be zeros of the zeta function (from (20.26)) and points $i\lambda_k$. At such a pole, the residue of the function under consideration is given by

$$\mathrm{Res}_{\nu=i\lambda_k}\,U_m(\,.\,, \frac{1-\nu}{2}) = -2\pi^{\frac{1}{2}}\,\frac{\Gamma(-i\lambda_k)}{\Gamma(\frac{1-i\lambda_k}{2})}\,(4\pi m)^{\frac{i\lambda_k}{2}}$$

$$\sum_\ell b_{m;k,\ell}\,\mathcal{M}_{k,\ell}(z) \tag{20.44}$$

if

$$\mathcal{M}_{k,\ell}(z) = y^{\frac{1}{2}} \sum_{n \neq 0} b_{n;k,\ell}\,K_{\frac{i\lambda_k}{2}}(2\pi|n|y)\,e^{2i\pi nx}. \tag{20.45}$$

All this, concerning the function $U_m(\,.\,, \frac{1-\nu}{2})$, is well known.

Taking the q-th ρ–related Fourier coefficient at $\theta = \frac{\pi}{2}$ (this only involves integrating on some finite path contained in Π) and using (20.43), we get that, for $m \geq 1$, $c_m^+(\nu, q)$ extends as a meromorphic function for $\mathrm{Re}\,\nu < 1$, $\nu \notin \mathbf{Z}$, with possible poles only as mentioned in the theorem, and that

$$\mathrm{Res}_{\nu=i\lambda_k}\,c_m^+(\nu, q) = -2^{\frac{3}{2}(1+i\lambda_k)}\,\pi^{\frac{1}{2}}\,(\cosh\frac{\pi\lambda_k}{2}) \times \frac{\Gamma(-i\lambda_k)\,\Gamma(\frac{i\lambda_k}{2})}{\Gamma(\frac{3-i\lambda_k}{4})\,\Gamma(\frac{1+i\lambda_k}{4})} \times$$

$$\frac{4\kappa\,\log\eta}{\Gamma(\frac{1-i\lambda_k}{4} - \frac{i\pi q}{2\kappa\,\log\eta})\,\Gamma(\frac{1-i\lambda_k}{4} + \frac{i\pi q}{2\kappa\,\log\eta})} \sum_\ell b_{m;k,\ell}\,a_q(\rho; k, \ell; \tfrac{\pi}{2}), \tag{20.46}$$

an expression that can be transformed using the duplication formula for the gamma function so as to get (20.42).

Now, from (20.41), $\sum_\ell b_{m;k,\ell}\,a_q(\rho; k, \ell; \frac{\pi}{2})$ is just the m-th Fourier coefficient of $\mathcal{M}_{\rho,q;k}$. All that precedes only holds for $m \geq 1$, since $U_m(\,.\,, \frac{1-\nu}{2})$

is not defined otherwise. However, from (19.82) and (20.30) respectively, one sees that

$$c^+_{-m}(\nu, q) = [c^+_m(\nu^*, -q)]^* \tag{20.47}$$

and

$$\tilde{c}_q(-m, \nu, \frac{\pi}{2}) = [\tilde{c}_{-q}(m, \nu^*, \frac{\pi}{2})]^*. \tag{20.48}$$

Then, from (20.40),

$$[a_{-q}(\rho; k, \ell; \frac{\pi}{2})]^* = \pm a_q(\rho; k, \ell; \frac{\pi}{2}), \tag{20.49}$$

also

$$b_{-m;k,\ell} = \pm b_{m;k,\ell} \tag{20.50}$$

according to whether $\mathcal{M}_{k,\ell}$ is even or odd so that (20.46) is still valid when $m < 0$, which shows that the two cusp–forms on both sides of (20.42) have the same sets of Fourier coefficients. □

Just like the Eisenstein series, the Γ–invariant functions $f^+_\nu = f^+_{\nu,0}$ satisfy a functional equation: recall that f^-_ν is really the same as $f^+_{-\nu}$.

THEOREM 20.6. *The functions f^+_ν and f^-_ν extend as meromorphic functions of ν throughout the complex plane. Moreover, one has the functional equation*

$$f^+_\nu + f^-_\nu = c^+_0(\nu) E_{\frac{1+\nu}{2}} \tag{20.51}$$

with

$$c^+_0(\nu) = 4\kappa \log \eta \times \frac{\pi^{\frac{3}{2}} \Gamma(\frac{\nu}{2}) \Gamma(-\frac{\nu}{2})}{\Gamma(\frac{3+\nu}{4}) \Gamma(\frac{3-\nu}{4}) (\Gamma(\frac{1-\nu}{4}) \Gamma(\frac{1+\nu}{4}))^2} \times b_0(\nu, \frac{\pi}{2}), \tag{20.52}$$

where $b_0(\nu, \frac{\pi}{2})$, expressed by (20.9) when $\mathrm{Re}\ \nu < -1$, is in general the constant ρ–related Fourier coefficient, evaluated at $\theta = \frac{\pi}{2}$, of the Eisenstein series $E_{\frac{1-\nu}{2}}$.

Recall that, in the case when the discriminant D (cf. proof of lemma 19.1) of the second degree equation for ρ is also the discriminant of the field $\mathbb{Q}(\rho)$, $b_0(\nu, \frac{\pi}{2})$ agrees, up to some coefficient, with the Hecke zeta function of a certain ideal class for that field.

PROOF. In the strip $-1 < \mathrm{Re}\ \nu < 1$, f^+_ν and f^-_ν both make sense if $\nu \neq 0$, $\zeta(1 \pm \nu) \neq 0$ and $\pm \nu \neq i\lambda_k$ for all k. On the other hand, theorem 19.6 shows that $(\Delta - \frac{1-\nu^2}{4})(f^+_\nu + f^-_\nu) = 0$ in the distribution sense, since $C(-\nu) = -C(\nu)$ in (19.69), then in (19.68) too. Thus $f^+_\nu + f^-_\nu$ can only be a multiple of the Eisenstein series, which we normalize by making this combination explicit up to an error term lying in $L^2(\Gamma \backslash \Pi)$. For this computation we may well assume that $\mathrm{Re}\ \nu = 0$, $\nu \neq i\lambda_k$ for all k. We already took advantage (using lemma 19.2) of the fact that the difference

$|\phi(z;g) - \psi(z;g)|$ made explicit in (19.76) was $\leq C\,(|m_1| + |m_2|)^{-1}$: now we use the fact that it is actually $\leq C\,y^{-1}\,(|m_1| + |m_2|)^{-1}$, while $|\psi(z;g)| \geq C\,y\,(|m_1| + |m_2|)$. We can then improve proposition 19.8 by adding that, for large y (then $z \notin \Sigma$ in view of (19.55), (19.56)), one has $|f_\nu^+(z) - \tilde{f}_\nu^+(z)| \leq C\,y^{\frac{\mathrm{Re}\ \nu - 3}{2}}$ for $\mathrm{Re}\ \nu < 1$, $\nu \notin \mathbf{Z}$. It then follows from theorem 19.9 that, when $\mathrm{Re}\ \nu = 0$, one has $f_\nu^+(z) \sim c_0^+(\nu)\,y^{\frac{1+\nu}{2}}$ up to some error term lying in $L^2(\Gamma\backslash\Pi)$, with $c_0^+(\nu)$ given as the continuation of the series (convergent when $\mathrm{Re}\ \nu < -1$, $\nu \notin \mathbf{Z}$)

$$c_0^+(\nu) = \frac{2^{\frac{\nu-1}{2}}\,\pi\,\Gamma(\frac{\nu}{2})\,\Gamma(\frac{-\nu}{2})}{\Gamma(\frac{1-\nu}{2})\,\Gamma(\frac{1+\nu}{2})\,\Gamma(\frac{3-\nu}{4})\,\Gamma(\frac{1+\nu}{4})}\,(\rho - \bar{\rho})^{\frac{1-\nu}{2}}$$
$$\sum_{\left(\begin{smallmatrix} n_1 & n_2 \\ m_1 & m_2 \end{smallmatrix}\right)\in\Gamma_\infty\backslash\Gamma/\Gamma_\rho} |N(m_1\rho + m_2)|^{\frac{\nu-1}{2}}. \quad (20.53)$$

Using (20.9) to make this continuation explicit, one may also write, using the duplication formula for the gamma function,

$$f_\nu^+(z) \sim c_0^+(\nu)\,y^{\frac{1+\nu}{2}} \quad (20.54)$$

with

$$c_0^+(\nu) = 4\kappa\,\log\eta \times \frac{\pi^{\frac{3}{2}}\,\Gamma(\frac{\nu}{2})\,\Gamma(-\frac{\nu}{2})}{\Gamma(\frac{3+\nu}{4})\,\Gamma(\frac{3-\nu}{4})\,(\Gamma(\frac{1-\nu}{4})\,\Gamma(\frac{1+\nu}{4}))^2} \times b_0(\nu, \frac{\pi}{2}), \quad (20.55)$$

where $b_0(\nu, \frac{\pi}{2})$ is the constant ρ-related Fourier coefficient, evaluated at $\theta = \frac{\pi}{2}$, of the Eisenstein series $E_{\frac{1-\nu}{2}}$. Consequently, $f_\nu^+ + f_\nu^-$ can only be (from (3.29)) $c_0^+(\nu)\,E_{\frac{1+\nu}{2}}$: but, for the peace of the mind, we then must check that

$$\frac{c_0^+(-\nu)}{c_0^+(\nu)} = \pi^\nu\,\frac{\Gamma(\frac{1-\nu}{2})}{\Gamma(\frac{1+\nu}{2})}\,\frac{\zeta(1-\nu)}{\zeta(1+\nu)}, \quad (20.56)$$

which reduces to

$$\frac{b_0^+(-\nu, \frac{\pi}{2})}{b_0^+(\nu, \frac{\pi}{2})} = \pi^\nu\,\frac{\Gamma(\frac{1-\nu}{2})}{\Gamma(\frac{1+\nu}{2})}\,\frac{\zeta(1-\nu)}{\zeta(1+\nu)}, \quad (20.57)$$

a functional equation already observed in the statement of theorem 20.1. □

Remark. As mentioned in the beginning of section 19, our starting point for the last two sections was trying to get some kind of Eisenstein-like *distribution* that would be concentrated on some quadratic orbit of $PSL(2, \mathbf{Z})$ in $P_1(\mathbf{R})$. The results obtained can be rephrased as follows, in the modular distribution setting.

For any ν with $\operatorname{Re} \nu < -1$, and $u \in C^{\infty}_{-\nu}$, set

$$
< \mathfrak{T}_{\nu}, u > = \frac{2^{\frac{\nu-1}{2}} \pi^{\frac{2-\nu}{2}} \Gamma(\frac{\nu}{2})}{\Gamma(\frac{1-\nu}{2}) \Gamma(\frac{1+\nu}{2}) \Gamma(\frac{3-\nu}{4}) \Gamma(\frac{1+\nu}{4})} (\rho - \bar{\rho})^{\frac{1-\nu}{2}} \times
$$
$$
\sum_{g=\left(\begin{smallmatrix} n_1 & n_2 \\ m_1 & m_2 \end{smallmatrix}\right) \in \Gamma/\Gamma_{\rho}} |N(m_1\rho + m_2)|^{\frac{\nu-1}{2}} u(\frac{n_1\rho + n_2}{m_1\rho + m_2}).
$$

$$(20.58)$$

Indeed, $\mathfrak{T}_{\nu}$, as so defined, lies in $\mathcal{D}'_{\nu}$ when $\operatorname{Re} \nu < -1$. It is not a modular distribution. Its image under the map Θ_{ν} introduced in definition 2.4, in other words the function $z \mapsto < \mathfrak{T}_{\nu}, \phi^{-\nu}_z >$ with $\phi^{-\nu}_z$ as given by (2.6), is just the same as $\tilde{f}^+_{\nu} = \tilde{f}^+_{\nu,0}$ as defined in (19.73), except that $\psi(z;g)$ (cf. (19.68)) has to be replaced by

$$
\tilde{\psi}(z;g) = \frac{y^2 + [x - \frac{n_1\rho+n_2}{m_1\rho+m_2}]^2}{y \frac{\rho-\bar{\rho}}{(m_1\rho+m_2)(m_1\bar{\rho}+m_2)}}.
$$

$$(20.59)$$

Using the equation (20.37) with g changed to g^{-1}, one can easily see that the difference $< \mathfrak{T}_{\nu}, \phi^{-\nu}_z > - \tilde{f}^+$ extends as a holomorphic function of ν for $\operatorname{Re} \nu < 1, \nu \notin \mathbf{Z}$. It is not difficult, either, to see, as a consequence of what has been done in the last two sections, that the distribution $\mathfrak{T}'_{\nu}$ extends as a meromorphic function of ν for $\operatorname{Re} \nu < 1$, and that its residues on the pure imaginary line are modular distributions.

A problem. The collection consisting of Δ together with the set of all Hecke operators T_n, finally the symmetry operator $f \mapsto f_1$, with $f_1(z) = f(-z^*)$, is complete in the sense that a joint eigenfunction of all these operators, i.e. a Maass–Hecke form $\mathcal{M}_{k,\ell}$, is determined (up to the multiplication by a constant) if the corresponding set of eigenvalues is known. As a consequence, it should be possible to express the ρ–related Fourier coefficients $a_q(\rho; k, \ell; \frac{\pi}{2})$ given as integrals in (20.40) in terms of the L–function $L(., \mathcal{M}_{k,\ell})$. Indeed, using some inverse Mellin transform, one can find such an expression as the integral, on a line $\operatorname{Re} z = c > \frac{1}{2}$, of the function $L(z, \mathcal{M}_{k,\ell})$ against some rather complicated function of z, q, itself an integral. But this is an analyst's answer. It seems likely that a better policy would be to try to understand the Hecke operators themselves in ρ–related terms. Whatever be the case, this seems to be an interesting problem, somewhat related in spirit to that of finding the connexion coefficients linking bases of solutions of some ordinary linear differential equation characterized by their behaviour near various singular points.

On the geometrical side, however, we may clarify things a little bit by moving from Π to $\Gamma\backslash\Pi$. Consider the path

$$t \mapsto z = \frac{\rho + \bar\rho}{2} + \frac{\rho - \bar\rho}{2} e^{it}, \qquad 2\arctan(\eta^{-2\kappa}) \le t \le \frac{\pi}{2}. \tag{20.60}$$

First, it is immediate that $\frac{dt}{\sin t}$, the measure which occurs in (20.40), is nothing but the length element with respect to the canonical Riemannian structure on Π. Next, call $\mathfrak{K}$ the image of this path in the quotient space $\Gamma\backslash\Pi$. This is a *closed* geodesic since, as a consequence of (20.39) and (19.22), the initial point z corresponding to the value $2\arctan(\eta^{-2\kappa})$ of t is the image of the end point $(t = \frac{\pi}{2})$ by the fractional linear transformation associated (in the notation of (19.22)) with $A^\kappa \in \Gamma_\rho \subset \Gamma$: thus, the two endpoints of the path are identical when considered in $\Gamma\backslash\Pi$. However, the closed geodesic $\mathfrak{K}$ may be covered more than once by the path under consideration. Indeed, in the case when, with the notations of lemma 20.1, k is odd (so that $\kappa = 2$) and $\alpha^2 \equiv -1 \bmod \gamma$, set

$$C = (\rho - \bar\rho)^{-1} \begin{pmatrix} \rho & \bar\rho \\ 1 & 1 \end{pmatrix} \begin{pmatrix} 0 & \eta \\ \bar\eta & 0 \end{pmatrix} \begin{pmatrix} 1 & -\bar\rho \\ -1 & \rho \end{pmatrix}$$

$$= \begin{pmatrix} -\alpha & \frac{\alpha^2 + 1}{\gamma} \\ -\gamma & \alpha \end{pmatrix}, \tag{20.61}$$

a matrix in Γ, and

$$C_1 = (\rho - \bar\rho)^{-1} \begin{pmatrix} \rho & \bar\rho \\ 1 & 1 \end{pmatrix} \begin{pmatrix} \eta & 0 \\ 0 & -\bar\eta \end{pmatrix} \begin{pmatrix} 1 & -\bar\rho \\ -1 & \rho \end{pmatrix}. \tag{20.62}$$

Since $\begin{pmatrix} 0 & \eta \\ \bar\eta & 0 \end{pmatrix} . i = \begin{pmatrix} \eta & 0 \\ 0 & -\bar\eta \end{pmatrix} . i$, and i is the ζ-value (*cf.* (20.6)) corresponding to the point attained on the path (20.60) for $t = \frac{\pi}{2}$, one sees that in this case the middle point on the path (20.60), corresponding to $t = 2\arctan(\eta^{-2})$, is actually Γ-equivalent to the two endpoints.

We denote as dm the measure on the closed geodesic $\mathfrak{K}$ which is the product of the Riemannian length element by the number of times $\mathfrak{K}$ is covered by the path (20.60): thus $\int_{\mathfrak{K}} dm = 2\kappa \log \eta$.

PROPOSITION 20.7. *For* $\operatorname{Re} \nu < -1$, $\nu \notin \mathbb{Z}$, *the Roelcke–Selberg decomposition of* f_ν^+ *is given as*

$$f_\nu^+(z) = \Phi^0 + \frac{1}{8\pi} \int_{-\infty}^{\infty} \Phi(\lambda)\, E_{\frac{1-i\lambda}{2}}(z)\, d\lambda + \sum_{k,\ell} c_{k,\ell}\, \mathcal{M}_{k,\ell}(z) \tag{20.63}$$

with

$$\Phi^0 = \frac{24\, C(\nu)}{\pi(\nu^2 - 1)}\, \kappa \log \eta, \tag{20.64}$$

$$\Phi(\lambda) = 2\kappa \, \log \eta \times \frac{4 \, C(\nu)}{\nu^2 + \lambda^2} \times b_0(-i\lambda, \frac{\pi}{2}) \tag{20.65}$$

and

$$c_{k,\ell} = \frac{4 \, C(\nu)}{\nu^2 + \lambda_k^2} \int_{\mathcal{R}} \mathcal{M}_{k,\ell}^* \, dm, \tag{20.66}$$

where

$$C(\nu) = 2\pi^{\frac{1}{2}} \frac{\Gamma(\frac{\nu}{2}) \, \Gamma(\frac{2-\nu}{2})}{\Gamma(\frac{1-\nu}{2}) \, \Gamma(\frac{1+\nu}{2}) \, \Gamma(\frac{1-\nu}{4}) \, \Gamma(\frac{1+\nu}{4})} \tag{20.67}$$

and $b_0(-i\lambda, \frac{\pi}{2})$ is defined by analytic continuation from the equation (20.9).

PROOF. One has

$$c_{k,\ell} = \int_{\Gamma \backslash \Pi} f_\nu^+(z) \, \mathcal{M}_{k,\ell}^*(z) \, d\mu(z). \tag{20.68}$$

Now, using a rephrased formulation of theorem 19.6,

$$0 = \int_{\Gamma \backslash \Pi} [(\Delta f_\nu^+) \cdot \mathcal{M}_{k,\ell}^* - f_\nu^+ \cdot (\Delta \mathcal{M}_{k,\ell}^*)] \, d\mu$$

$$= \frac{1 - \nu^2}{4} \int_{\Gamma \backslash \Pi} f_\nu^+ \cdot \mathcal{M}_{k,\ell}^* \, d\mu + C(\nu) \int_{\mathcal{R}} \mathcal{M}_{k,\ell}^* \, dm - \frac{1 + \lambda_k^2}{4} \int_{\Gamma \backslash \Pi} f_\nu^+ \cdot \mathcal{M}_{k,\ell}^* \, d\mu \tag{20.69}$$

with

$$C(\nu) = \frac{\pi(1 - \nu^2)}{4} \frac{\Gamma(\nu) \, \Gamma(1 - \nu)}{(\Gamma(\frac{1-\nu}{2}) \, \Gamma(\frac{1+\nu}{2}))^2} \, {}_2F_1(\frac{3 - \nu}{2}, \frac{3 + \nu}{2}; 2; \frac{1}{2}), \tag{20.70}$$

which can be expressed as in (20.67) thanks to the formula [27, p. 40]

$${}_2F_1(\frac{3 - \nu}{2}, \frac{3 + \nu}{2}; 2; \frac{1}{2}) = \frac{\pi^{\frac{1}{2}}}{\Gamma(\frac{5-\nu}{4}) \, \Gamma(\frac{5+\nu}{4})} \tag{20.71}$$

and a new application of the duplication formula for the Gamma function.

Concerning $\Phi(\lambda)$, (19.81) and what has been said in the proof of theorem 20.6 show that, when Re $\nu < -1$, $f_\nu^+(z)$ is square integrable in the fundamental domain even after having been multiplied by $y^{\frac{1}{2}}$: this permits to compute the continuous coefficient of the Roelcke–Selberg decomposition of $f_\nu^+(z)$ by the integral formula

$$\Phi(\lambda) = \int_{\Gamma \backslash \Pi} f_\nu^+(z) \, E_{\frac{1+i\lambda}{2}}(z) \, d\mu(z), \tag{20.72}$$

from which one can conclude, using the same method of evaluation of this integral as the one used in (20.69), with the help of (20.38): indeed, this latter formula permits to interpret the integral $\int_{\mathcal{R}} E_{\frac{1+i\lambda}{2}} \, dm$ as a constant times the constant ρ–related Fourier coefficient of $E_{\frac{1+i\lambda}{2}}$. Finally, using the like of (20.69) again with the constant 1 substituted for $\mathcal{M}_{k,\ell}$, we get the value of Φ^0. $\qquad \square$

Remark. Using the duplication formula twice, it is an easy matter, computing the residue of $c_{k,\ell}$, as a function of ν, at $\nu = i\lambda_k$, to see that it agrees with the one which can be obtained from (20.42), (20.41) and (20.38).

THEOREM 20.8. *With $C(\nu)$ defined in (20.67), one has if* Re $\nu < -1$ *the expression*

$$f_\nu^+(z) = C(\nu) \int_{\mathcal{R}} G_{\frac{1-\nu}{2}}(z; z')\, dm(z')\,, \tag{20.73}$$

where the automorphic Green function $G_{\frac{1-\nu}{2}}(z; z')$ was defined in (18.117).

PROOF. It suffices to compare the Roelcke–Selberg expansion (20.63) to the expansion

$$G_{\frac{1-\nu}{2}}(z; z') = \frac{12}{\pi}(\nu^2 - 1)^{-1} + 4\sum_{k,\ell} \frac{\mathcal{M}_{k,\ell}(z)\,\mathcal{M}_{k,\ell}^*(z')}{\nu^2 + \lambda_k^2}$$

$$+ \frac{1}{2\pi} \int_{-\infty}^{\infty} E_{\frac{1-i\lambda}{2}}(z)\, E_{\frac{1+i\lambda}{2}}(z')\, \frac{d\lambda}{\nu^2 + \lambda^2} \tag{20.74}$$

(*cf.* [**36**, p. 271]).   □

Remarks. The last theorem, together with the fact that $(\Delta - \frac{1-\nu^2}{4})(z \mapsto G_{\frac{1-\nu}{2}}(z; z'))$ is the Dirac mass at z', explains the formula (19.68) from theorem 19.6.

From theorem 20.8 and from (18.118), it follows that in the estimate

$$f_\nu^+(z) = c_0^+(\nu)\, y^{\frac{1+\nu}{2}} + \text{error term}$$

quoted in (20.53) (with Re $\nu < -1$), the error term is actually a $O(e^{-2\pi y})$ as $y \to \infty$. In the case when D, as defined in the beginning of the proof of lemma 19.1, is the discriminant of the field $\mathbb{Q}(\rho)$, the coefficient $c_0^+(\nu)$ is, according to (20.53) and to [**50**, p. 281], the value at $\frac{1-\nu}{2}$, up to some coefficient, of the (Hecke) zeta–function of some ideal class in $\mathbb{Q}(\rho)$.

Index of notations: page 2

Subject Index

Bibliography

[1] F.A.Berezin, *Quantization in Complex Symmetric Spaces*, Math.U.S.S.R.Izvestija **9**(2) (1975), 341-379.

[2] F.A.Berezin, *A connection between co- and contravariant symbols of operators on classical complex symmetric spaces*, Soviet Math.Dokl. **19** (1978), 786-789.

[3] D.Bump, *Automorphic Forms and Representations*, Cambridge Series in Adv.Math. **55**, Cambridge, 1996.

[4] A.Cerezo, *Hyperfonctions à une variable*, in *Hyperfunctions and Theoretical Physics* (F.Pham, ed.), Lecture Notes in Math. 449, Springer–Verlag, Berlin (1975), 5–20.

[5] H.Cohen, *Sums involving the values at negative integers of L–functions of quadratic characters*, Math.Ann. **217** (1975), 271-295.

[6] P.B.Cohen, Y.Manin, D.Zagier, *Automorphic pseudodifferential operators*, in *Algebraic aspects of integrable systems*, Progress Nonlinear Diff.Equ.Appl. **26**, Birkhäuser (1996), 17–47.

[7] J.M.Deshouillers, H.Iwaniec, *Kloosterman sums and Fourier coefficients of cusp forms*, Inv.Math. **70** (1982), 219-288.

[8] H.M.Edwards, *Riemann's zeta function*, Acad. Press, New York-London, 1974.

[9] S.Gelbart, *Automorphic forms on Adele groups*, Ann.of Math.Studies **83** Princeton Univ.Press, Princeton, 1975.

[10] I.M.Gelfand, M.I.Graev, I.I.Pyatetskii-Shapiro, *Representation theory and automorphic functions*, W.B.Saunders Co, Philadelphia–London–Toronto, 1969.

[11] I.M.Gelfand, M.I.Graev, N.Ya.Vilenkin, *Generalized Functions*, vol.5, Academic Press, London–New York, 1966.

[12] D.Goldfeld, P.Sarnak, *Sums of Kloosterman sums*, Inv.Math. **71**(2) (1983), 243-250.

[13] I.S.Gradstein, I.M.Ryshik, *Tables of Series, Products and Integrals*, vol.2, Verlag Harri Deutsch, Thun–Frankfurt/M, 1981.

[14] K.I.Gross, R.A.Kunze, *Fourier Bessel transforms and holomorphic discrete series*, Lecture Notes in Math. **266** (1972), 79-122.

[15] K.I.Gross, R.A.Kunze, *Bessel functions and representation theory I,II*, J.Funct.Anal. **22** (1976), 73-105; *ibid.*(1977), 1-49.

[16] G.H.Hardy, E.M.Wright, *An Introduction to the Theory of Numbers*, fourth edition, Oxford Univ.Press, London, 1962.

[17] S.Helgason, *Geometric Analysis on Symmetric Spaces*, Math.Surveys and Monographs **39**, A.M.S., Providence, 1994.

[18] D.A.Hejhal, *Some observations concerning eigenvalues of the Laplacian and Dirichlet L–series*, in *Recent Progress in Analytic Number Theory* (H.Halberstam and C.Hooley, eds), vol.2, Acad.Press, London–New York (1981), 95–110.

[19] L.K.Hua, *Introduction to Number Theory*, Springer–Verlag, Berlin, 1982.

[20] H.Iwaniec, *Introduction to the spectral theory of automorphic forms*, Revista Matemática Iberoamericana, Madrid, 1995.

[21] H.Iwaniec, *Topics in Classical Automorphic Forms*, Graduate Studies in Math. **17**, A.M.S., Providence, 1997.

[22] A.W.Knapp, *Representation Theory of Semi–Simple Groups*, Princeton Univ.Press, Princeton, 1986.

[23] T.Kubota, *Elementary Theory of Eisenstein Series*, Kodansha Ltd, Tokyo; Halsted Press, New York, 1973.

[24] S.Lang, *Algebraic Number Theory*, Addison–Wesley, Reading (Mass.), 1970.

[25] P.D.Lax, R.S.Phillips, *Scattering Theory for Automorphic Functions*, Ann. Math. Studies **87**, Princeton Univ. Press, 1976.

[26] J.B.Lewis, *Eigenfunctions on symmetric spaces with distribution-valued boundary forms*, J.Funct.Anal. **29** (1978), 287-307.

[27] W.Magnus, F.Oberhettinger, R.P.Soni, *Formulas and theorems for the special functions of mathematical physics*, 3rd edition, Springer–Verlag, Berlin, 1966.

[28] D.Niebur, *A class of nonanalytic automorphic functions*, Nagoya Math. J. **52** (1973), 133-145.

[29] H.Rademacher, *Topics in Analytic Number Theory*, Springer–Verlag, Berlin, 1973.

[30] R.A.Rankin, *The construction of automorphic forms from the derivatives of a given form*, J.Indian Math.Soc. **20** (1956), 103-116.

[31] R.A.Rankin, *Modular Forms and Functions*, Cambridge Univ. Press, Cambridge–London–New York–Melbourne, 1977.

[32] L.Schwartz, *Théorie des distributions*, vol.1, Hermann, Paris, 1959.

[33] A.Selberg, *On the Estimation of Fourier Coefficients of Modular Forms*, Proc.Symp.Pure Math. **8** (1963), 1-15.

[34] C.L.Siegel, *Lectures on advanced analytic number theory*, Tata Institute of Fundamental Research, Bombay, 1961.

[35] R.A.Smith, *The L^2-norm of Maass wave functions*, Proc.Amer.Math.Soc. **82** (1981), 179-182.

[36] A.Terras, *Harmonic analysis on symmetric spaces and applications 1*, Springer–Verlag, New York–Berlin–Heidelberg–Tokyo, 1985.

[37] G.Tenenbaum, *Introduction à la théorie analytique et probabiliste des nombres*, Cours spécialisés Soc.Math.France, Paris, 1995.

[38] A.Unterberger, *Pseudodifferential operators and applications: an introduction*, Lecture Notes **46**, Aarhus Universitet, Aarhus (Denmark), 1976.

[39] A.Unterberger, J.Unterberger, *La série discrète de $SL(2,\mathbb{R})$ et les opérateurs pseudo-différentiels sur une demi-droite*, Ann.Sci.Ecole Norm.Sup. **17** (1984), 83-116.

[40] A.Unterberger, *Symbolic calculi and the duality of homogeneous spaces*, Contemp.Math. **27** (1984), 237-252.

[41] A.Unterberger, *L'opérateur de Laplace–Beltrami du demi-plan et les quantifications linéaire et projective de $SL(2,\mathbb{R})$* in *Colloque en l'honneur de L.Schwartz*, Astérisque **131** (1985), 255-275.

[42] A.Unterberger, J.Unterberger, *Quantification et analyse pseudodifférentielle*, Ann.Sci.Ecole Norm.Sup. **21** (1988), 133-158.

[43] A.Unterberger, *Quantification relativiste*, Mémoires Soc. Math. de France **44–45**, 1991.

[44] A.Unterberger, J.Unterberger, *Representations of $SL(2,\mathbb{R})$ and symbolic calculi*, Integr. Equ. Oper. Theory **18** (1994), 303-334.

[45] A.Unterberger, J.Unterberger, *Algebras of Symbols and Modular Forms*, J.d'Analyse Math. **68** (1996), 121-143.

[46] A.Unterberger, H.Upmeier, *The Berezin Transform and Invariant Differential Operators*, Comm.Math.Phys. **164** (1994), 563-597.

[47] A.B.Venkov, *Spectral theory of automorphic functions and its applications*, Kluwer Acad. Pub., Dordrecht–Boston–London, 1990.

[48] A.Weil, *On some exponential sums*, Proc.Nat.Acad.Sci. USA **34** (1948), 204-207.

[49] D.Zagier, *Introduction to modular forms*, in *From Number Theory to Physics* (M.Waldschmidt, P.Moussa, J.M.Luck and C.Itzykson, eds), Springer–Verlag, Berlin (1992), 238–291.

[50] D.Zagier, *Automorphic forms, representation theory and arithmetic*, Bombay Colloquium 1979, Tata Institute of Fund.Res.Stud.in Math, (1981), 275–301.

[51] D.Zagier, *Modular forms whose coefficients involve zeta–functions of quadratic fields* in *Modular Functions of one variable VI*, Lecture Notes in Math. **627**, Springer–Verlag, Berlin, 1977.

Vol. 1654: R. W. Ghrist, P. J. Holmes, M. C. Sullivan, Knots and Links in Three-Dimensional Flows. X, 208 pages. 1997.

Vol. 1655: J. Azéma, M. Emery, M. Yor (Eds.), Séminaire de Probabilités XXXI. VIII, 329 pages. 1997.

Vol. 1656: B. Biais, T. Björk, J. Cvitanic, N. El Karoui, E. Jouini, J. C. Rochet, Financial Mathematics. Bressanone, 1996. Editor: W. J. Runggaldier. VII, 316 pages. 1997.

Vol. 1657: H. Reimann, The semi-simple zeta function of quaternionic Shimura varieties. IX, 143 pages. 1997.

Vol. 1658: A. Pumarino, J. A. Rodrıguez, Coexistence and Persistence of Strange Attractors. VIII, 195 pages. 1997.

Vol. 1659: V, Kozlov, V. Maz'ya, Theory of a Higher-Order Sturm-Liouville Equation. XI, 140 pages. 1997.

Vol. 1660: M. Bardi, M. G. Crandall, L. C. Evans, H. M. Soner, P. E. Souganidis, Viscosity Solutions and Applications. Montecatini Terme, 1995. Editors: I. Capuzzo Dolcetta, P. L. Lions. IX, 259 pages. 1997.

Vol. 1661: A. Tralle, J. Oprea, Symplectic Manifolds with no Kähler Structure. VIII, 207 pages. 1997.

Vol. 1662: J. W. Rutter, Spaces of Homotopy Self-Equivalences – A Survey. IX, 170 pages. 1997.

Vol. 1663: Y. E. Karpeshina; Perturbation Theory for the Schrödinger Operator with a Periodic Potential. VII, 352 pages. 1997.

Vol. 1664: M. Väth, Ideal Spaces. V, 146 pages. 1997.

Vol. 1665: E. Giné, G. R. Grimmett, L. C. Saloff-Coste, Lectures on Probability Theory and Statistics 1996. Editor: P. Bernard. X, 424 pages, 1997.

Vol. 1666: M. van der Put, M. F. Singer, Galois Theory of Difference Equations. VII, 179 pages. 1997.

Vol. 1667: J. M. F. Castillo, M. González, Three-space Problems in Banach Space Theory. XII, 267 pages. 1997.

Vol. 1668: D. B. Dix, Large-Time Behavior of Solutions of Linear Dispersive Equations. XIV, 203 pages. 1997.

Vol. 1669: U. Kaiser, Link Theory in Manifolds. XIV, 167 pages. 1997.

Vol. 1670: J. W. Neuberger, Sobolev Gradients and Differential Equations. VIII, 150 pages. 1997.

Vol. 1671: S. Bouc, Green Functors and G-sets. VII, 342 pages. 1997.

Vol. 1672: S. Mandal, Projective Modules and Complete Intersections. VIII, 114 pages. 1997.

Vol. 1673: F. D. Grosshans, Algebraic Homogeneous Spaces and Invariant Theory. VI, 148 pages. 1997.

Vol. 1674: G. Klaas, C. R. Leedham-Green, W. Plesken, Linear Pro-p-Groups of Finite Width. VIII, 115 pages. 1997.

Vol. 1675: J. E. Yukich, Probability Theory of Classical Euclidean Optimization Problems. X, 152 pages. 1998.

Vol. 1676: P. Cembranos, J. Mendoza, Banach Spaces of Vector-Valued Functions. VIII, 118 pages. 1997.

Vol. 1677: N. Proskurin, Cubic Metaplectic Forms and Theta Functions. VIII, 196 pages. 1998.

Vol. 1678: O. Krupková, The Geometry of Ordinary Variational Equations. X, 251 pages. 1997.

Vol. 1679: K.-G. Grosse-Erdmann, The Blocking Technique. Weighted Mean Operators and Hardy's Inequality. IX, 114 pages. 1998.

Vol. 1680: K.-Z. Li, F. Oort, Moduli of Supersingular Abelian Varieties. V, 116 pages. 1998.

Vol. 1681: G. J. Wirsching, The Dynamical System Generated by the 3n+1 Function. VII, 158 pages. 1998.

Vol. 1682: H.-D. Alber, Materials with Memory. X, 166 pages. 1998.

Vol. 1683: A. Pomp, The Boundary-Domain Integral Method for Elliptic Systems. XVI, 163 pages. 1998.

Vol. 1684: C. A. Berenstein, P. F. Ebenfelt, S. G. Gindikin, S. Helgason, A. E. Tumanov, Integral Geometry, Radon Transforms and Complex Analysis. Firenze, 1996. Editors: E. Casadio Tarabusi, M. A. Picardello, G. Zampieri. VII, 160 pages. 1998

Vol. 1685: S. König, A. Zimmermann, Derived Equivalences for Group Rings. X, 146 pages. 1998.

Vol. 1686: J. Azéma, M. Émery, M. Ledoux, M. Yor (Eds.), Séminaire de Probabilités XXXII. VI, 440 pages. 1998.

Vol. 1687: F. Bornemann, Homogenization in Time of Singularly Perturbed Mechanical Systems. XII, 156 pages. 1998.

Vol. 1688: S. Assing, W. Schmidt, Continuous Strong Markov Processes in Dimension One. XII, 137 page. 1998.

Vol. 1689: W. Fulton, P. Pragacz, Schubert Varieties and Degeneracy Loci. XI, 148 pages. 1998.

Vol. 1690: M. T. Barlow, D. Nualart, Lectures on Probability Theory and Statistics. Editor: P. Bernard. VIII, 237 pages. 1998.

Vol. 1691: R. Bezrukavnikov, M. Finkelberg, V. Schechtman, Factorizable Sheaves and Quantum Groups. X, 282 pages. 1998.

Vol. 1692: T. M. W. Eyre, Quantum Stochastic Calculus and Representations of Lie Superalgebras. IX, 138 pages. 1998.

Vol. 1694: A. Braides, Approximation of Free-Discontinuity Problems. XI, 149 pages. 1998.

Vol. 1695: D. J. Hartfiel, Markov Set-Chains. VIII, 131 pages. 1998.

Vol. 1696: E. Bouscaren (Ed.): Model Theory and Algebraic Geometry. XV, 211 pages. 1998.

Vol. 1697: B. Cockburn, C. Johnson, C.-W. Shu, E. Tadmor, Advanced Numerical Approximation of Nonlinear Hyperbolic Equations. Cetraro, Italy, 1997. Editor: A. Quarteroni. VII, 390 pages. 1998.

Vol. 1698: M. Bhattacharjee, D. Macpherson, R. G. Möller, P. Neumann, Notes on Infinite Permutation Groups. XI, 202 pages. 1998.

Vol. 1699: A. Inoue,Tomita-Takesaki Theory in Algebras of Unbounded Operators. VIII, 241 pages. 1998.

Vol. 1700: W. A. Woyczyński, Burgers-KPZ Turbulence, XI, 318 pages. 1998.

Vol. 1701: Ti-Jun Xiao, J. Liang, The Cauchy Problem of Higher Order Abstract Differential Equations. XII, 302 pages. 1998.

Vol. 1702: J. Ma, J. Yong, Forward-Backward Stochastic Differential Equations and Their Applications. XIII, 270 pages. 1999.

Vol. 1703: R. M. Dudley, R. Norvaiša, Differentiability of Six Operators on Nonsmooth Functions and p-Variation. VIII, 272 pages. 1999.

Vol. 1704: H. Tamanoi, Elliptic Genera and Vertex Operator Super-Algebras. VI, 390 pages. 1999.

Vol. 1705: I. Nikolaev, E. Zhuzhoma, Flows in 2-dimensional Manifolds. XIX, 294 pages. 1999.

Vol. 1706: S. Yu. Pilyugin, Shadowing in Dynamical Systems. XVII, 271 pages. 1999.

Vol. 1707: R. Pytlak, Numerical Methods for Optimal Control Problems with State Constraints. XV, 215 pages. 1999.

Vol. 1708: K. Zuo, Representations of Fundamental Groups of Algebraic Varieties. VII, 139 pages. 1999.

Vol. 1709: J. Azéma, M. Émery, M. Ledoux, M. Yor (Eds), Séminaire de Probabilités XXXIII. VIII, 418 pages. 1999.

Vol. 1710: M. Koecher, The Minnesota Notes on Jordan Algebras and Their Applications. IX, 173 pages. 1999.

Vol. 1711: W. Ricker, Operator Algebras Generated by Commuting Projections: A Vector Measure Approach. XVII, 159 pages. 1999.

Vol. 1712: N. Schwartz, J. J. Madden, Semi-algebraic Function Rings and Reflectors of Partially Ordered Rings. XI, 279 pages. 1999.

Vol. 1713: F. Bethuel, G. Huisken, S. Müller, K. Steffen, Calculus of Variations and Geometric Evolution Problems. Cetraro, 1996. Editors: S. Hildebrandt, M. Struwe. VII, 293 pages. 1999.

Vol. 1714: O. Diekmann, R. Durrett, K. P. Hadeler, P. K. Maini, H. L. Smith, Mathematics Inspired by Biology. Martina Franca, 1997. Editors: V. Capasso, O. Diekmann. VII, 268 pages. 1999.

Vol. 1715: N. V. Krylov, M. Röckner, J. Zabczyk, Stochastic PDE's and Kolmogorov Equations in Infinite Dimensions. Cetraro, 1998. Editor: G. Da Prato. VIII, 239 pages. 1999.

Vol. 1716: J. Coates, R. Greenberg, K. A. Ribet, K. Rubin, Arithmetic Theory of Elliptic Curves. Cetraro, 1997. Editor: C. Viola. VIII, 260 pages. 1999.

Vol. 1717: J. Bertoin, F. Martinelli, Y. Peres, Lectures on Probability Theory and Statistics. Saint-Flour, 1997. Editor: P. Bernard. IX, 291 pages. 1999.

Vol. 1718: A. Eberle, Uniqueness and Non-Uniqueness of Semigroups Generated by Singular Diffusion Operators. VIII, 262 pages. 1999.

Vol. 1719: K. R. Meyer, Periodic Solutions of the N-Body Problem. IX, 144 pages. 1999.

Vol. 1720: D. Elworthy, Y. Le Jan, X-M. Li, On the Geometry of Diffusion Operators and Stochastic Flows. IV, 118 pages. 1999.

Vol. 1721: A. Iarrobino, V. Kanev, Power Sums, Gorenstein Algebras, and Determinantal Loci. XXVII, 345 pages. 1999.

Vol. 1722: R. McCutcheon, Elemental Methods in Ergodic Ramsey Theory. VI, 160 pages. 1999.

Vol. 1723: J. P. Croisille, C. Lebeau, Diffraction by an Immersed Elastic Wedge. VI, 134 pages. 1999.

Vol. 1724: V. N. Kolokoltsov, Semiclassical Analysis for Diffusions and Stochastic Processes. VIII, 347 pages. 2000.

Vol. 1725: D. A. Wolf-Gladrow, Lattice-Gas Cellular Automata and Lattice Boltzmann Models. IX, 308 pages. 2000.

Vol. 1726: V. Marić, Regular Variation and Differential Equations. X, 127 pages. 2000.

Vol. 1727: P. Kravanja, M. Van Barel, Computing the Zeros of Analytic Functions. VII, 111 pages. 2000.

Vol. 1728: K. Gatermann, Computer Algebra Methods for Equivariant Dynamical Systems. XV, 153 pages. 2000.

Vol. 1729: J. Azéma, M. Émery, M. Ledoux, M. Yor, Séminaire de Probabilités XXXIV. VI, 431 pages. 2000.

Vol. 1730: S. Graf, H. Luschgy, Foundations of Quantization for Probability Distributions. X, 230 pages. 2000.

Vol. 1731: T. Hsu, Quilts: Central Extensions, Braid Actions, and Finite Groups,. XII, 185 pages. 2000.

Vol. 1732: K. Keller, Invariant Factors, Julia Equivalences and the (Abstract) Mandelbrot Set. X, 206 pages. 2000.

Vol. 1733: K. Ritter, Average-Case Analysis of NumericalProblems. IX, 254 pages. 2000.

Vol. 1734: M. Espedal, A. Fasano, A. Mikelić, Filtration in Porous Media and Industrial Applications. Cetraro 1998. Editor: A. Fasano. 2000.

Vol. 1735: D. Yafaev, Scattering Theory: Some Old and New Problems. XVI, 169 pages. 2000.

Vol. 1736: B. O. Turesson, Nonlinear Potential Theory and Weighted Sobolev Spaces. XIV, 173 pages. 2000.

Vol. 1737: S. Wakabayashi, Classical Microlocal Analysis in the Space of Hyperfunctions. VIII, 367 pages. 2000.

Vol. 1738: M. Emery, A. Nemirovski, D. Voiculescu, Lectures on Probability Theory and Statistics. XI, 356 pages. 2000.

Vol. 1739: R. Burkard, P. Deuflhard, A. Jameson, J.-L. Lions, G. Strang, Computational Mathematics Driven by Industrial Problems. Martina Franca, 1999. Editors: V. Capasso, H. Engl, J. Periaux. VI, 414 pages. 2000.

Vol. 1740: B. Kawohl, O. Pironneau, L. Tartar, J.-P. Zolesio, Optimal Shape Design. Troia 1999. Editors: A. Cellina, A. Ornelas. IX, 393 pages. 2000.

Vol. 1741: E. Lombardi, Oscillatory Integrals and Phenomena Beyond all Algebraic Orders. XV, 413 pages. 2000.

Vol. 1742: A. Unterberger, Quantization and Non-holomorphic Modular Forms.VIII, 253 pages. 2000.

4. Lecture Notes are printed by photo-offset from the master-copy delivered in camera-ready form by the authors. Springer-Verlag provides technical instructions for the preparation of manuscripts. Macro packages in T_EX, L^AT_EX2e, $L^AT_EX2.09$ are available from Springer's web-pages at

http://www.springer.de/math/authors/b-tex.html.

Careful preparation of the manuscripts will help keep production time short and ensure satisfactory appearance of the finished book.

The actual production of a Lecture Notes volume takes approximately 12 weeks.

5. Authors receive a total of 50 free copies of their volume, but no royalties. They are entitled to a discount of 33.3 % on the price of Springer books purchase for their personal use, if ordering directly from Springer-Verlag.

Commitment to publish is made by letter of intent rather than by signing a formal contract. Springer-Verlag secures the copyright for each volume. Authors are free to reuse material contained in their LNM volumes in later publications: A brief written (or e-mail) request for formal permission is sufficient.

Addresses:

Professor F. Takens, Mathematisch Instituut,
Rijksuniversiteit Groningen, Postbus 800,
9700 AV Groningen, The Netherlands
E-mail: F.Takens@math.rug.nl

Professor B. Teissier
Université Paris 7
UFR de Mathématiques
Equipe Géométrie et Dynamique
Case 7012
2 place Jussieu
75251 Paris Cedex 05
E-mail: Teissier@math.jussieu.fr

Springer-Verlag, Mathematics Editorial, Tiergartenstr. 17,
D-69121 Heidelberg, Germany,
Tel.: *49 (6221) 487-701
Fax: *49 (6221) 487-355
E-mail: lnm@Springer.de

9 783540 678618